CALIFORNIA NATURAL HISTORY GUIDES

**FIELD GUIDE TO
BEETLES OF CALIFORNIA**

California Natural History Guides

Phyllis M. Faber and Bruce M. Pavlik, General Editors

Field Guide to
BEETLES of
CALIFORNIA

Arthur V. Evans
James N. Hogue

UNIVERSITY OF CALIFORNIA PRESS
Berkeley   Los Angeles   London

We dedicate this book to our parents, Edwin and Lois Evans
and Charles and Barbara Hogue. Much of the success of our
entomological endeavors is due to their support and
encouragement.

University of California Press, one of the most distinguished university presses in
the United States, enriches lives around the world by advancing scholarship in the
humanities, social sciences, and natural sciences. Its activities are supported by
the UC Press Foundation and by philanthropic contributions from individuals
and institutions. For more information, visit www.ucpress.edu.

California Natural History Guide Series No. 88

University of California Press
Berkeley and Los Angeles, California

University of California Press, Ltd.
London, England

© 2006 by the Regents of the University of California

Library of Congress Cataloging-in-Publication Data

Evans, Arthur V. 1956–
 Field guide to beetles of California / Arthur V. Evans and James N. Hogue.
  p. cm. — (California natural history guides ; 88)
 Includes bibliographical references and index.
 ISBN-13 978-0-520-24655-3 (cloth : alk. paper)—ISBN-10 0-520-24655-1
(cloth : alk. paper)
 ISBN-13 978-0-520-24657-7 (pbk. : alk. paper)—ISBN-10 0-520-24657-8
(pbk. : alk. paper)
 1. Beetles—California—Identification. I. Hogue, James N., 1961– II. Title. III.
Series.
QL584.C2E93 2006
595.7609794—dc22   2005034493

Manufactured in China
10 09 08 07 06
10 9 8 7 6 5 4 3 2 1

The paper used in this publication meets the minimum requirements of
ANSI/NISO Z39.48–1992 (R 1997) (*Permanence of Paper*).

Cover: Ten-lined June Beetle *(Polyphylla decemlineata)*. Photograph by
Arthur V. Evans.

The publisher and authors gratefully acknowledge the generous contributions to this book provided by

the Gordon and Betty Moore Fund
in Environmental Studies
and
the General Endowment Fund of the
University of California Press Foundation.

CONTENTS

Acknowledgments	xi

INTRODUCTION   1

Getting to Know California Beetles	3
Activities with Beetles	21
How to Use This Book	25

ILLUSTRATED KEY TO FAMILIES OF CALIFORNIA BEETLES   29

FAMILY ACCOUNTS   49

Reticulated Beetles (Cupedidae)	51
Wrinkled Bark Beetles (Rhysodidae)	52
Ground Beetles and Tiger Beetles (Carabidae)	54
Whirligig Beetles (Gyrinidae)	63
Crawling Water Beetles (Haliplidae)	65
Trout-stream Beetles (Amphizoidae)	68
Predaceous Diving Beetles (Dytiscidae)	70
Water Scavenger Beetles (Hydrophilidae)	75
Clown Beetles (Histeridae)	80
Primitive Carrion Beetles (Agyrtidae)	83
Carrion Beetles (Silphidae)	85
Rove Beetles (Staphylinidae)	88
Stag Beetles (Lucanidae)	91
Hide Beetles (Trogidae)	94

Rain Beetles (Pleocomidae)	96
Earth-boring Scarab Beetles (Geotrupidae)	100
Bumblebee Scarab Beetles (Glaphyridae)	103
Scarab Beetles (Scarabaeidae)	105
Soft-bodied Plant Beetles (Dascillidae)	119
Cedar Beetles or Cicada Parasite Beetles (Rhipiceridae)	121
Schizopodid Beetles or False Jewel Beetles (Schizopodidae)	123
Metallic Wood-boring Beetles or Jewel Beetles (Buprestidae)	125
Riffle Beetles (Elmidae)	134
Long-toed Water Beetles (Dryopidae)	137
Variegated Mud-loving Beetles (Heteroceridae)	140
Water Penny Beetles (Psephenidae)	142
False Click Beetles (Eucnemidae)	144
Click Beetles (Elateridae)	146
Net-winged Beetles (Lycidae)	151
Glowworms (Phengodidae)	153
Fireflies and Glowworms (Lampyridae)	156
Soldier Beetles (Cantharidae)	159
Skin Beetles (Dermestidae)	162
Bostrichid Beetles (Bostrichidae)	166
Deathwatch Beetles and Spider Beetles (Anobiidae)	170
Bark-gnawing Beetles (Trogossitidae)	174
Checkered Beetles (Cleridae)	177
Soft-winged Flower Beetles (Melyridae)	182
Sap Beetles (Nitidulidae)	184
Silvanid Flat Bark Beetles (Silvanidae)	188
Flat Bark Beetles (Cucujidae)	190
Pleasing Fungus Beetles (Erotylidae)	191
Bothriderid Beetles (Bothrideridae)	194
Handsome Fungus Beetles (Endomychidae)	196
Lady Beetles (Coccinellidae)	198
Tumbling Flower Beetles (Mordellidae)	204
Ripiphorid Beetles (Ripiphoridae)	206
Zopherid Beetles (Zopheridae)	208
Darkling Beetles (Tenebrionidae)	212

False Blister Beetles (Oedemeridae)	221
Blister Beetles (Meloidae)	224
Antlike Flower Beetles (Anthicidae)	231
Longhorn Beetles (Cerambycidae)	234
Leaf Beetles and Seed Beetles (Chrysomelidae)	254
Leaf-rolling Weevils (Attelabidae)	264
Weevils or Snout Beetles (Curculionidae)	267
Checklist of North American Beetle Families	279
California's Sensitive Beetles	285
Collections, Societies, and Other Resources	289
Glossary	291
Selected General References	295
Art Credits	299
Index	301

ACKNOWLEDGMENTS

As with our first book, *Introduction to California Beetles* (2004), this work is the result of a cooperative effort. It has benefited greatly from the generous assistance given to us by our friends and colleagues working in both the public and private sectors.

The following supplied copies of pertinent literature, identifications, and information that contributed substantially to the taxonomic, biological, and distribution content of this book: Nancy Adams, Steve Lingafelter, and Warren Steiner (National Museum of Natural History, Smithsonian Institution, Washington, D.C.); John Chemsak (University of California, Berkeley); Lee Herman (American Museum of Natural History, New York); Margaret Thayer (Field Museum of Natural History, Chicago); John Pinto (University of California, Riverside); Charles Bellamy (Department of Food and Agriculture, Sacramento); and Rick Westcott (Salem, Oregon).

Cheryl Barr (University of California, Berkeley), Roberta Brett (California Academy of Sciences, San Francisco), Brian Harris and Weiping Xie (Natural History Museum of Los Angeles County, Los Angeles), and Steve Heydon (University of California, Davis) kindly provided us with species records of California beetles from collections in their care.

We would also like to thank the following institutions for providing much-needed resources and access to libraries and insect collections: National Museum of Natural History, Smithsonian Institution; Natural History Museum of Los Angeles County; and California State University, Northridge.

The following people generously supplied unpublished state checklists generated from their personal databases of North American beetles: Larry Bezark and Richard Penrose, California Department of Food and Agriculture, Sacramento (Cerambycidae); Ed Riley, Texas A & M University, College Station (Chrysomelidae); and Robert Rabaglia, Maryland Department of Agriculture, Baltimore (Curculionidae, Scolytinae). Yves Alarie and Jennifer Babin, Laurentian University, Sudbury, Ontario, Canada, clarified taxonomic questions regarding the Gyrinidae and verified records for the state.

The following friends and colleagues reviewed various family sections of the book, updated species names, and provided numerous suggestions that improved the usefulness and readability of this field guide: Rolf Aalbu, Sacramento (Tenebrionidae, Zopheridae); Robert Allen, Mission Viejo (Coccinellidae); Robert Anderson, Canadian Museum of Nature, Ottawa, Canada (Curculionidae); Charles Bellamy (Buprestidae, Schizopodidae); Larry Bezark (Cerambycidae); Michael Caterino, Santa Barbara Museum of Natural History, Santa Barbara (Histeridae); Gill Challet, Foothill Ranch, California (Dytiscidae); Donald Chandler, University of New Hampshire, Durham (Anthicidae); Andrew Cline, Department of Food and Agriculture, Sacramento (Nitidulidae); Zachary Falin, University of Kansas, Lawrence (Ripiphoridae); Arthur Gilbert, Clovis, California (Chrysomelidae); Frank Hovore, Canyon Country, California (Pleocomidae); Michael Ivie, University of Montana, Bozeman (Rhysodidae); John Jackman, Texas A & M, College Station (Mordellidae); Paul Johnson, South Dakota State University (Eucnemidae, Elateridae); John Kingsolver, Florida State Collection of Arthropods, Gainesville (Chrysomelidae, Bruchinae); Nadine Kriska, University of Wisconsin, Madison (Oedemeridae); Adriean Mayor, Great Smoky Mountains National Park, Gatlinburg, Tennessee (Melyridae); Matthew Paulsen, University of Nebraska, Lincoln (Lucanidae); Keith Philips, Western Kentucky University, Bowling Green (Anobiidae, Bothrideridae); John Pinto (Meloidae); Alistair Ramsdale, Bishop Museum, Honolulu, Hawai'i (Cantharidae, Lycidae); Jacques Rifkind, Valley Village, California (Cleridae); William Shepard, California State University, Sacramento (Dryopidae, Elmidae, Gyrinidae, Haliplidae, Heteroceridae, Hydrophilidae, Psephenidae); Paul Skelley, Florida State Collection of Arthropods, Gainesville (Endomychidae, Erotylidae); Andrew Smith, Canadian Museum of Nature, Ottawa, Canada (Geotrupidae, Pleocomidae, Scarabaeidae, Trogidae); Margaret Thayer, Field Museum of Natural History, Chicago (Staphylinidae); Michael Thomas, Florida State Collection of Arthropods, Gainesville (Cucujidae, Silvanidae); Natalia Vandenberg, Smithsonian Institution, Washington, D.C. (Coccinellidae).

We are greatly indebted to the following individuals and institutions for allowing us to collect and photograph beetles on their properties, or for logistical support for field trips in California. Deanne Rushall and Sam Dakin, Bouverie Audubon Preserve, Glen Ellen, California, provided Evans with marvelous accommodations at the preserve while photographing beetles in the northern Coast Ranges in May of 2002. Jeff Brown, Station Manager of the Sagehen Creek Field Station, University of California at Berkeley, secured accommodations for Evans in July of 2002. During this time Philip Ward and Alex Wild, University of California at Davis, and the students of his Entomology 109 course at the field station, enthusiastically wrangled live beetles for Evans to photograph before immortalizing them in their collections. Edwin and Lois Evans, Juniper Hills, California, provided logistical support and housing while Evans collected and photographed beetles in the Mojave Desert and Transverse Ranges during the summers of 2000 to 2004. Cameron Barrows, Coachella Valley Preserve, made arrangements for the authors to visit several Colorado Desert sites in his care. Rosser and Jo Garrison, Azusa, California, provided numerous kindnesses to Evans during his spring and summer visits to California. Steve Prchal and Chip Hedgecock, Sonoran Arthropod Studies, Tucson, Arizona, provided housing, logistical support, and excellent company during Evans's spring trip to the Sonoran Desert in April of 2003.

Robert Allen, Ron Alten, Diana Andres, James Dilley, Ann Dittmer, Robert Espinoza, Michael Gaddis, Frank Hovore, Tuynh Huynh, Brenda Kanno, and Paula Schiffman provided live specimens for photography. William Shepard graciously supplied specimens of Elmidae. We express our gratitude to Weiping Xie, who provided invaluable assistance to Hogue in generating some of the digital images of pinned specimens. Steve Lingafelter, Charyn Micheli, and Lisa Roberts, all of the Smithsonian Institution, graciously and patiently assisted Evans with the AutoMontage images that appear in this book.

We especially thank Doris Kretschmer, Jenny Wapner, Scott Norton, and Kate Hoffman at the University of California Press for their enthusiasm, encouragement, and support during all stages of this book's development. It was their vision and hard work that transformed the manuscript and hundreds of photographs and illustrations into this handsome and unique book.

We are particularly grateful to the reviewers assigned by the University of California Press to critique and correct the penultimate draft of the book. Rosser Garrison, California Department of Food and Agriculture, Sacramento, applied his broad knowledge of California insects to the work. Michael Caterino also painstakingly reviewed the manuscript. Although his vision for the work differed from our own, he added substantially to the accuracy of the species descriptions, distribution accounts, and natural history notes. His diligent and thorough reading contributed immeasurably to the overall utility of the book. We were fortunate to benefit from his vast knowledge of California's beetles.

Chuck Bellamy tracked down numerous references and locality records, and checked the key to families. He also accompanied Evans on a photography trip through the Mojave Desert in the spring of 2003. As always, his friendship, humor, and unflagging support of this project added enormously to the overall quality of this work.

We share the success of *Field Guide to Beetles of California* with all the aforementioned individuals, but the responsibility for any and all of its shortcomings, misrepresentations, inaccuracies, and omissions is entirely our own.

Arthur V. Evans, Richmond, Virginia, and
James N. Hogue, Los Angeles, California
June 2006

INTRODUCTION

CALIFORNIA IS HOME to perhaps 8,000 species of beetles, making them the largest group of animals in the state. In comparison there are only 5,800 species of vascular plants, 547 birds, 186 mammals, 240 butterflies, and 108 dragonflies and damselflies known to occur in California. We still don't have a complete listing of all the beetles in the state. The sheer number of species, coupled with vast and still unexplored regions awaiting surveys by professional and amateur beetle specialists, continues to present catalogers with considerable challenges.

Still, beetles are a prominent part of California life, and people who take the time to observe them want to know "What are they? What do they do? Will they hurt my pets or me, or damage my belongings?" This book, the first of its kind for California beetles, provides a means for amateur naturalists, students, and professional biologists to recognize more than 500 of the state's most conspicuous species of beetles found in the home and garden, or out in the field.

Most of the families and species of beetles covered in this book are those that are most likely to be seen and arouse curiosity. The initial species list was generated from surveys of some of the largest and most extensive beetle collections in the state, including those at the California Academy of Sciences in San Francisco, Natural History Museum of Los Angeles County, and University of California at Davis. Relatively large (10 mm or more) or colorful species were selected, along with smaller or seldom seen species with unusual or otherwise noteworthy habits that would make them conspicuous to beetle enthusiasts, observant naturalists, or even the casual passerby. Particular attention was given to genera and species that are broadly distributed in the state. We have also included additional species that are seldom stumbled upon by chance, but would still generate interest if encountered. The final selection reflects our combined 60 years as collectors and photographers of beetles throughout California's mountains, islands, valleys, and deserts.

The species presented here represent only a fraction of the thousands of species of beetles known to live in the state. Beetle specialists will certainly find that many species familiar to them are not included. And no doubt many amateur naturalists will find that some locally abundant species are not represented either. Still, we expect that the majority of species commonly encountered throughout California are found among these pages.

Getting to Know California Beetles

Identifying a beetle in the field is essentially no different from identifying a bird, mammal, flowering plant, or butterfly. The identification of any organism requires some knowledge of the group's distinguishing characteristics and being able to locate and see them clearly. Identifying California beetles can be trying at times simply because of their overwhelming numbers and because their distinguishing features are often small and difficult to see. Read through the following introductory sections of the book. Get to know beetles, their bodies, distinguishing characteristics, classification, and what makes them unique among insects and other animals. This book will also help you gain an understanding of where beetles live and which habitats support a greater diversity of species. For a more comprehensive treatment of beetle morphology and the natural history of the state's beetles, we suggest you refer to the companion book, *Introduction to California Beetles* (Evans and Hogue 2004). For other photos of California beetle larvae and additional selected references visit www.sbnature.org/calbeetles/fieldguidesuppl.

What Is a Beetle?

Beetles are insects with uniquely hard or leathery wing covers called elytra. The elytra nearly always meet in a straight line down the middle of the back. The membranous flight wings, if present, are generally longer than the elytra and are carefully folded beneath them when not in use. As with all insects, beetles have three body regions. However, the "midsection" of a beetle is actually only part of its thorax. The remaining two sections of the thorax, along with most or all of the abdomen, are covered by the elytra. Beetles are further distinguished from other insects by having chewing mouthparts and a developmental process known as complete metamorphosis. Other beetlelike insects (cockroaches, some true bugs, earwigs) do not have this combination of characteristics.

The Beetle Body Plan

Beetles are encased in an external skeleton, or exoskeleton, that functions as both skin and skeleton. The exoskeleton protects the

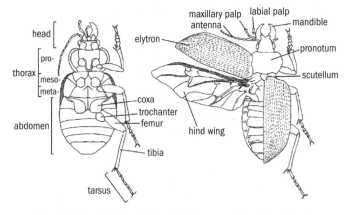

Figure 1. Ventral and dorsal view of *Calosoma semilaeve* (Carabidae), illustrating the major body parts of an adult beetle.

delicate internal organs and serves as a foundation for muscle attachment. It also provides a platform for important receptors that equip beetles to find food, mates, and egg-laying sites. The exoskeleton is subdivided into three major body regions: head, thorax, and abdomen. Each body region is formed by a series of distinct or seamlessly fused plates known as sclerites (fig. 1).

Mouthparts

The head possesses chewing mouthparts variously modified to cut flesh, grind leaves, or strain fluids. The structure of the mouthparts of all beetles is founded on the same basic plan: an upper lip, or labrum; two pairs of chewing appendages, the mandibles and maxillae; and a lower lip, or labium. The maxilla and labium usually possess delicate, flexible structures, called palps, that function like fingers to sense and manipulate food and bring it to the mouth (fig. 2). Protecting the mouthparts from above is a broad plate of cuticle formed by the leading edge of the head known as the clypeus. Below the head and behind the mouthparts are two sclerites known as the mentum and the gula.

The mouthparts of some beetles are prognathous, directed forward and parallel to the long axis of the body. Prognathous mouthparts are typical of predators such as ground beetles

(Carabidae) and whirligig beetles (Gyrinidae), as well as some wood-boring beetles. Hypognathous mouthparts are directed downward and are typical of most plant feeders, including chafers (Scarabaeidae), many longhorn beetles (Cerambycidae), leaf beetles (Chrysomelidae), and weevils (Attelabidae and Curculionidae). The hypognathous mouthparts of some net-winged beetles (Lycidae) and many weevils are drawn out into a beak.

Antennae

Each beetle antenna consists of three basic parts: scape, pedicel, and flagellum (fig. 3). Although the antennae appear to be segmented, the "segments" of the flagellum lack internal musculature. Without muscles the segments of the flagellum are not true segments and are properly called flagellomeres. Taken together, the segments of the flagellum, along with the scape and pedicel, are properly called antennomeres. However, we believe that the term "antennal segments" is more familiar to most readers and use it throughout this book in place of antennomeres. The usual number of antennal segments for beetles is 11, but 10 or fewer segments is common.

Thorax

The thorax is divided into a series of dorsal (upper) and ventral (lower) plates that join with lateral plates to form three rings or

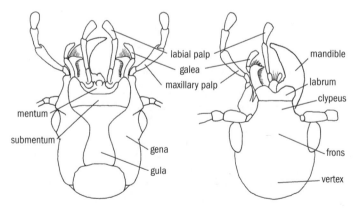

Figure 2. Ventral and dorsal view of *Calosoma semilaeve* (Carabidae) head, illustrating the major parts of an adult beetle head and mouthparts.

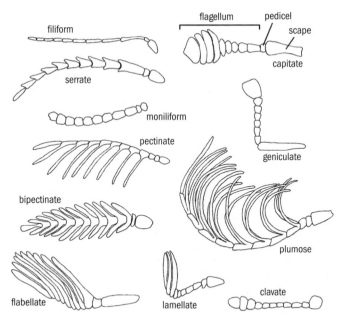

Figure 3. Examples of types of antennae possessed by California beetles:
   Filiform or threadlike (*Cicindela oregona*, Carabidae)
   Capitate or clubbed (*Nicrophorus nigrita*, Silphidae)
   Serrate or saw-toothed (*Prionus californicus*, Cerambycidae)
   Moniliform or beadlike (*Nyctoporis carinata*, Tenebrionidae)
   Geniculate or elbowed (*Scyphophorus yuccae*, Curculionidae)
   Pectinate (*Euthysanius lautus*, Elateridae)
   Bipectinate (*Pleotomus nigripennis*, Lampyridae)
   Plumose or feathery (*Zarhipis integripennis*, Phengodidae)
   Flabellate or fanlike (*Sandalus cribricollis*, Rhipiceridae)
   Lamellate (*Amblonoxia palpalis*, Scarabaeidae)
   Clavate (*Ostoma pippingskoeldi*, Trogossitidae).

thoracic segments: prothorax, mesothorax, and metathorax. Each thoracic segment has a pair of legs, but only the meso- and metathorax possess wings (fig. 1).

The top sclerite of the prothorax is called the pronotum and is variable in shape and texture. In some antlike flower beetles (Anthicidae) the pronotum is hornlike, extending forward over the head. The pronota of some earth-boring scarab beetles (Geotrupidae) and twig borers (Bostrichidae) are scooped out and used like a bulldozer to clear their excavations of soil or sawdust. The sculpturing of the pronotum—such as the size and placement of horns, pits, and bumps—is important in the identification of some beetle species. The pronotum is either margined, with a distinct ridge or keel running along most or all of each side, or evenly rounded.

Elytra

The most prominent and characteristic feature of nearly all beetles is their thick and leathery wing covers, known collectively as elytra, or individually as an elytron. The elytra meet in a distinctive line that runs down the middle of the back and is known as the elytral suture. In some blister beetles (Meloidae) the elytra partially overlap one another at their bases. At the base of the suture, just behind the pronotum, if visible, is an oval, triangular, or hatchet-shaped sclerite called the scutellum.

The elytra cover most or all of the abdominal segments but are typically short in some rove beetles (Staphylinidae), sap beetles (Nitidulidae), clown beetles (Histeridae), soft-winged flower beetles (Melyridae), adult male glowworms (Phengodidae), and ripiphorid beetles (Ripiphoridae). Some longhorn beetles, blister beetles, and click beetles (Elateridae) also have short elytra. Adult females of both glowworms and some fireflies (Lampyridae) lack both elytra and flight wings and resemble larvae.

Legs

Beetle legs consist of six segments. Beginning at the base, the trochantin is partially or completely hidden by the coxa when the leg is in its normal position. The coxae firmly anchor the legs into the coxal cavities of the thorax, yet allow for the horizontal to and fro movement of the legs. Usually small, the coxae of ground beetles and their relatives are enlarged and extend backward across part of their abdomen, an important diagnostic feature. In the crawling water beetles (Haliplidae) the coxae form broad plates that conceal nearly the entire abdomen. The next four segments are the trochanter, femur (thigh), tibia (shin), and tarsus (foot).

Each tarsus (plural: tarsi) consists of five or fewer segments. Like the antennae, the tarsi lack internal musculature and their segments are technically called tarsomeres. The number of segments on each foot is often characteristic of certain beetle fami-

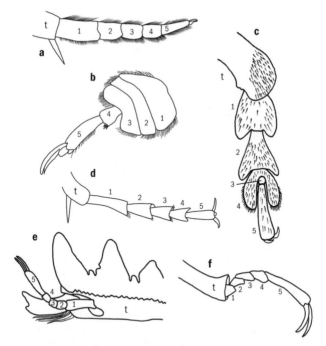

Figure 4. Examples of tarsi possessed by California beetles showing variation of tarsal segments (numbered) and adjacent tibia (t). **a**—*Dytiscus marginicollis* (Dytiscidae), hind tarsus. **b**—*Dytiscus marginicollis* (Dytiscidae), front tarsus of male. **c**—*Chrysochus cobaltinus* (Chrysomelidae), front tarsus. **d**—*Calosoma prominens* (Carabidae), hind tarsus. **e**—*Canthon simplex* (Scarabaeidae), front tarsus. **f**—*Postelclichus immsi* (Dryopidae), hind tarsus.

lies or groups of families and is thus a useful characteristic for beetle identification (fig. 4). The tarsal formula referred to in each family description, such as 5-5-5, 5-5-4, or 4-4-4, indicates the number of segments in the front, middle, and hind feet, respectively. Some segments are difficult to see without careful examination from all angles under high magnification. For example, in cases where each tarsus possesses a hidden segment, the tarsal formula may be referred to as appearing 4-4-4, but actually 5-5-5. The tarsus usually bears a pair of claws attached to the last segment. Claws lacking notches, teeth, or saw-toothed edges on their undersides are referred to as "simple."

Beetle Classification

The naming and arrangement of beetles and other organisms into categories is called taxonomy. Taxonomists use categories known as taxa (singular: taxon) and arrange them in a hierarchical system. The species taxon is the most exclusive of all animal taxa. A beetle species is generally considered to be a group of actually or potentially interbreeding individuals capable of producing successive generations of reproductively viable offspring. Each beetle species is defined by a set of physical and behavioral characteristics wrought by its unique evolutionary history and biological niche within a particular geographical range. On the basis of shared features that imply a shared evolutionary history, beetle species are grouped into genera (singular: genus), which are then arranged into tribes, which in turn are organized into families. All beetle families are placed in the order of insects known as Coleoptera. The thickened elytra of beetles inspired the name of this group, derived from the Greek words *koleos*, meaning "sheath," and *pteron*, meaning "wing."

Beetle Life Cycles

Unlike most vertebrates that grow and develop with relatively little outward change in bodily form, beetles, like most other insects, undergo dramatic changes in body shape during the course of their lives. Like butterflies and moths, a beetle's life consists of a series of successive changes called complete metamorphosis, a transformation marked by four distinct stages: egg, larva, pupa, and adult. Each of these stages allows beetles to exploit different aspects of their environment and gives them considerable flexibility to adjust to ever-changing conditions. Parental care is rare among beetles, so adult and immature stages are seldom found together. By living apart, adults and their offspring rarely compete with one another for food or space.

Reproductive Strategies

Although some beetles, such as those infesting stored grain products, are literally wallowing in "target-rich environments," others must disperse over wide distances and search through tangled vegetation or layers of leaf litter and soil to find a mate. With their relatively short adult lives coming to an end in a matter of weeks or months, beetles have little time to waste and have thus developed various channels of communication to maximize their efforts at finding a mate. They locate and seduce potential mates using sight, sound, or scent. Sometimes these strategies are remarkably effective, luring in numerous eager mates from considerable distances.

Defense

Beetles are attacked and eaten by mammals, birds, reptiles, amphibians, fish, and other arthropods. The large size of some longhorn beetles, backed up by powerful mandibles, may be enough to deter some predators, but most beetles must rely on various behavioral and chemical methods of defense to avoid being attacked by predators and pathogens. Feigning death is a behavioral strategy employed by some beetles. Others hide from their enemies or disguise themselves to look inedible. Some species resemble insects that are difficult to capture or capable of inflicting painful stings, whereas others possess an arsenal of noxious chemicals to discourage predators.

The Role of Beetles

Beetles are a major component of the "F.B.I."—fungi, bacteria, and insects—the primary agents of decomposition. Equipped with powerful chewing mouthparts, both larval and adult beetles are capable of cutting, grinding, or boring through all kinds of plant and animal materials. Living or dead, there is a good chance that most plants and animals will eventually be consumed by some kind of beetle. Their role in the physical and metabolic breakdown of organic matter is critical for recycling nutrients in almost every terrestrial ecosystem on the planet.

Distribution of California Beetles

Even the most casual traveler can see that the state is divided into four major zones based on topography: mountains, islands, valleys, and deserts (maps 1, 2). Each of these physiogeographic regions serves as a corridor of dispersal for some beetles, while acting as a barrier to others. Their unique and shared vegetational communities, soil types, and climates act in concert to create

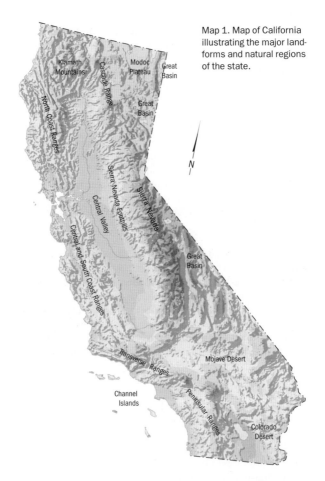

Map 1. Map of California illustrating the major landforms and natural regions of the state.

seemingly infinite habitats that support a breathtaking diversity of beetles.

Mountains

The mountains of California are part of an extensive system of highlands covering much of western North America and are divided into six major regions: the Klamath Mountains, Cascade Range, Sierra Nevada, and Coast, Transverse, and Peninsular

Ranges. The San Francisco Bay divides the Coast Ranges into their northern and southern regions. The Transverse Ranges, which include the Santa Ynez, Santa Monica, San Gabriel, San Bernardino, and Tehachapi Mountains, are unique to the state in that they run east to west. The higher elevations of the San Gabriel and San Bernardino Mountains, as well as the San Jacinto Mountains of the Peninsular Ranges, share many species with similar habitats in the Sierra Nevada.

The respective locations of these mountain ranges, combined with their exposure to sunlight, amount of precipitation, soils, and vegetation, determine the unique or shared nature of their beetle faunas. They are particularly rich in ground beetles, predaceous diving beetles (Dytiscidae), metallic wood-boring beetles (Buprestidae), and longhorn beetles.

Deserts

California has three deserts: the Great Basin, Mojave, and Colorado. All are located in the rain shadows of mountain chains lying to their west or south. Each receives less than 10 inches of rain annually and experiences extreme seasonal temperatures. High winds and dry, clear air enhance the intensity of light reaching the ground, increasing evaporation rates. Nutrient-poor, alkaline soils, often a by-product of dry climates, are sparsely vegetated. Still, California's deserts, particularly their sand dunes, are home to an amazing diversity of beetles.

Great Central Valley

The Great Central Valley is composed of the Sacramento Valley to the north and the San Joaquin Valley to the south. They combine to form a single north-south valley separating the Coast Ranges and the Sierra Nevada. More than any other region of the state, the Great Central Valley has been greatly altered by human activity. Nearly all of the original grasslands are gone, replaced by farmland and introduced grasses. Most of the low-lying marshes are also gone, as well as the gallery forest lining the major rivers and their tributaries.

Like the deserts, the Great Central Valley is one of the California's driest regions. Positioned in the rain shadow of the Coast Ranges, the winter rainfall ranges from moderate in the north to light in the south. Many of the low-lying areas in the southern portion of the valley support desertlike communities and share

Map 2. County map of California.

several genera and species of beetles, especially darkling beetles (Tenebrionidae) and weevils, with the state's deserts.

California Islands

Grazing, agriculture, and introduced species have caused significant alteration and damage to the native flora and fauna of California's islands. In spite of this, these islands continue to serve as refuges for beetles once broadly distributed but now absent, or nearly so, from the mainland. The beetle fauna of California's Channel, Farrallon, Año Nuevo, and San Francisco Bay Islands are poorly known. Of the more than 300 species of beetles recorded from the islands thus far, about 40 are thought to be endemic, occurring nowhere else.

California Beetles and Humans

Some beetles capture the attention of researchers, naturalists, and homeowners, not because of what they look like, but because of what they do. For example, the remains of beetles thousands of years old can reveal particular details of ancient environments far better than the bones of large extinct mammals. Highly specialized species with limited distributions, some of which are under federal protection as threatened or endangered, serve as red flags for entire habitats in need of protection.

In the wild, the activities of wood-boring beetles are essential to the health of forest systems, accelerating the decay of wood and recycling its basic components for use by other plants and animals. However, in forests that are managed for timber production, the activities of these beetles can kill thousands of trees annually, resulting in significant monetary losses. Severe outbreaks of pest beetles and other insects also increase the hazard of fire, damage watersheds and wildlife habitats, pollute stream systems by increased erosion due to flooding, and reduce the aesthetic value of California's national parks and forests.

Closer to home, pantry pests, such as Confused Flour Beetles *(Tribolium confusum)*, rice weevils *(Sitophilus oryzae* and *S. zeamais)*, and Saw-toothed Grain Beetles *(Oryzaephilus surinamensis)*, are particularly annoying to homeowners because of their small size and ability to infest dried foods and spices that are not properly stored. They fly into buildings from outside sources such as nearby rodent, bird, or insect nests. They also hitchhike into homes on old furniture, rugs, drapes, bedding, and other materials of plant or animal origin. Once inside they take up residence in dark secluded corners, living on organic debris accumulating in cracks and crevices. Carefully tucked away in their hiding places, pantry beetles are poised to attack uninfested items as soon as they are brought in from the grocery store.

Other pests from as far away as Australia and Asia have found their way to the Golden State, sometimes via circuitous routes. Some of these exotic beetles readily settled in urban landscapes filled with similarly exotic plants, while aggressive detection and quarantine programs thwarted the establishment of others. Not all beetle immigrants are considered pests, at least not yet. Many of California's dairies, equestrians, and dog owners apparently benefit from the activities of African and European dung beetles (Scarabaeidae), purposely introduced to bury animal waste. And several species of metallic wood-boring beetles, leaf beetles, and weevils have been imported as biological control agents of introduced weeds.

What Is the Best Season to Find California Beetles?

In short, beetles are found throughout the year. It is the arrival of precipitation that usually determines their activity period. Much of the state has a mediterranean climate, with a wet season that runs from September through April, the heaviest precipitation occurring in December through February. Although September marks the beginning of the wet season in northern California, the southern part of the state usually must wait another month or more. The amount of rainfall generally decreases from north to south and increases with elevation. Good winter rains stimulate plant growth from late March through early June, which in turn triggers a flush of insect activity, especially of plant-feeding beetles.

Spring

The sudden profusion of spring flowers in the Colorado and Mojave Deserts in March and April creates a haven for pollen- and flower-feeding beetles, as well as those that scour the desert floor for dried bits of plants or animal remains. Fresh piles of soil known as "push ups" mark the entrances of burrowing beetles as they prepare nests for their young or simply for escape from ever-increasing daytime temperatures. The steady succession of flowering annuals and perennials from April through June in the coastal sage communities of the Coast, Transverse, and Peninsular Ranges provides beetles with not only a source of food, but also a place to find mates.

Spring is also the time when ground squirrels and pack rats remove plant refuse and waste from their underground nest chambers, inadvertently displacing beetles that live deep within those homes. Upon their eviction these beetles take to the air, sometimes by the hundreds, to search for mates and new burrows to occupy.

Aquatic beetles also begin to emerge from overwintering chambers beneath objects along streambeds, migrating to fresh pools to search for food and to reproduce. The shimmering surfaces of swimming pools and shiny tin-covered buildings and car rooftops often attract these would-be aquatic colonists.

Summer

As plants of the deserts and foothills brace for the dry season (May through August), the mountains above 5,000 ft in elevation begin to warm up. Mountain flowers attract different kinds of pollen- and flower-feeding beetles from those of the lowlands, while various species of wood-boring beetles take wing in search of mates and egg-laying sites. Isolated and intense summer thundershowers, typical of the montane regions of the Sierra Nevada and the Transverse and Peninsular Ranges, spark the sudden emergence of still more beetles. Storm runoff feeds cold, fast-moving mountain streams, creating new habitats for montane species of aquatic beetles. An entirely different set of aquatic species prefers the warmer, slower waters in the canyons and valleys below. In late summer (August and September), strong tropical storms occasionally push their way northward from Mexico to drench isolated patches of parched desert and the surrounding foothills, initiating a second flush of flower and beetle activity.

Autumn

Fall can be a slow time of year for beetle enthusiasts in California. The long summer drought that began five or six months earlier is nearing the end of its reign. To avoid the drying heat of late summer and the impending cooler temperatures of fall and winter, most California beetles have already sought refuge deep in the soil or beneath stones and logs. Stands of fall-blooming native plants are uncommon in California. However, expanses of rabbit brush *(Chrysothamnus)* sprinkled along the flats of the Mojave Desert and Great Basin or dry desert streambeds lined with scale-broom *(Lepidospartum squamatum)* attract insects of all kinds with their pungent yellow blossoms. These plants serve as the last gathering places for California beetles before the onset of the first frost. By late fall, most California beetles are decidedly less conspicuous.

Winter

Most beetles spend the winter as eggs, larvae, or pupae. These life stages are better equipped to wait out shorter days and freezing temperatures. However, many species do overwinter as adults, hibernating beneath stones or the loose bark of trees, sometimes congregating by the hundreds or thousands. Aquatic beetles seek shelter, sometimes by the hundreds, in the moist mud or gravel beneath stones in otherwise dry streambeds.

For a surprising number of California beetles, winter is their season of activity. While other parts of the country are in the grips of bone chilling temperatures, much of coastal California remains frost free. Daytime temperatures in some inland valleys and the Colorado Desert can be quite mild and pleasant during the winter, permitting further opportunities to observe and collect beetles. Coastal and desert dune-dwelling beetles move closer to the sand surface as temperatures drop and moisture levels increase. Even at higher elevations, with nighttime temperatures hovering around freezing, winter-active beetles may be found crawling slowly on the ground or across patches of snow well into the chilly night. In fact, any habitat that is not covered with snow is probably a good place to look for beetles in winter.

Where Are the Best Places to Look for Beetles?

Beetles are found everywhere in California, from rocky and sandy beaches upward to barren, windswept mountaintops thousands of feet above sea level. Although many beetles are habitat generalists with broad distributions throughout the state, others are specialists, inhabiting a single coastal dune, desert spring, or rocky mountaintop. Listed below are some of the more productive microhabitats to search for beetles. Checking these habitats throughout the day and into the evening will produce a stunning array of species.

Flowers and Vegetation

Plant-feeding beetles use all parts of plants not only for food, but also as places to mate and lay eggs. Check all parts of shrubs and trees, especially flowers, fruit, cones, branches, leaves, and needles. A source of sweet nectar, high-protein pollen, and succulent petals, flowers are especially attractive to scarabs, tumbling flower beetles (Mordellidae), metallic wood-boring beetles, sap beetles, blister beetles, and longhorns. Members of the sunflower family, such as brittlebush *(Encelia farinosa)* and coreopsis *(Coreopsis)*, are particularly attractive to flower-visiting beetles in spring. In summer, the blossoms of California-lilac *(Ceanothus)*, California buckwheat *(Eriogonum fasciculatum)*, lupine *(Lupinus)*, and other flowers sprinkled over the mountains and their foothills often teem with beetles. Rabbit brush and scale-broom are magnets for fall species, such as blister and longhorn beetles in the valleys and deserts.

Many flowers, shrubs, and trees are also attractive to nocturnal beetles. Oaks *(Quercus)*, chamise *(Adenostoma fasciculatum)*, mountain-mahogany *(Cercocarpus)*, California-lilac, mesquite *(Prosopis)*, buckwheat *(Eriogonum)*, pines *(Pinus)*, and firs *(Abies)* are all excellent plants to search after dark.

Dead Trees, Logs, and Stumps

Moist wood is usually more productive than completely dry wood. The quality of the wood changes as it decomposes and so will its desirability as a habitat. Checking the same stumps and logs over a period of years will reveal a succession of beetle species that use the wood for mating, feeding, and egg laying. Night collecting on dead wood, particularly on warm evenings in the spring and summer, is also productive. Look for tumbling flower beetles, darkling beetles, click beetles, longhorn beetles, and others emerging from their tunnels or wandering about in search of mates. During the day these beetles are found hiding underneath the bark. When pulling bark off of trees, be sure to examine the exposed area very carefully, checking both the log and the inner surface of the bark. Always try to replace the bark when you are finished looking for specimens. Tie or nail the bark back to the log and it will continue to attract and support beetles and other insects.

Fresh-Cut and Burned Wood

The smell of fresh cut or recently burned wood will attract beetles, especially metallic wood-boring beetles and longhorn beetles looking for mates and egg-laying sites. Boards or wood chips laid out across freshly cut stumps will provide shelter for visiting beetles and should be checked frequently for new arrivals throughout the day and evening.

Fungi, Mushrooms, Mosses, and Lichens

Several families of small beetles may be found in good numbers on fungi, mosses, and lichens and nowhere else. As with logs and stumps, inspect the plants and fungi carefully with a hand lens and leave them in good condition so that they continue to lure new beetles. If abundant, take home samples of fungi, mosses, and lichens to extract beetles with a Berlese funnel.

Stream Banks and Ocean- and Lakeshores

Floating debris on the surface of streams and lakes contains flying and crawling beetles trapped by flowing waters. The flumes that criss-cross the western foothills of the Sierra Nevada regularly produce rare and unusual specimens. Ground beetles and rove beetles often hide during the day beneath plant debris that has been washed up on the shores of lakes and oceans. Burrowing species, such as variegated mud-loving beetles (Heteroceridae), can be flushed from their burrows by splashing water across flat mud banks. Fast-moving adult tiger beetles (Carabidae) hunt and fly along the shore searching for insect prey.

Freshwater Pools, Streams, and Lakes

Although a few beetles prefer cold, fast streams, most favor ponds or slow-moving streams. Look for whirligig beetles on the surface of pools. Predaceous diving beetles are often found on gravelly bottoms or beneath submerged objects, whereas water scavengers (Hydrophilidae) and long-toed water beetles (Dryopidae) are found swimming near aquatic plants or crawling among mats of algae. Pick up and search rocks from flowing water for water penny beetle larvae (Psephenidae) and riffle beetles (Elmidae).

Coastal and Desert Sand Dunes

Small scarabs, clown beetles, and weevils often hide in the sand among the roots of dune grasses and other plants, usually in the moisture layers, or in accumulations of plant material beneath the surface. In fall and winter, the moisture layer and the beetles that live within it move closer to the surface and are easier to collect by hand or by sifting.

Carcasses

Dead animals provide food for many kinds of adult and larval beetles. In order to secure adequate food supplies for their young, burying beetles conceal smaller carcasses before other carrion-feeding insects arrive. Carrion beetles (Silphidae) feed on fresh, juicy tissues, but skin beetles (Dermestidae) consume dried flesh. Hide beetles (Trogidae) gnaw on hair, feathers, hooves, and horns. Predatory clown beetles, rove beetles, and ham beetles (*Necrobia,* Cleridae) search for the eggs of other carrion-feeding insects and mites. Other families of beetles are simply attracted by the moisture and shelter afforded to them by a dead body. Be sure to pick through or sift the soil directly beneath the body for all stages of beetles.

Dung

California's native dung beetle (Scarabaeidae) fauna is small when compared to other regions such as Arizona or states farther east. However, large numbers of a few exotic species imported from Europe and Africa may be found in cattle, horse, and dog dung throughout the state. Native dung beetles are often specialists, preferring the dung of burrowing rodents or deer. In the foothills of coastal southern California, hide beetles are commonly found during the winter wet season gnawing on the hair-filled scats of coyotes. Relatives of water scavenger beetles are also commonly found on fresh cattle dung, especially in the central and northern parts of the state.

Beneath Stones and Other Objects

Ground beetles, rove beetles, and darkling beetles may be found beneath stones, boards, and other objects lying on the ground. Stones in grass, along streams, or in other wet habitats are most productive. When finished searching the exposed area, always return the stone or board back to its original place for the benefit of the organisms living there, the aesthetic appearance of the area, and to maintain the productivity of the site for future collecting trips.

Leaf Litter and Compost

Layers of leaves and needles that gather beneath trees, accumulate along canyon bottoms, or wash up on beaches and lakeshores harbor numerous beetles. Backyard compost heaps and other accumulations of decomposing vegetation are particularly productive. Some beetles, particularly some rove beetles, are found only under decomposing piles of seaweed along the beach.

Nests

Nests of birds and burrowing mammals may produce rare or poorly known species of beetles. Many small dung beetles prefer the waste of burrowing rodents. These beetles are active in spring and are sometimes found flying low over fields. Be careful when searching these nests, as there is a chance of contracting flea-borne diseases. Never disturb occupied nests, especially those of sensitive, threatened, or endangered bird or mammal species.

Lights

Beetles of many families are attracted to lights at night. Check porch lights, mercury vapor streetlights, and storefronts, especially those in undeveloped areas. Although many species settle on the ground directly beneath the light, others are typically found climbing on nearby walls or plants. Others seldom if ever come to lights, preferring instead to remain in the nearby shadows.

At Home

Beetles trapped in buildings will usually fly to windows and other light sources while attempting to escape. Look for living and dead beetles on windowsills and light fixtures located inside houses, garages, sheds, and warehouses. High numbers of pest species may indicate an infestation of beetles attacking stored foods, skins, plant materials, or wood products.

Activities with Beetles

One of the most appealing aspects of California beetles is the fact that they can be found virtually anywhere, anytime. Whether exploring your own backyard, visiting a state park, or hiking in a national forest, you will discover something new and exciting about beetles on every outing throughout the year. The study of beetles through observing and photographing them in the field, making a collection, or rearing them in the classroom, can provide a lifetime of exploration and pleasure. In California the chances of discovering new species or revealing unknown aspects of beetles' lives are virtually guaranteed. For more detailed information on collecting, studying, and observing California beetles, refer to the companion book, *Introduction to California Beetles* (Evans and Hogue 2004).

Collecting Beetles

Little in the way of specialized or expensive equipment is required to observe and collect beetles. Some basic pieces of collecting equipment include a net, beating sheet, large kitchen strainer, headlamp, containers of various sizes, and a killing jar. Attracting beetles with baited and unbaited pitfall traps or a black light and sheet will also maximize your efforts in the field. Forceps made of spring aluminum, known as "featherweights," are extremely useful for picking up small beetles, while camel-hair brushes are used to probe for and dislodge beetles from their resting and hiding places. Aspirators are used suck up small beetles into a glass or plastic vial. Hand lenses are small and compact devices for revealing beetle anatomy and other details that might otherwise escape notice by the naked eye. Magnifications of 8× or 10× are ideal, with some units employing several lenses used in concert to increase magnification.

Always record the date, place, and collector for your specimens. These data will become the basis for labels that accompany each specimen in your collection. Beetles are mounted on special, rustproof pins and kept in boxes with tight-fitting lids to prevent damage from pests.

Since dead specimens reveal little of their lives, keep careful notes on time of day, temperature and humidity, plant or animal associations, behavior, or other important facts in a notebook. Anything that catches your eye is worth recording and may easily prove to be new to science. Whenever possible, record your observations in the field as they are happening. Never trust your memory for long; it is all too easy to confuse bits of information in time and place.

Responsible collectors are mindful of beetle populations and their habitats, never taking more specimens then they need. The purpose of responsible beetle collecting is to develop and maintain a reference collection that can be used for both scientific and educational purposes. Carefully prepared and documented beetle collections are important tools of scientific advancement, environmental education, and ultimately, conservation.

When on a collecting trip, take only the specimens you need. Because of the great numbers of most beetles, you need not worry that your collecting activities will adversely affect most populations. Field collecting should be selective and minimize trampling or other damage to the habitat and food plants. Collecting a large number of the same species at the same time or place seldom makes for a good reference collection.

Always ask for permission to collect on private and Native American lands, and be respectful of other naturalists and their activities. When visiting public lands (county, state, and national parks, state and national forests, monuments, recreational areas, etc.) it is advisable and often required to obtain written permission to collect beetles and other insects. Make sure all of your collecting activities are in compliance with regulations relating to the public and to individual species and their habitats.

Watching Beetles with Binoculars

Beetles, especially conspicuous flower-feeding and predatory species, even those living in waters with a still surface, can be easily observed through binoculars. Binoculars with close-focusing capability that allow you to focus on scenes 6 feet away or less are ideal for observing the details of beetle behavior. Even though many beetles are easily approached, viewing their magnified images through binoculars can be awe inspiring.

Keeping Beetles in Captivity

Captive beetles make excellent educational displays and are a sure-fire way of bringing life to a discovery room, classroom, or nature center. Captive beetles also provide a means of acquiring pictures of species or behaviors that are difficult or impossible to photograph in the field.

Captive beetles exposed to sudden increases in temperature die quickly, especially aquatic species. Never leave them in direct sunlight or unprotected in a closed car for any period of time. Carry an ice chest with one or more 2-liter bottles filled with frozen water for transporting live beetles from the field. Half-pint and pint-size plastic deli containers, available at supermarkets and restaurant supply chains, are ideal containers for housing beetles because they are inexpensive, light weight, and stack easily for packing. It is not necessary to punch air holes in the lid for a day trip, especially if the containers are kept cool. A piece of paper towel, some leaf litter, or a piece of moss, slightly moistened with water, will provide a bit of comfort for your animals and protect them from the jostling of travel. Never transport aquatic beetles in water; they will drown. Instead, put wet moss or moistened paper towels in the container.

To keep beetles in captivity it is important to know what they need in terms of food, water, and substrate. With these basic requirements in mind, you will have a better chance of duplicating their environment as much as possible and ensuring that they thrive and behave as naturally as possible.

Rearing Beetles

Late winter and early spring are good times to gather fungi, dead limbs, and rotten logs and stumps that contain beetle larvae and pupae. A large-mouth glass jar can serve as a practical and inexpensive rearing chamber for smaller beetles. Sprinkle the collected material with a bit of water from time to time to keep moisture levels up. Use cheesecloth secured with a rubber band, or window screen glued to a dome lid ring of a canning jar as a top to allow excess moisture to escape. Be sure to place your jar away from outside doors, windows, and heating and cooling ducts to avoid exposing your animals to extreme temperatures. The warmer indoor temperatures will accelerate their development.

Place larger dead limbs in sturdy plastic sweater boxes or inside large square metal cans. Seal the containers tight enough to prevent the escape of beetles when they emerge from the wood, while permitting air to enter and exit the container. Wrap the plastic boxes with black garbage bags to keep out light. Then cut a hole into the end of each container and insert a small jar or vial. Emerging beetles are drawn to the light and will find their way into the bottle. Be sure to add a bit of water from time to time so the rearing material does not completely dry out.

Conserving California Beetles

Habitat loss, not beetle collecting, is the greatest threat to California's beetle populations. Urban development, grazing, agriculture, pollution, and the introduction of exotic species all contribute to the local disappearance or extinction of the state's beetles. Indiscriminate use of pesticides in both urban and rural situations also unnecessarily kills millions of California beetles annually.

California is particularly susceptible to species loss because of the sheer number of beetles specifically adapted to living in its abundance of unique habitats. In a state as large, diverse, and unexplored as California, it is very likely that we have already lost species that we didn't even know existed. At least three species or subspecies of California beetles are already known to have become extinct in the last century and a half: the Oblivious Tiger Beetle (*Cicindela latesignata obliviosa,* Carabidae), the San Joaquin Valley Tiger Beetle (an undescribed subspecies of *C. tranquebarica*), and the Mono Lake Hygrotus Diving Beetle (*Hygrotus artus,* Dytiscidae).

Only four California beetles have been afforded governmental protection and are listed as threatened or endangered by the U.S. Fish and Wildlife Service: the Ohlone Tiger Beetle *(C. ohlone),* the Delta Green Ground Beetle (*Elaphrus viridis,* Carabidae), the Mount Hermon June Beetle (*Polyphylla barbata,* Scarabaeidae), and the Valley Elderberry Longhorn Beetle (*Desmocerus californicus dimorphus,* Cerambycidae). Nearly 60 additional species have been submitted for consideration. These and many other species living in sensitive habitats, especially those inhabiting coastal and desert sand dunes, require further study to determine if they are in need of state or federal protection.

How to Use This Book

To get the most out of your *Beetles of California*, carry it with you in your car or in your pack. When encountering a species of interest, most people flip through the illustrations in search of the closest match to what is in front of them. Since field guides covering birds, reptiles and amphibians, or butterflies usually cover all of the species within a given region, this method of use usually results in reliable identification to species. However, for reasons already mentioned, there is a real possibility that the beetle before you may not appear in the book. Don't be discouraged. You still have the opportunity to use the illustrated key to place your beetle to family, review the family accounts section, and then refer to the reference section for additional resources.

Although some distinctive beetles are readily identified on the basis of a photograph alone, many others will require the careful examination of antennae, mouthparts, male genitalia, and other features not easily seen without a hand lens or some other means of magnification. Serious students requiring identifications of these beetles are encouraged to consult the references and resources offered at the back of the book for further information.

The order of beetle families presented here follows that of *American Beetles* (Arnett and Thomas 2001; Arnett et al. 2002), which is based on interpretations of their evolutionary relationships as revealed through the physical features they have in common.

Family Identificaton

One hundred and fifteen of North America's 127 beetle families are known to occur in California. In addition to the illustrated key to families of California beetles, many beetles can also be placed to family based on the essays presenting 56 of the state's most conspicuous and commonly encountered families. Each essay includes information on natural history and physical features to help distinguish a family's adults and larvae from those of other families. Features of interest include color, overall body shape, length, orientation of mouthparts (prognathous or hypognathous), number of antennal segments, and antennal type(s). Also included is the width of the thorax in relation to the base of the elytra, whether or not the scutellum is visible, length and surface sculpturing of the elytra, tarsal formula, and number of abdominal segments. A similar diagnosis for the larvae is given. The presence or absence of projections, or urogomphi, on the ninth abdominal segment is noted. For supplementary images of California beetle larvae visit www.sbnature.org/calbeetles/fieldguidesuppl.

These diagnoses are intended to provide an overall impression of each family's more conspicuous California members based on characteristics easily seen in the field, and not to provide a detailed and definitive description for all species in the family. This is especially true for the larvae, for which no more than 5 percent have been reliably associated with an adult beetle.

Next is a listing of similar California families, a bulleted section that presents pertinent characteristics of similar families with which the family under discussion might be easily confused. Check the characteristics listed here against those of your specimen to verify that you have placed your beetle in the correct family.

The numbers of known California species and genera in each family are given not only to highlight the family's diversity in the state, but also to caution the reader about making snap species identifications. Next come paragraphs featuring different genera, each with one or more species. The diagnoses of each species consist of brief information on length, distribution, and season of adult activity. Natural history notes are given, if known.

ILLUSTRATED KEY TO FAMILIES OF CALIFORNIA BEETLES

The following key is designed as a guide to help place some of the more conspicuous species of California beetles in their proper families. By determining the family, or narrowing down the possibilities, you can turn to the appropriate section(s) of the book and compare your specimen with the photos and descriptions. Note that some beetle families are incredibly diverse in form and appear more than once in the key. Also note that it is beyond the scope of this book to provide a comprehensive key that will place all of California's 8,000 or so beetle species to family. If you would like more detailed information, consult the "Key to Families" in volume 2 of *American Beetles* (Arnett et al. 2002).

Families followed by * are not covered in this book.

1a Body small; abdominal segments partially covered by plates (hind coxae)
Crawling water beetles (Haliplidae)

1b Body variable in size; abdominal segments clearly visible
.... go to 2

2a First abdominal segment divided by the hind coxae; trochanter large and offset from the femur
Ground beetles, tiger beetles (Carabidae)
Predaceous diving beetles (Dytiscidae)
Whirligig beetles (Gyrinidae)
Wrinkled bark beetles (Rhysodidae)
Trout-stream beetles (Amphizoidae)
False ground beetles (Trachypachidae)*
Burrowing water beetles (Noteridae)*

2b First abdominal segment not divided by hind coxae
.... go to 3

3a Head with snout (a distinctly elongated region in front of eyes)
Weevils (Curculionidae)
Net-winged beetles (Lycidae)
Seed beetles (Bruchinae, Chrysomelidae)
Leaf-rolling weevils (Attelabidae)
Narrow-waisted bark beetles (Salpingidae)*
Pine weevils (Nemonychidae)*
Fungus weevils (Anthribidae)*
Straight-snouted weevils (Brentidae)*

3b Head without snout .... go to 4

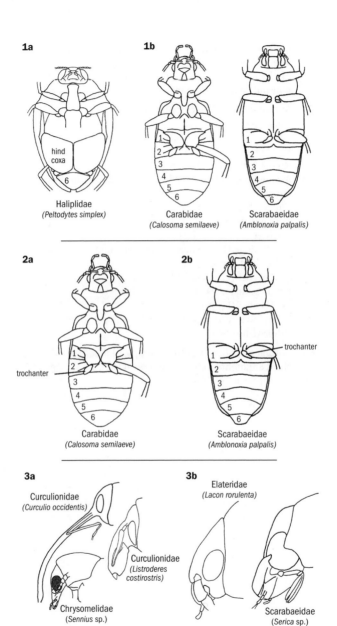

4a   Elytra short, exposing one or more complete segments, or elytra absent .... go to 5
4b   Elytra completely covering the abdomen or exposing only part of last segment .... go to 6

5a   Antennae comblike or feathery
Click beetles (Elateridae)
Glowworms (Phengodidae)
Fireflies and glowworms (Lampyridae)
Ripiphorid beetles (Ripiphoridae)

5b   Antennae beadlike or threadlike to slightly or distinctly clubbed
Carrion beetles (Silphidae)
Rove beetles (Staphylinidae)
Sap beetles (Nitidulidae)
Short-winged flower beetles (Kateretidae)*
Clown beetles (Histeridae)
Seed beetles (Chrysomelidae)
Fireflies and glowworms (Lampyridae)
Skiff beetles (Hydroscaphidae)*
False clown beetles (Sphaeritidae)*
Palmetto beetles (Smicripidae)*

6a   Antennae distinctly swollen at tips, clubbed, or with very short segments extended to one side .... go to 7
6b   Antennae not clubbed .... go to 16

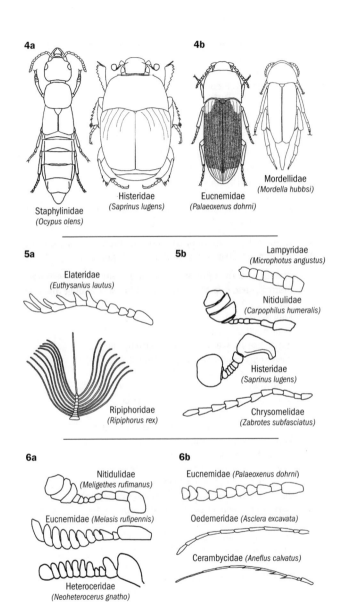

7a Last five or more antennal segments distinctly extended to one side
Scarab beetles (Scarabaeidae)
Rain beetles (Pleocomidae)
Cedar beetles (Rhipiceridae)
Long-toed water beetles (Dryopidae)
Variegated mud-loving beetles (Heteroceridae)
Water penny beetles (Psephenidae)
False click beetles (Eucnemidae)
Net-winged beetles (Lycidae)
Deathwatch beetles (Anobiidae)
Seed beetles (Chrysomelidae)

7b No more than four antennal segments extended to one side, if at all. . . . . go to 8

8a Antennal club asymmetrical or lopsided . . . . . . go to 9
8b Antennal club symmetrical, not lopsided . . . . . go to 10

9a Antennal club segments platelike, sometimes with segment at base cup-shaped and partially enveloping the next
Scarab beetles (Scarabaeidae)
Stag beetles (Lucanidae)
Hide beetles (Trogidae)
Earth-boring scarab beetles (Geotrupidae)
False stag beetles (Diphyllostomatidae)*
Enigmatic scarab beetles (Glaresidae)*
Sand-loving scarab beetles (Ochodaeidae)*
Scavenger scarab beetles (Hybosoridae)*
Bumblebee scarab beetles (Glaphyridae)

9b Antennal club without platelike segments
Water scavenger beetles (Hydrophilidae)
Bostrichid beetles (Bostrichidae)
Primitive carrion beetles (Agyrtidae)
Burying beetles, carrion beetles (Silphidae)
Deathwatch beetles, spider beetles (Anobiidae)
Bark-gnawing beetles (Trogossitidae)
Checkered beetles (Cleridae)
Handsome fungus beetles (Endomychidae)
Throscid beetles (Throscidae)*
Silken fungus beetles (Cryptophagidae)*
False skin beetles (Biphyllidae)*
Polypore fungus beetles (Tetratomidae)*
Jugular-horned beetles (Prostomidae)*
Conifer bark beetles (Boridae)*

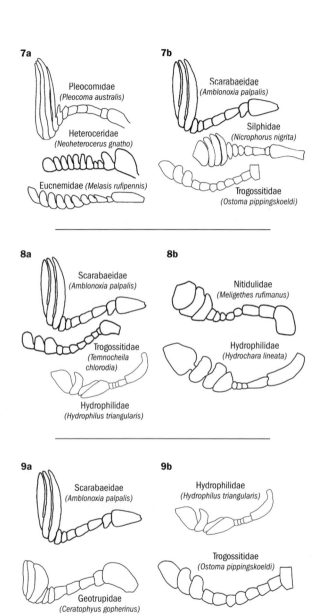

10a Head hidden from above or nearly so . . . . **go to 11**
10b Head clearly visible from above . . . . **go to 12**

11a Body compact, almost humpbacked in profile, with a strongly oval to nearly circular outline when viewed from above

Water scavenger beetles (Hydrophilidae)
Skin beetles (Dermestidae)
Lady beetles (Coccinellidae)
Darkling beetles (Tenebrionidae)
Leaf beetles (Chrysomelidae)
Clown beetles (Histeridae)
Primitive carrion beetles (Agyrtidae)
Burying beetles, carrion beetles (Silphidae)
Deathwatch beetles (Anobiidae)
Handsome fungus beetles (Endomychidae)
Round fungus beetles (Leiodidae)*
Minute beetles (Clambidae)*
Pill beetles, moss beetles (Byrrhidae)*
Minute fungus beetles (Corylophidae)*
Shining flower beetles (Phalacridae)*

11b Body more elongate

Skin beetles (Dermestidae)
Sap beetles (Nitidulidae)
Darkling beetles (Tenebrionidae)
Leaf beetles (Chrysomelidae)
Water scavenger beetles (Hydrophilidae)
Riffle beetles (Elmidae)
Long-toed water beetles (Dryopidae)
Bostrichid beetles (Bostrichidae)
Short-winged flower beetles (Kateretidae)*
Zopherid beetles (Zopheridae)
Antlike flower beetles (Anthicidae)
Bark beetles (Scolytinae, Curculionidae)
Throscid beetles (Throscidae)*
Cryptic slime mold beetles (Sphindidae)*
Shining flower beetles (Phalacridae)*
Fruitworms (Byturidae)*
Hairy fungus beetles (Mycetophagidae)*
Minute tree-fungus beetles (Ciidae)*
False darkling beetles (Melandryidae)*
False flower beetles (Scraptiidae)*

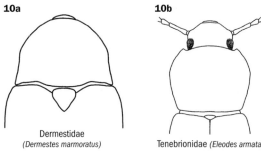

10a Dermestidae *(Dermestes marmoratus)*

10b Tenebrionidae *(Eleodes armata)*

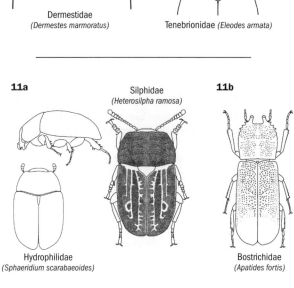

11a Hydrophilidae *(Sphaeridium scarabaeoides)*

Silphidae *(Heterosilpha ramosa)*

11b Bostrichidae *(Apatides fortis)*

12a Body long and slender
Bostrichid beetles (Bostrichidae)
Bark-gnawing beetles (Trogossitidae)
Checkered beetles (Cleridae)
Silvanid flat bark beetles (Silvanidae)
Flat bark beetles (Cucujidae)
Lizard beetles (Languriinae, Erotylidae)
Bothriderid beetles (Bothrideridae)
Zopherid beetles (Zopheridae)
Antlike flower beetles (Anthicidae)
Bark beetles (Scolytinae, Curculionidae)
False longhorn beetles (Stenotrachelidae)*
Lined flat bark beetles (Laemophloeidae)*
False antlike flower beetles (Pyrochroidae)*
Root-eating beetles (Monotomidae)*

12b Body not long and slender . . . . **go to 13**

13a Body compact, almost humpbacked in profile with a strongly oval to nearly circular outline when viewed from above
Water scavenger beetles (Hydrophilidae)
Sap beetles (Nitidulidae)
Darkling beetles (Tenebrionidae)
Leaf beetles (Chrysomelidae)
Short-winged flower beetles (Kateretidae)*
Pleasing fungus beetles (Erotylidae)
Minute bog beetles (Microsporidae)*
Round fungus beetles (Leiodidae)*

13b Body not humpbacked, more elongate. . . . . **go to 14**

12a

Trogossitidae
(Temnocheila chlorodia)

12b

Trogossitidae
(Ostoma pippingskoeldi)

13a

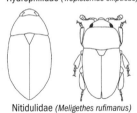

Hydrophilidae *(Tropisternus ellipticus)*

Nitidulidae *(Meligethes rufimanus)*

13b

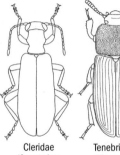

Cleridae
(Cymatodera pseudotsugae)

Tenebrionidae
(Uloma longula)

14a Pronotum at base narrower than base of elytra
　　Water scavenger beetles (Hydrophilidae)
　　Leaf beetles (Chrysomelidae)
　　Bostrichid beetles (Bostrichidae)
　　Checkered beetles (Cleridae)
　　Silvanid flat bark beetles (Silvanidae)
　　Flat bark beetles (Cucujidae)
　　Zopherid beetles (Zopheridae)
　　Antlike flower beetles (Anthicidae)
　　Minute moss beetles (Hydraenidae)*
　　Toothed-neck fungus beetles (Derodontidae)*
　　Lined flat bark beetles (Laemophloeidae)*
　　Minute brown scavenger beetles (Latridiidae)*
　　Narrow-waisted bark beetles (Salpingidae)*
14b Pronotum at base equal to base of elytra, or nearly so
　　............................... **go to 15**

15a Mandible directed downward (head hypognathous)
　　Lady beetles (Coccinellidae)
　　Darkling beetles (Tenebrionidae)
　　Leaf beetles (Chrysomelidae)
　　Minute moss beetles (Hydraenidae)*
　　Feather-winged beetles (Ptiliidae)*
　　Antlike stone beetles (Scydmaenidae)*
　　Minute marsh-loving beetles (Limnichidae)*
15b Mandibles directed forward (head prognathous)
　　Sap beetles (Nitidulidae)
　　Darkling beetles (Tenebrionidae)
　　Bark-gnawing beetles (Trogossitidae)
　　Handsome fungus beetles (Endomychidae)
　　Short-winged flower beetles (Kateretidae)*
　　Hyporhagus gilensis (Zopheridae)
　　Nosodendrid beetles (Nosodendridae)*
　　Minute bark beetles (Cerylonidae)*

16a Antennae saw-toothed, comblike, or feathery ....
　　.................................. **go to 17**
16b Antennae threadlike or beadlike .... **go to 19**

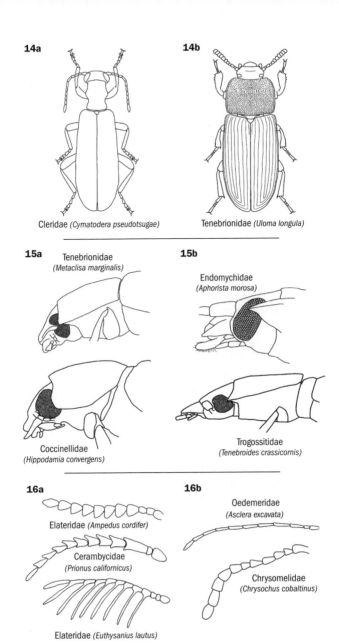

17a Head visible from above
 Click beetles (Elateridae)
 Longhorn beetles (Cerambycidae)
 Leaf beetles (Chrysomelidae)
 Cedar beetles (Rhipiceridae)
 Metallic wood-boring beetles (Buprestidae)
 False click beetles (Eucnemidae)
 Checkered beetles (Cleridae)
 False darkling beetles (Melandryidae)*
 Ripiphorid beetles (Ripiphoridae)
 Comb-clawed beetles (Tenebrionidae)
 Forest-stream beetles (Eulichadidae)*
 Armatopodid beetles (Artematopodidae)*
 Rare click beetles (Cerophytidae)*
 Texas beetles (Brachypsectridae)*
 False soldier beetles, false firefly beetles (Omethidae)*
 Fire-colored beetles (Pyrochroidae)*
 Palm beetles (Mycteridae)*

17b Head mostly or completely hidden from above . . . .
. . . . . go to 18

18a Elytra faintly to distinctly grooved or ridged
 Soft-bodied plant beetles (Dascillidae)
 Cedar beetles (Rhipiceridae)
 Schizopodid beetles (Schizopodidae)
 False click beetles (Eucnemidae)
 Net-winged beetles (Lycidae)
 Deathwatch beetles (Anobiidae)
 Comb-clawed beetles (Alleculinae, Tenebrionidae)
 Seed beetles (Chrysomelidae)

18b Elytra smooth, not grooved or ridged
 Leaf beetles (Chrysomelidae)
 Tumbling flower beetles (Mordellidae)
 Metallic wood-boring beetles (Buprestidae)
 Water penny beetles (Psephenidae)
 Fireflies and glowworms (Lampyridae)
 Soldier beetles (Cantharidae)
 Deathwatch beetles (Anobiidae)
 Soft-winged flower beetles (Melyridae)
 False darkling beetles (Melandryidae)*
 Marsh beetles (Scirtidae)*
 Throscid beetles (Throscidae)*

17a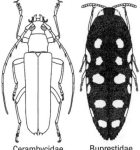

Cerambycidae
(Aneflus calvatus)

Buprestidae
(Acmaeodera gibbula)

17b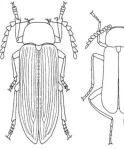

Dascillidae
(Dascillus plumbeus)

Melyridae
(Malachius sp.)

18a

Dascillidae
(Dascillus plumbeus)

18b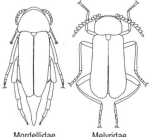

Mordellidae
(Mordella hubbsi)

Melyridae
(Malachius sp.)

19a Head visible from above . . . . go to 20
19b Head mostly or completely hidden from above . . . .
. . . . go to 22

20a Body antlike
  Rove beetles (Staphylinidae)
  Antlike flower beetles (Anthicidae)
  Antlike stone beetles (Pyrochroidae)*
  Antlike leaf beetles (Aderidae)*
20b Body not antlike . . . . go to 21

21a Pronotum ridged or keeled on sides
  Rove beetles (Staphylinidae)
  Click beetles (Elateridae)
  Darkling beetles (Tenebrionidae)
  Longhorn beetles (Cerambycidae)
  Leaf beetles (Chrysomelidae)
  Reticulated beetles (Cupedidae)
  Soft-bodied plant beetles (Dascillidae)
  Riffle beetles (Elmidae)
  Long-toed water beetles (Dryopidae)
  Water penny beetles (Psephenidae)
  False click beetles (Eucnemidae)
  Soldier beetles (Cantharidae)
  Silvanid flat bark beetles (Silvanidae)
  Flat bark beetles (Cucujidae)
  False soldier beetles, false firefly beetles (Omethidae)*
  Lined flat bark beetles (Laemophloeidae)*
  False darkling beetles (Melandryidae)*
  Dead log beetles (Pythidae)*
  Orsodacnid leaf beetles (Orsodacnidae)*
21b Pronotum rounded on sides, not ridged or keeled
  Rove beetles (Staphylinidae)
  Leaf beetles (Chrysomelidae)
  Checkered beetles (Cleridae)
  False blister beetles (Oedemeridae)
  Blister beetles (Meloidae)
  Darkling beetles (Tenebrionidae)
  False longhorn beetles (Stenotrachelidae)*
  Seed beetles (Chrysomelidae)
  False darkling beetles (Melandryidae)*
  Narrow-waisted beetles (Salpingidae)*
  Magalopodid leaf beetles (Megalopodidae)*
  Orsodacnid leaf beetles (Orsodacnidae)*

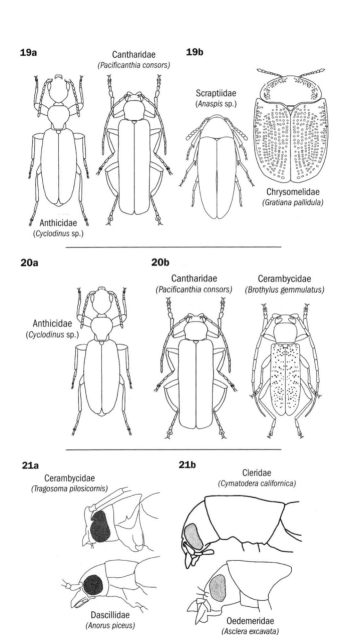

22a Body compact, oval to round in outline when viewed from above
   Leaf beetles, seed beetles (Chrysomelidae)
   Water penny beetles (Psephenidae)
   Deathwatch beetles, spider beetles (Anobiidae)
   Pill beetles, moss beetles (Byrrhidae)*
   Marsh beetles (Scirtidae)*

22b Body more elongate and straight-sided
   Rove beetles (Staphylinidae)
   Longhorn beetles (Cerambycidae)
   Leaf beetles (Chrysomelidae)
   Soft-bodied plant beetles (Dascillidae)
   Riffle beetles (Elmidae)
   Long-toed water beetles (Dryopidae)
   False click beetles (Eucnemidae)
   Net-winged beetles (Lycidae)
   Fireflies and glowworms (Lampyridae)
   Soldier beetles (Cantharidae)
   False flower beetles (Scraptiidae)*

22a   **22b**

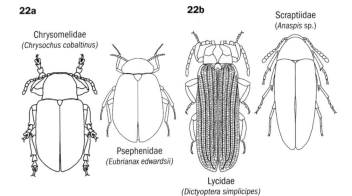

Chrysomelidae
(*Chrysochus cobaltinus*)

Psephenidae
(*Eubrianax edwardsii*)

Lycidae
(*Dictyoptera simplicipes*)

Scraptiidae
(*Anaspis* sp.)

FAMILY ACCOUNTS

Antlike Flower Beetles (Anthicidae)	231
Bark-gnawing Beetles (Trogossitidae)	174
Blister Beetles (Meloidae)	224
Bostrichid Beetles (Bostrichidae)	166
Bothriderid Beetles (Bothrideridae)	194
Bumblebee Scarab Beetles (Glaphyridae)	103
Carrion Beetles (Silphidae)	85
Cedar Beetles or Cicada Parasite Beetles (Rhipiceridae)	121
Checkered Beetles (Cleridae)	177
Click Beetles (Elateridae)	146
Clown Beetles (Histeridae)	80
Crawling Water Beetles (Haliplidae)	65
Darkling Beetles (Tenebrionidae)	212
Deathwatch Beetles and Spider Beetles (Anobiidae)	170
Earth-boring Scarab Beetles (Geotrupidae)	100
False Blister Beetles (Oedemeridae)	221
False Click Beetles (Eucnemidae)	144
Fireflies and Glowworms (Lampyridae)	156
Flat Bark Beetles (Cucujidae)	190
Glowworms (Phengodidae)	153
Ground Beetles and Tiger Beetles (Carabidae)	54
Handsome Fungus Beetles (Endomychidae)	196
Hide Beetles (Trogidae)	94

continued ➤

Lady Beetles (Coccinellidae)	198
Leaf Beetles and Seed Beetles (Chrysomelidae)	254
Leaf-rolling Weevils (Attelabidae)	264
Longhorn Beetles (Cerambycidae)	234
Long-toed Water Beetles (Dryopidae)	137
Metallic Wood-boring Beetles or Jewel Beetles (Buprestidae)	125
Net-winged Beetles (Lycidae)	151
Pleasing Fungus Beetles (Erotylidae)	191
Predaceous Diving Beetles (Dytiscidae)	70
Primitive Carrion Beetles (Agyrtidae)	83
Rain Beetles (Pleocomidae)	96
Reticulated Beetles (Cupedidae)	51
Riffle Beetles (Elmidae)	134
Ripiphorid Beetles (Ripiphoridae)	206
Rove Beetles (Staphylinidae)	88
Sap Beetles (Nitidulidae)	184
Scarab Beetles (Scarabaeidae)	105
Schizopodid Beetles or False Jewel Beetles (Schizopodidae)	123
Silvanid Flat Bark Beetles (Silvanidae)	188
Skin Beetles (Dermestidae)	162
Soft-bodied Plant Beetles (Dascillidae)	119
Soft-winged Flower Beetles (Melyridae)	182
Soldier Beetles (Cantharidae)	159
Stag Beetles (Lucanidae)	91
Trout-stream Beetles (Amphizoidae)	68
Tumbling Flower Beetles (Mordellidae)	204
Variegated Mud-loving Beetles (Heteroceridae)	140
Water Penny Beetles (Psephenidae)	142
Water Scavenger Beetles (Hydrophilidae)	75
Weevils or Snout Beetles (Curculionidae)	267
Whirligig Beetles (Gyrinidae)	63
Wrinkled Bark Beetles (Rhysodidae)	52
Zopherid Beetles (Zopheridae)	208

RETICULATED BEETLES   Cupedidae

A small and unusual family of primitive beetles, the Cupedidae contains only 30 species worldwide. They are considered to be relicts of a once more diverse group of beetles dating back to the Triassic more than 200 million years ago. Their elongate elytra are sculpted with square punctures, a feature that distinguishes this family from all other families of beetles.

Although uncommon in collections, a few species have been found in numbers in old logs, on dead branches, swarming in the sunlight near dense woods, at lights, or attracted to laundry bleach containing sodium hypochlorite. Both adults and larvae are found together boring into rotten wood beneath the bark of hard- and softwood logs or branches, preferring firm and moist wood previously attacked by fungi. Reticulated beetles also occur in fairly dry habitats and will attack basement timbers of old houses, but are not generally considered to be of economic importance.

IDENTIFICATION: Reticulated beetles are usually brown or tan, ranging in length from 10.0 to 20.0 mm. Their elongate, parallel-sided, and somewhat flattened body is covered with broad scale-like setae. The head is prognathous, with protruding eyes and a distinct neck. The antennae are thick and threadlike, consisting of 11 segments. The pronotum is narrower than the elytra, rough, parallel-sided, and more or less distinctly margined. The scutellum is not visible. The elytra are elongate, parallel-sided, and completely conceal the abdomen. The elytral surface is strongly ribbed and covered with rows of square punctures. The tarsal formula is 5-5-5; the claws are not toothed. The abdomen has five segments visible from below.

The known mature larvae are pale, long, slender, and cylindrical in shape, with short legs. The head is partially withdrawn into the prothorax and is prognathous. The antennae are four-segmented and one or more pairs of simple eyes are present or absent. The thoracic segments are less than one-fourth the length of the abdomen in later instars. The legs are six-segmented, including the claw. The last abdominal segment is enlarged and extends into a sharp conical single or double process.

SIMILAR CALIFORNIA FAMILIES:
- hispine leaf beetles (Chrysomelidae)—antennae short and clubbed; head narrower than pronotum
- net-winged beetles (Lycidae)—head not visible from above

CALIFORNIA FAUNA: Two species in two genera.

The adult *Prolixocupes lobiceps* (8.0 to 11.0 mm) (pl. 1) has been found under the bark of western sycamore *(Platanus racemosa)*, while all stages of development are known to occur in the decaying stumps of coast live oak *(Quercus agrifolia)*. The antennae of this species extend beyond half the length of the body. It is found in the foothills of southern California and also occurs in Arizona.

Priacma serrata (10.0 to 22.0 mm) (pl. 2) is found on dead branches of white fir *(Abies concolor)* and is distributed throughout the Sierra Nevada and the Transverse and Peninsular Ranges. The antennae are scarcely half the length of the body. The adult male is attracted to clothes freshly laundered with bleach in areas of mixed montane forest. The highly excited nature of the males lured to this bait suggests that bleach is chemically similar to the female's sexual pheromone. This species also occurs in Idaho, Montana, Oregon, Washington, and British Columbia.

COLLECTING METHODS: Both California species are active in late spring and summer. Adult beetles are collected by chopping into old logs and stumps, netting in flight near infested wood, beating from dead branches, or by attraction to lights at night. Males of *Priacma serrata* are attracted to sheets freshly laundered with bleach. *Prolixocupes lobiceps* has been collected in the early evening by sweeping low vegetation.

WRINKLED BARK BEETLES   Rhysodidae

Wrinkled bark beetles are represented by eight species in North America. Both species occurring in the western part of the continent occur in California. Adults and larvae are found in moist fallen logs, stumps, roots, and dead limbs infested with slime molds, especially on beech *(Fagus)*, ash *(Fraxinus)*, elm *(Ulmus)*, Douglas-fir *(Pseudotsuga menziesii)*, and pine *(Pinus)*.

The larvae are slow and grublike and feed only on liquids inside rotten logs. They do not create galleries. Pupation occurs in the wood. Adults also push their way through rotten wood and sometimes hibernate in groups over the winter.

The mandibles of adults are incapable of chewing and are configured in such a way that they protect the other mouthparts. Instead they use part of their maxillae to feed, presumably on slime molds. They are not considered to be of economic importance since they attack only rotten wood.

Some researchers consider wrinkled bark beetles to be highly specialized members of the ground beetle and tiger beetle family (Carabidae).

IDENTIFICATION: California wrinkled bark beetles are reddish brown or black and range up to 10.0 mm. Their bodies are elongate, narrow, and parallel sided. The head is prognathous and distinctively grooved, with a conspicuous neck. The antennae are beadlike and consist of 11 segments. The pronotum has one or three distinct lengthwise grooves. The scutellum is not visible. The elytra are distinctly grooved or punctured and completely conceal the abdomen. The abdomen has five segments visible from below. The tarsal formula is 5-5-5, with all claws equal in size and simple.

The larva is yellowish, grublike, and somewhat flattened, with very short legs. The head lacks simple eyes and the mouthparts are directed forward. The legs are six-segmented, including the claw. Both the thorax and abdomen lack well-defined, darkened plates. The abdomen is 10-segmented, with the last segment forming a false foot. The segments are nearly equal in length and posses a transverse row of spines.

SIMILAR CALIFORNIA FAMILIES:
- cylindrical bark beetles (Colydiinae, Zopheridae)—antennae clubbed; tarsi 4-4-4
- bothriderid beetles (*Deretaphrus,* Bothrideridae)—antennae clubbed; single interrupted pronotal groove in middle; tarsi 4-4-4

CALIFORNIA FAUNA: Two species in two genera.

The Wrinkled Bark Beetle *(Omoglymmius hamatus)* (6.2 to 6.8 mm) (pl. 3) is reddish brown to black and is known from the Sierra Nevada, Klamath Mountains, and Transverse Ranges, where it lives in Douglas-fir and pine logs. The elytra are not deeply grooved but are covered in longitudinal lines of deep pits. The grooves on both sides of the pronotum are long, extending nearly its entire length. The adult can be found virtually year-round. It also occurs in Washington, Idaho, and Oregon.

Clinidium calcaratum (5.8 to 8.1 mm) lives in the northern Coast Ranges and Sierra Nevada, where it is found on rotten logs of Douglas-fir and other conifers. The adult can be found virtually year-round. The elytra are deeply grooved throughout their length, while only the middle pronotal groove is long. This species also occurs in the humid forests of coastal British Columbia, Washington, and Oregon.

COLLECTING METHODS: Look for adults under the loose bark of dead conifer logs and stumps. Wrinkled bark beetles prefer wood that is moist and partially decayed, either as logs resting on the soil or in the roots or lower parts of dead stumps. They are usually not taken in wood that is crumbling or filled with the tunnels of other wood-boring insects. Look for larvae in soft, somewhat spongy wood under slime mold fruiting bodies. The presence of yellow slime-mold plasmodia is a good indicator of places to search. A hatchet or screwdriver may be necessary to split firm, but rotten wood.

GROUND BEETLES and TIGER BEETLES Carabidae

Ground beetles and tiger beetles are among the most common and frequently encountered beetles. They are quite diverse, comprising the third largest family of beetles in the world, behind weevils (Curculionidae) and rove beetles (Staphylinidae). With long legs and powerful jaws, they are superbly adapted for hunting and killing insect prey. Both larvae and adults are considered largely beneficial because they prey on caterpillars, grubs, grasshoppers, snails, and other small invertebrates considered pests. Most tiger beetles hunt during the warmest part of the day, while many ground beetles hunt at dusk or at night. A few species may occasionally consume fruits and seeds. Both ground beetles and tiger beetles run swiftly when disturbed and often release a stream of noxious fluid from their anus, sometimes with amazing accuracy. Pygidial glands located at the tip of their abdomens produce a variety of hydrocarbons, aldehydes, phenols, quinones, esters, and acids that combine to produce smelly and sometimes caustic fluids.

Many species of ground beetles lay individual eggs in small divots dug in the soil. The females of several genera *(Chlaenius, Brachinus, Pterostichus,* and *Calosoma)* lay their eggs in specially constructed cells made of mud, twigs, and leaves. The larvae molt three times before pupating. Ground beetle larvae are generally active predators that crawl on the ground or lurk beneath bark.

Tiger beetles, with their distinct shape and bulging eyes, were once placed in their own family, the Cicindelidae, but are now considered a subfamily of the Carabidae. Their bright, shiny colors and distinct markings help them blend into their background, making it difficult for potential predators to track them. The body patterns of some species may mimic the appearance of blistering or stinging insects, or help them to regulate their body temperature. They are opportunistic daytime hunters that will eat anything they can capture, including woodlice, moths, ants, and flies. In spite of their speed, tiger beetles do fall prey to robber flies and dragonflies, as well as hawks, Killdeer *(Charadrius vociferus)*, toads, lizards, shrews, and Raccoons *(Procyon lotor)*. They are also parasitized by bee flies *(Anthrax)* and erythraeid mites.

Tiger beetle larvae live in vertical burrows they dig themselves in fine or sandy soils. The long cylindrical body of a tiger beetle larva is anchored inside the burrow with abdominal hooks. Sensing the vibrations of an insect near the burrow's entrance, the larva lunges outside the burrow toward its target, often exposing up to half its body length. If successful, it grabs its prey with large scythelike jaws and drags it back into the burrow, where it can dine in relative safety.

IDENTIFICATION: Ground beetles are small to large, elongate beetles with a flattened, generally hairless body tapered at both ends. They are usually uniformly dark and shiny, but some species are brownish or pale. A few species are metallic, bi- or tricolored, or have elytra brightly marked with patterns of yellow or orange. Tiger beetles often have bold white or yellow markings on otherwise metallic elytra. California ground beetles range in length up to 34.0 mm, whereas the largest tiger beetle reaches 28.0 mm. The head is prognathous, with 11-segmented threadlike or beadlike antennae. The pronotum is usually narrower than the elytra. In ground beetles the sides of the pronotum are sharply margined, less so in tiger beetles. The scutellum is visible except in round sand beetles *(Omophron)*. The elytra are always present, completely covering the abdomen, and are fused in flightless species. The elytral surface is smooth, pitted, or grooved. Flight wings range from fully developed to absent. The tarsal formula is 5-5-5, with claws equal and simple, saw-toothed, or comblike. The abdomen has six segments visible from below, except for the genus

Brachinus, which has seven or eight. The first abdominal segment is partially divided by the hind coxae, and the trochanter is large and offset from the femur.

The yellow, brown, reddish brown, or black larvae are almost cylindrical or slightly flattened. The head is distinct and often has prominent mandibles directed forward. The antennae are as long as or longer than the mandibles and are usually four-segmented. Most species have six simple eyes on each side of the head. The thoracic segments have a thickened plate above, with a distinct groove running down the middle. The legs are long and usually six-segmented, each with one or two claws. The 10-segmented abdomen has a pair of dorsal projections on segment nine, except in the tiger beetles *(Amblycheila, Cicindela, Omus,* and *Tetracha).* The tenth segment is prolonged and usually functions as a foot.

SIMILAR CALIFORNIA FAMILIES:
- false tiger beetles (Othniinae, Salpingidae)—small (5.0 to 9.0 mm); antennae clubbed
- darkling beetles (Tenebrionidae)—tarsi 5-5-4; antennae sometimes gradually clubbed
- comb-clawed beetles (Alleculinae, Tenebrionidae)—tarsi 5-5-4; claws saw-toothed
- narrow-waisted bark beetles (Salpingidae)—tarsi 5-5-4

CALIFORNIA FAUNA: Approximately 677 species in 91 genera.

The round sand beetles *(Omophron),* are distinctively oval and convex, with dark markings on a yellowish background. They are found worldwide, with six species in California. Both the larvae and adults live in damp sand along lakes, rivers, and streams. Round sand beetles run swiftly but do not fly. *Omophron dentatus* (5.1 to 7.1 mm) (pl. 4) is the most widespread of the California species, known from southern California and Arizona to just north of San Francisco. It is somewhat more elongate than other species in the genus. The head has a pale, weakly M-shaped patch. *Omophron tesselatus* (5.4 to 7.0 mm) has a distinctive M-shaped patch on its head. One of the most widespread species in the genus, it is found in southeastern Canada and the northeastern, central, and southwestern United States, including northern California.

Twenty-three species of *Calosoma* occur in California. In the deserts they are often attracted to lights by the hundreds, where they feed on crushed insects. When disturbed, *Calosoma* produces a strong, disagreeable odor. The common Black Calosoma *(C. semilaeve)* (20.0 to 27.0 mm) (pl. 5) is common throughout western United States and can be extremely numerous in some years in the Great Central Valley and Mojave Desert in spring. Another common black desert species, *C. parvicollis* (20.0 to 28.0 mm), has a faint bluish luster at the base of the pronotum and the elytral markings near the humeri. It is active in spring, hunting night and day for caterpillar prey. In the Coachella Valley this species attacks the larvae of the White-lined Sphinx Moth *(Hyles lineata)* as they feed on sand verbena *(Abronia)*. *Calosoma latipenne* (14.0 to 20.0 mm) occurs in the Central Valley and western Mojave Desert. The prothorax of the Fiery Searcher *(C. scrutator)* (30.0 to 35.0 mm) (pl. 6) is steel blue, and the grooved elytra are metallic green. The head, prothorax, and elytra are all trimmed in red or gold. This striking species is found throughout the United States and southern Canada. In California it appears to be most abundant in the Great Central Valley and the southern coastal plain in late spring and summer. This caterpillar hunter runs on the ground or climbs shrubs in search of its prey.

Seventeen species of snail eaters *(Scaphinotus)* live in California. They have powerful jaws, a long narrow head and prothorax, and large abdomen. Their narrow heads are apparently adapted for feeding on snails through the shell opening. Most California snail eaters live in coastal woods and wet mountain canyons under cool and moist conditions. They sometimes form large aggregations beneath the bark of dead trees or under logs in summer, apparently in an effort to remain in favorable conditions. The shiny black *S. punctatus* (14.0 to 25.0 mm) (pl. 7) lives in chaparral and pine forests of the Transverse and Peninsular Ranges, where it feeds on berries, moth larvae, and other ground beetles. *Scaphinotus cristatus* (11.0 to 24.0 mm) is black or brown with a small crest on the middle of its head and a deep furrow behind each eye. It lives along coastal California from Del Norte County south to Point Sur, Monterey County. *Scaphinotus striatopunctatus* (14.0 to 24.0 mm) is very dark brown or black, with 18 or fewer grooves running the length of its elytra. It occurs in the coastal counties of California but is most abundant in the vicinity of San Francisco Bay. *Scaphinotus ventricosus* (15.5 to 20.0 mm) has elytra with more than 20 grooves and is found in the Coast Ranges, especially around Alameda, San Francisco, and San Mateo Counties, as well as the central Sierra Nevada. These and other species in northern California feed on the small Gray Garden Slug *(Agriolimax agrestis).*

The genera *Amblycheila* and *Omus* are small eyed and flightless. These crepuscular and nocturnal hunters are black, have fused elytra, and lack hind wings. The Mojave Giant Tiger Beetle *(A. schwarzi)* (21.0 to 28.0 mm) (pl. 8) is the only species of the genus in California. Described from a single Arizona specimen in 1893, only two or three more specimens were known until 1939. Encountered from spring through late summer, it prefers habitats dotted with large granite boulders or traversed by dry sandy washes in the Great Basin and Mojave Desert mountains of southern and eastern California. Individuals are particularly active on nights following midafternoon rains. It is also known from northwestern Arizona, southern Nevada, and extreme southeastern Utah.

Three species and subspecies of *Omus* occur in forested areas of California. The California Night-stalking Tiger Beetle *(O. californicus)* (12.0 to 21.0 mm) (pl. 9) is black with a shiny, wrinkled pronotum and domed, rough elytra. It is active in winter and spring and frequents redwood groves along streams in the Coast Ranges and also occurs in the western foothills of the Sierra Nevada. This species thrives in steep, ivy-covered slopes in the cities of San Francisco and Monterey. Auduoin's Night-stalking Tiger Beetle *(O. auduoini)* (13.0 to 18.0 mm) is a summer species that occurs from British Columbia to California where it is restricted to the extreme northern Coast Ranges. It seems to prefer open and grassy, low coastal terrain and coastal bluffs. This dull black species also has domed elytra, but the corners of the pronotum are directed forward and distinctly curved downward. The Lustrous Night-stalking Tiger Beetle *(O. submetallicus)* (13.0 to 18.0 mm) is distinguished from all other species of *Omus* by having a row of black setae along the side of the pronotum. It is active in summer and is known only from Warthan Canyon on the eastern slope of the Diablo Range, where Monterey and Fresno Counties meet. It lives in open live oak *(Quercus)* woodland and gray pine *(Pinus sabiniana)* forest. The adult is sometimes found under leaf litter on the forest floor and under logs.

The robust Pan-American Big-headed Tiger Beetle *(Tetracha carolina)* (12.0 to 20.0 mm) (pl. 10), formerly in the genus *Megacephala,* is easily recognized by its bright metallic green colors with gold reflections and pale brownish yellow markings on the tips of the elytra. The legs, antennae, and the tip of the abdomen are creamy yellow, while the body is pinkish and iridescent. Broadly distributed throughout the southern United States, this tiger beetle is restricted in California to the Colorado Desert, especially in the vicinity of the Colorado River. The Pan-American Big-headed Tiger Beetle is active during summer at dusk and in the evening and is attracted to lights. During the day it rests under rocks, boards, and other objects. Like many other ground and tiger beetles, this species secretes noxious defensive compounds, including hydrogen cyanide, but this chemical arsenal does not pose a threat to humans or pets. This species ranges across the southern United States, reaching as far north as Virginia, and south to Chile.

Members of the genus *Cicindela* are most often found along sandy or gravelly streamsides, lakeshores, dry lake beds, or on coastal beaches. The white or cream-colored elytral markings are quite variable. The "normal" pattern consists of a crescent-shaped marking on each shoulder, called the humeral lunule; a middle band; and an apical lunule, a crescent-shaped marking at the apex of the elytra. The lunules and bands may or may not be connected by a marginal line. Twenty-six species and 17 subspecies occur in California. The Western Tiger Beetle *(C. oregona)* (9.0 to 14.0 mm) (pl. 11) is the most widespread species of *Cicindela* in western North America, ranging from Alaska to Baja California and east to Colorado and New Mexico, and is quite variable. In California it is typically bronze brown, with the transverse portion of the middle band concave toward the head. It is common during the warmer months on gravelly or sandy banks of streams throughout the state, except in the Colorado Desert. It also occurs along the coastal beaches of southern California and on the dark mud of estuaries and tidal mudlflats, occasionally venturing onto nearby dry saline flats. The California Tiger Beetle *(C. californica)* (11.0 to 17.0 mm) (pl. 12) has a broad marginal band ending in a narrow or rounded point. The state has two subspecies, both of which are commonly attracted to lights. The coppery or reddish brown *C. c. mojavi* occupies eastern Kern County and a narrow band of desert from the northwestern reaches of the Mojave Desert south through the Colorado Desert to the mouth of the Colorado River, with discontinuous populations in coastal salt marshes and adjacent beaches on both sides of the Gulf of California. The subspecies *C. c. pseudoerronea* is dark green to blue and is restricted to damp salt flats and river and stream basins in Death Valley. The White-striped Tiger Beetle *(C. lemniscata)* (7.0 to 9.0 mm) (pl. 13) is widespread in southwestern United States and northern Mexico but in California is restricted to the Colorado Desert. This species is nocturnal and lives along the banks of streams, ponds, dry lakebeds, or well away from water among grasses in sandy or clay soils, and is frequently attracted to lights. The upper surface of the body has a bright coppery sheen, with each elytron bearing a long white stripe running lengthwise along the entire length and curving inward to meet the suture at the wing tips. The Boreal Long-lipped Tiger Beetle *(C. longilabris)* has a boreal distribution, ranging from southeastern Alaska across to southeastern Canada, the Great Lakes states, and New England. In western North America it is found in the Rockies and Washington south through the Cascades of Oregon. In California the bright metallic green or bronze subspecies *C. l. perviridis* (12.0 to 13.0 mm) is found at higher elevations in the northern Sierra Nevada and Cascades along paths and in open areas during late spring and summer.

The Big-headed Ground Beetle *(Scarites subterraneus)* (16.0 to 25.0 mm) (pl. 14) is widespread throughout North America. This distinctive species is shiny black and elongate. The combined length of the head and pronotum is nearly equal to the length of the elytra. The pronotum is attached to the rest of the body by a thin, cylindrical "waist." The strong, rakelike front legs are adapted for digging. It is usually found in moist areas beneath objects in cultivated fields and other disturbed areas and is occasionally attracted to lights.

The genus *Harpalus* occurs worldwide, except in South America and Australia, with more than 110 species cataloged thus far. Ten species are known from California. The Murky Ground Beetle *(H. caliginosus)* (17.5 to 35.5 mm) has a black body with shining elytra that are deeply grooved, and reddish yellow to reddish brown mouthparts and antennae. This species is widely distributed in North America and throughout California. Adults are found in a wide variety of habitats. It is commonly attracted to light and is recorded to feed on both plant and animal tissues.

Members of the genus *Pterostichus* are similar in body form to those of *Harpalus* and are among the most frequently encoun tered ground beetles in California. *Pterostichus lama* (18.0 to 30.0 mm) (pl. 15) is among the largest ground beetles in the state and is found in northern California, the Sierra Nevada, and the Transverse and Peninsular Ranges. The other *Pterostichus* in California are much smaller. For example, *P. californicus* (12.0 to 17.5 mm) is similar in color and shape but is only half the size.

Frequently encountered members of two other related genera bear a superficial resemblance to *Pterostichus*. *Anchomenus funebris* (8.1 to 9.8 mm) (pl. 16) is extremely common throughout the state, where it is found under rocks along the shores of streams and rivers. *Laemostenus complanatus* (14.0 mm) (pl. 17) is also widespread and is found under rocks in fields along the west coast from southern British Columbia to southern California. Both of these species lack the folded appearance to the apical portion of the lateral margin of the elytra that is found on *Pterostichus*.

The brightly colored and foul-smelling genus *Chlaenius* is distributed worldwide. These beetles feed on a wide range of plant and animal materials. Thirteen species and two subspecies are known from California. The elytra of the False Bombardier Beetle *(C. cumatilis)* (11.9 to 14.5 mm) (pl. 18) range from blue or blue black to violet tinged with bluish green. The slender legs are reddish or orange. The False Bombardier lives throughout California, especially along the coast, and is also known from Arizona and Baja California. *Chlaenius sericeus viridifrons* (11.4 to 16.1 mm) is bright metallic blue tinged with green. It lives throughout much of southern California, the Coast Ranges north to San Francisco, the San Joaquin Valley, and the Sierra Nevada north to Sequoia National Park. Additional subspecies and forms occur in other parts of northern California.

The Tule Beetle *(Tanystoma maculicolle)* (9.0 mm) (pl. 19) is common from western Oregon to Baja California. In California it is most abundant west and south of the Sierra Nevada. The wing covers are distinctively marked with pale brown borders and a dark brown center. It occupies a variety of habitats ranging from grassland to chaparral. Reproduction occurs in winter, and the larvae develop during the wet season. By late summer the populations may become large enough to become a nuisance in the Great Central Valley and elsewhere.

Of the 40 species of *Brachinus* occuring in the United States and Canada, nine live in California. They are called bombardier beetles because of their explosive defense system. A mixture of hydroquinones and hydrogen peroxide is produced by a pair of abdominal glands and stored in two reservoirs near the end of the abdomen. When disturbed, the beetle releases these chemicals into a reaction chamber at the tip of the abdomen, where they mix with enzymes that catalyze a violent chemical reaction that explodes out of the body with an audible "pop," producing a small yet potent cloud of acrid spray of benzoquinones heated to the boiling point. This noxious mist is delivered with startling accuracy with the aid of a flexible abdominal turret. The sound and "smoke" both serve as warnings to predators. The liquid will burn human skin for just an instant, leaving a dark brown color that persists for a few days. *Brachinus* are seldom found far from water, preferring to live beneath objects along streams, rivers, and lakes. All species have amber or reddish bodies with blue elytra. *Brachinus costipennis* (5.0 to 8.0 mm) is widely distributed in the state and lacks dense pubescence on the elytra. Of the species with elytral pubescence, *B. mexicanus* (7.0 to 9.0 mm) (formerly known as *B. fidelis*) is also widely distributed in California and has antennal segments three to five black. *Brachinus pallidus* (7.0 to 9.0 mm) has entirely pale reddish antennae and is found primarily in northern California. The pronotal surface of *B. favicollis* (9.5 to 10.5 mm) (pl. 20) is deeply punctured and rough. It is restricted to the coastal plain and adjacent canyons of southern California. The larvae of both *B. mexicanus* and *B. pallidus* have been found under stones feeding on the pupae of water scavenger beetles (Hydrophilidae).

COLLECTING METHODS: Ground beetles are often attracted to lights at night. Search for them during the day or night beneath rocks, boards, logs, and leaves, especially along the banks of ponds, streams, and rivers. Species living along the edges of streams and rivers can be flushed from their hiding places in mud and sand by pouring or splashing water on the shore. Beating and sweeping vegetation, especially at night, along with pitfall traps baited with meat, may also produce some specimens.

Tiger beetles of the genus *Cicindela* are alert and can be a challenge to collect. They are extremely fast runners and take flight quickly when threatened, always landing to face their attacker. *Cicindela* species are active on hot, sunny days during spring, summer, and fall. Look for them in sunny, open places along sandy roads, ocean beaches, lakeshores, riverbanks, and mudflats. During the day, walk slowly through their habitat and watch for beetles taking to the air. Then approach them cautiously and quickly clap down the net on top of them and wait for them to climb up into the net. Carefully reach into the net and gently grab them with thumb and forefinger. Nets with larger diameters and thin, flexible rims reduce the chances of beetles escaping. On cloudy or rainy days, or at night, these species are found in burrows under stones, or beneath bark. A few tiger beetles are active at night and are attracted to lights, particularly in the deserts. *Omus* and *Amblycheila* are collected by hand while they hunt at dusk and at night or are captured in pitfall traps fitted with runners or placed along the edges of boulders. A 30 cm cardboard square supported 5 to 10 cm above a pitfall trap with 15 cm nails pushed completely through each corner of the square and anchored several centimeters into the ground will attract shade-seeking tiger beetles living on salt flats.

WHIRLIGIG BEETLES Gyrinidae

Whirligigs are readily distinguished from other beetles by their flattened bodies, short, paddlelike middle and hind legs, and extraordinary divided compound eyes. They are the only group of beetles that regularly uses the surface film of water for support. Whirligigs live on the surface of ponds or along the edges of slow-moving streams. They are sometimes found in cattle tanks, canals, swimming pools, even rain puddles. Moving about singly or in large groups, whirligigs scour the water's surface in search of food and mates. Recently emerged adults sometimes congregate by the dozens or hundreds in shady or sheltered spots in late summer and fall.

The rigid, streamlined bodies of whirligigs meet little resistance on the water. The divided compound eyes are especially adapted for life on the water's surface. The lenses of the upper and lower eye are specifically designed for gathering images in the air and water, respectively. The raptorial front legs of both sexes are used to grasp dead or dying insects floating on the water. In males, the raptorial front legs are also used to grapple with smooth and slippery mates. Propelled by their paddlelike middle and hind legs, whirligigs steer with their rudderlike abdomen. They are good fliers and are sometimes attracted to lights at night. Like many other aquatic beetles, whirligigs breathe air stored beneath the elytra.

The name "whirligig" was bestowed upon these animals by virtue of their wild gyrations when disturbed. They are equally at home under water and will occasionally dive beneath the surface for brief periods to escape from harm. Incredibly buoyant, whirligigs must cling to underwater objects to remain submerged. Adults also produce defensive secretions from the tip of the abdomen that have a pungent odor suggesting apples, inspiring the nicknames "apple bugs," "apple smellers," and "mellow bugs." The secretion is not only distasteful to fish, amphibians, birds, and other predators, but it may also serve as an alarm pheromone to alert nearby whirligigs of possible danger. This secretion also propels them through the water by lowering the surface tension. Whirligigs are thought to ride on the wave of water molecules recoiling from the chemical.

Eggs are laid in rows, end to end in clusters, on submerged plants. The predatory larvae crawl about the bottom debris of ponds and streams in search of immature insects and other small invertebrates. They breathe directly through their body wall with the aid of threadlike feathery gills attached to the sides of their bodies. Mature larvae pupate within a case constructed of sand and debris and are located on the shore *(Dineutus)* or attached to plants above the water surface *(Gyrinus)*. Parasitic wasps are known to attack the pupae.

IDENTIFICATION: Adult whirligigs are shiny or dull black beetles ranging in length from 3.0 to 15.0 mm. Their body is oval, flattened, and streamlined, with the combined margins of their head, thorax, and elytra forming a continuous outline. The head is prognathous, and the short, clubbed antennae consist of eight to 11 segments. The compound eyes are distinctly divided. The scutellum may or may not be visible. The elytra are smooth and do not completely conceal the abdomen. The front legs are normally developed for grasping. The middle and hind legs are flattened and paddlelike. The tarsal formula is 5-5-5. The abdomen has six segments visible from below.

The white or yellowish brown larvae are elongate and slightly flattened. The distinctive head is prognathous and bears a pair of four-segmented antennae and six pairs of simple eyes. A thick plate covers the first thoracic segment. The legs are six-segmented, each with two claws. The last two segments of the 10-segmented abdomen are relatively narrow. Segments one through eight each have a pair of threadlike or feathery gills projecting from the sides. The ninth segment lacks gills but has four thick, curved hooks.

SIMILAR CALIFORNIA FAMILIES: The eyes, legs, and habits of whirligigs distinguish them from all other beetles.

CALIFORNIA FAUNA: Nine species in three genera.

Dineutus solitarius (9.0 to 10.0 mm) (pl. 21), a species described from Mexico, was recorded in California from only four specimens collected in August 1934 from Mecca, in Riverside County just north of the Salton Sea. Species of *Dineutus* are distinguished from the other genera of California whirligigs by their large size and hidden scutellum.

Gyrinus is the most commonly encountered genus of whirligig in California, with seven species recorded from the state. They range in size from 3.4 to 7.5 mm and are distinguished from other California genera by the presence of a visible scutellum. *Gyrinus plicifer* (5.0 to 6.5 mm) (pl. 22) is widely distributed in California, where it is usually found in foothill and mountain streams.

Gyretes californicus (3.0 to 5.0 mm) (previously known as *G. sinuatus*) has been found on the Colorado River. Like *Dineutus,* the scutellum in *Gyretes* is also hidden. In addition to their small size, adult *Gyretes* are readily distinguished from those of *Dineutus* by the presence of thick pubescence on the side margins of the pronotum and elytra.

COLLECTING METHODS: Rapidly sweeping an aquatic net through groups of beetles on the surface of ponds and streams is the best way to collect specimens. Individuals may dive beneath the water to avoid capture but can be collected moments later as they resurface. Some whirligigs are readily attracted to lights at night.

CRAWLING WATER BEETLES   Haliplidae

Crawling water beetles resemble small, loosely built predaceous diving beetles (Dytiscidae). However, they are easily distinguished from all other aquatic beetles by their enlarged coxal plates that obscure most of the abdomen. Crawling water beetles are found along the edges of standing or slow-moving freshwater habitats. They prefer permanent sources of clean water, but some species are found in nutrient-rich ponds and lakes, slow backwaters of streams, and vernal pools. A few species are permanent residents of saline waters, but none are known to have adapted to hot springs.

Although known as crawling water beetles, they are fair swimmers, propelling themselves through the water by moving their legs alternately instead of in unison. The long hairs on the hind legs increase their effectiveness as oars. Adults are usually found crawling over and feeding on mats of stringy green algae, but they also eat insect eggs and freshwater sponges. In captivity they will scavenge dead insect larvae and crustaceans.

Crawling water beetles use the space under their elytra to trap a bubble of air and keep it in direct contact with their spiracles. The bubble becomes a physical gill and expands to include the space between the oversized coxal plates of the hind legs and abdomen. The bubble serves not only as an oxygen supply, but also as a means of regulating the beetle's buoyancy. Equipped with a bubble of air, the crawling water beetle can easily rise to the surface to obtain more air. Beetles without an air bubble must laboriously climb or swim to the surface and may have difficulty in breaking through the surface film.

Eggs are laid on or inside submerged plant tissues. The larvae are most commonly found feeding on algae. First- and second-instar larvae respire directly through their body wall, but later instars have spiracles in preparation for pupation on land. The larvae molt three times before reaching the pupal stage. Mature larvae crawl on shore to dig a pupal chamber, usually beneath stones or logs. Crawling water beetles are active year-round, and most probably overwinter as adults.

IDENTIFICATION: Crawling water beetles are yellowish or brownish, with black spots, and range in length from 1.7 to 4.5 mm. Their body is oval and broadly tapered at each end. The head is prognathous. The antennae are 11-segmented; the first two segments are short and broad, while the rest are longer and threadlike. The prothorax is widest at its point of contact with the elytra and keeled below. The scutellum is not visible. The elytra completely conceal the abdomen and are covered with rows of punctures. The abdomen below is at least partly concealed by the large, flattened coxal plates of the hind legs.

The yellowish or reddish brown larvae are long, slender, and almost cylindrical in shape. The body is coarsely sculptured with small bumps *(Apteraliplus, Brychius,* and *Haliplus)* or festooned with long threadlike tracheal gills *(Peltodytes)*. The head has the mouthparts either hypognathous or somewhat prognathous. There are six pairs of simple eyes, and the antennae are four-segmented. The thoracic and abdominal segments are thickly armored across the back. The legs are six-segmented, including the claw. The abdomen is 10-segmented. In *Peltodytes*, segments one through eight each have two pairs of elongate tracheal gills on the back; segments nine and 10 each have one pair. *Apteraliplus*, *Brychius*, and *Haliplus* lack gills, and the last segment is prolonged into an almost spinelike process that is either forked at the tip or not.

SIMILAR CALIFORNIA FAMILIES:
- burrowing water beetles (Noteridae)—abdominal segments all visible from below, not covered by expanded coxal plates
- predaceous diving beetles (Dytiscidae)—abdominal segments all visible from below, not covered by expanded coxal plates

CALIFORNIA FAUNA: 14 species in four genera.

The hind coxal plates of *Haliplus* (1.7 to 5.0 mm) are shorter, exposing the last three abdominal segments, and are not bordered by a groove. Species in this genus are found along the edges of small ponds, lakes, or slow-moving streams. Nine species are known in California.

Both species of California *Peltodytes* (3.0 to 4.0 mm) (pl. 23) have two distinct black spots on the pronotum and hind coxal plates bordered by a groove that reach the last abdominal segment. They occur in large numbers, preferring nonacidic ponds, sheltered parts of lakes, and pools along streams and rivers that support algae.

The tiny *Apteralipus parvulus* (1.5 to 2.5 mm) is a spindle-shaped, humpbacked species that lives in temporary ponds and vernal pools. The adult appears during spring in temporarily flooded areas that are usually dry for most of the year. This species is found in the northern half of California, including the region east of the Sierra Nevada. It is also known from Washington and Oregon. The wings of this flightless species are reduced to mere pads. Described on the basis of characteristics associated with reduced wings, some researchers believe that *A. parvulus* is a highly derived member of the genus *Haliplus*. Because of the ephemeral and disturbed nature of its habitat, it has been suggested that *A. parvulus* should be considered for federal listing as an endangered species.

The pronotum of *Brychius* (3.5 to 4.5 mm) has nearly parallel sides, giving the body a distinct bell shape. They prefer the clean waters of creeks, streams, and rivers with rocky or gravelly bottoms, and the wave-swept shores of lakes. Two species occur in California, both in the northern part of the state.

COLLECTING METHODS: *Haliplus* and *Peltodytes* normally occur in weedy ditches, ponds, and lakes, whereas *Apteraliplus* is found only in vernal pools. Sweeping with an aquatic net through shallow, vegetated areas of permanent ponds, lakes, and streams with lots of algae will produce the most specimens of crawling water beetles. Look for *Brychius* along the shores of creeks, streams, or lakes.

TROUT-STREAM BEETLES   Amphizoidae

This small and distinct family of aquatic beetles contains only one genus, *Amphizoa*. Six species are distributed in western North America and China. They have received considerable attention from researchers because they are thought to represent an intermediate evolutionary stage between the terrestrial ground beetles (Carabidae) and the aquatic predaceous diving beetles (Dytiscidae). Recent studies suggest that trout-stream beetles are more closely related to the latter.

Trout-stream beetles resemble small darkling beetles (Tenebrionidae) or ground beetles, but both adults and larvae are primarily aquatic. Lacking modified swimming legs, they are usually found clinging to driftwood and other debris that accumulates in backwaters and eddies. The larvae are also found in these situations and remain close to the surface to obtain oxygen. Both adults and larvae are sometimes found out of the water.

Eggs are laid underwater in the nooks and crannies of submerged wood and other debris. Both the larvae and adults are predators, but they will also scavenge dead insects. The larvae cling to floating debris with their head downward, ready to dive after insect prey. After molting three times, mature larvae pupate in mud or sand some distance away from the shoreline. Beetles freshly emerged from their pupae are frequently caked with mud.

During the day, adults and larvae are sluggish but become more active at night. When handled, adult beetles secrete a yellow, sticky fluid from their anus that smells to some like rotten wood or overripe cantaloupe.

IDENTIFICATION: Their relatively large size, body shape, and habitat distinguish trout-stream beetles from other aquatic families of beetles. Adults are black, elongate, and oval. The head is prognathous and has thick, threadlike, 11-segmented antennae. The pronotum is margined on the sides and narrower than the elytra. The tarsal formula is 5-5-5. The abdomen has six segments visible from below. The first segment is divided by the hind coxae.

Mature larvae are long and spindle shaped, broadest at middle, and somewhat flattened. The combined length of the head and thorax is nearly equal to that of the abdomen. The upper part of the body is covered with a tough, dark exoskeleton, and the underside is paler and softer. The head is distinct and flattened, with antennae apparently three-segmented, but the fourth segment is greatly reduced. There are three pairs of simple eyes. The thorax is cylindrical, but the abdominal segments are more flattened. The legs are moderately long and six-segmented, including the two claws. The abdomen has eight segments, each with an extended flange on both sides. The last segment is tipped with a pair of prominent, fleshy, one-segmented projections.

SIMILAR CALIFORNIA FAMILIES:
- some ground beetles (Carabidae)—elytra distinctly grooved
- some darkling beetles (Tenebrionidae)—tarsi 5-5-4

CALIFORNIA FAUNA: One species in one genus.

The Trout-stream Beetle *(Amphizoa insolens)* (10.9 to 15.0 mm) (pl. 24) is black, with appendages black or reddish black. It ranges throughout western North America, from southern Yukon Territory and southeastern Alaska, eastward to western Alberta, and southward to eastern Nevada and the Transverse Ranges of southern California. It is generally found in cold streams ranging from 1,400 to 8,800 ft in elevation and seems to be most common around 6,000 ft.

COLLECTING METHODS: Trout-stream Beetles are found most often at the edges of cold, fast streams, where they live under rocks, or in coarse gravel along the shore, or cling to roots exposed beneath undercut banks. Sift through foam-covered floating debris that accumulates in backwater eddies. They are also found in masses of pine needles and twigs washed up on shore. Adults are sometimes found in the greatest numbers at the bases of waterfalls. They are occasionally found in ponds and lakes, but these individuals were probably washed downstream.

PREDACEOUS DIVING BEETLES   Dytiscidae

Adult and larval diving beetles are predators and scavengers, consuming both invertebrate and vertebrate tissues. They occur in a wide variety of aquatic habitats, particularly along the edges of pools, ponds, and slow streams where emergent vegetation grows. A few species are specialists, preferring cold-water streams, seeps, and springs, or other specialized bodies of water. Clear, gravel-bottomed streams, brackish waters, and thermal pools all have their own distinctive predaceous diving beetle fauna. The diverse aquatic habitats of the wetter coastal and montane regions of northern and central California support the greatest diversity of these beetles.

Predaceous diving beetles and their relatives are descendents of terrestrial insects that evolved a number of adaptations for living in water. Their oval and flattened bodies are streamlined to reduce drag as they swim. The short, fringed hind legs move in unison, propelling the beetle through the water. The hind legs are placed well back on the body to increase speed and maneuverability. Diving beetles are quite awkward on land. Their legs are attached to plates tightly fused to the body and are incapable of moving up and down like those of their terrestrial counterparts.

Adults store air in a cavity beneath their wing covers, replacing it periodically by hanging head down from the water surface. This bubble also makes them quite buoyant. To control their buoyancy, diving beetles can reduce the size and carrying capacity of the cavity by expanding their abdomen into the space. Another method is to swallow water and store it as ballast in an expandable chamber at the end of their digestive system. The amount of water carried can be modified to regulate the beetle's position in the water.

While submerged, the exoskeleton is quite permeable to water. However, when the diving beetle has been out of the water for some time, it becomes somewhat waterproof. Beetles attempting to reenter the water often have difficulty in submerging and can become trapped on the surface film. Using the surface film as a foothold, struggling beetles usually manage to yank themselves below the water. To facilitate reentry into the water, glands located at the tip of the abdomen produce surfactants—chemicals that act as wetting agents. When spread over the body surface, the surfactants make the exoskeleton more permeable to water and aid the beetle's efforts to become submerged.

The life histories of most California diving beetles are unknown. In general, the eggs are laid on moist soil or attached to or inserted into aquatic plants. The larvae are predaceous, and those of the larger species are sometimes called water tigers. Most larvae are probably beneficial, consuming the larvae of mosquitoes, biting midges, and other biting insects. Birds, mammals, and other aquatic insects prey on them.

Like the adults, many larvae obtain their oxygen from the surface, and it may be stored internally in their tracheal trunks. Young larvae have closed spiracles along the sides of the body, but as they mature the spiracles open in preparation for breathing on land. Larvae of some species obtain oxygen from the water through their cuticle, and a few have gills. The larvae molt three times before pupation. Mature larvae leave the water to construct pupal chambers beneath streamside objects.

Adult diving beetles are quite capable of flight and take to the air to colonize new bodies of water, occasionally turning up in puddles of rainwater or swimming pools. They also fly to locate overwintering sites or to leave habitats that have dried up or become otherwise unsuitable.

The study of all water beetles is useful because they are long-lived, and their presence or absence can serve as an indicator of environmental quality. It is possible to calibrate the environmental tolerances of each species and use them to gauge a range of conditions. Predaceous diving beetles are of particular use as indicator species in small pools and springs.

IDENTIFICATION: Adult predaceous diving beetles are oval and streamlined beetles ranging in length up to 33 mm. They are usually reddish brown to black or pale, with or without distinct markings. The head is prognathous and has beadlike antennae consisting of 11 segments. The pronotum is widest at the base. The scutellum is visible or not. The elytra are usually smooth and polished but may be pitted, sparsely hairy, or grooved, and they completely conceal the abdomen. The tarsal formula is 5-5-5, sometimes appearing 4-4-4. The claws are equal or unequal in size and are not toothed. The abdomen has six segments visible from below.

The white, yellowish brown, or reddish brown to black larvae may be variously marked with yellow or black and are long, spindle shaped, or somewhat flattened. The head is distinct, prognathous, and marked with a distinctive Y-shaped seam. The antennae are four-segmented, and there are usually three or fewer simple eyes on either side of the head. The legs are long, slender, and six-segmented, including the two claws. The nine-segmented abdomen rarely has gills on the sides. The eighth abdominal segment is sometimes elongate, while the last is greatly reduced and tipped with one- or two-segmented projections.

SIMILAR CALIFORNIA FAMILIES:

- whirligig beetles (Gyrinidae)—eyes divided; antennae clubbed
- crawling water beetles (Haliplidae)—head small; hind coxae greatly expanded
- burrowing water beetles (Noteridae)—scutellum not visible; hind tarsus with two similar claws
- water scavenger beetles (Hydrophilidae)—antennae clubbed; mouthparts (maxillary palps) long; underside flat, sometimes with a spinelike keel

CALIFORNIA FAUNA: 157 species in 27 genera.

Of the nine species of *Stictotarsus* found in California, *S. striatellus* (3.85 to 4.70 mm) (pl. 25) is the most widespread and one of the most commonly encountered. It is variable in color, ranging from nearly all black, to highly patterned, to nearly all pale. It is usually found in cool, clear mountain streams and ponds with sandy, muddy, or gravelly and rocky bottoms but may also be found in stagnant pools and even in hot springs. It is found throughout the state and north into British Columbia and Alberta and south into Baja California and southern Mexico.

Beetles of the genus *Agabus* are probably the most widespread and conspicuous predaceous diving beetles in California, where 33 species are known to occur. Usually black or pale with dark brown wing covers, *Agabus* species usually lack any markings on their elytra. They prefer permanent sources of water. At high elevations they are found along barren shorelines among gravel or rocks. At lower elevations *Agabus* species are found among the stems and roots of emergent vegetation. Some species emit a strong odor when handled. *Agabus strigulosus* (5.7 to 7.1 mm) ranges widely throughout western North America and is abundant in the Sierra Nevada. *Agabus lutosus* (6.4 to 8.4 mm) is black with narrow reddish margins on the pronotum and brownish yellow margins all around the elytra. It prefers the margins of ponds and slow, warm streams, swimming among emergent plants, and ranges all along the Pacific Coast, from southern British Columbia to northern Baja California. It is widespread throughout California. *Agabus disintegratus* (6.3 to 8.0 mm) has pale yellow elytra with black stripes. This species spends its summers in the Great Central Valley in a state of diapause, buried in dry pond bottom debris, and resumes activity with the onset of fall rains that refill the ponds. It is found across the United States and is widespread in California. *Agabus regularis* (9.2 to 11.3 mm) (pl. 26) is large and very dark brown. It is found in streams throughout much of the state, but primarily in the Coast Ranges from Mendocino County southward into Baja California.

The genus *Rhantus* is found throughout the world, including many oceanic islands. Of the 10 species known from North America, six occur in California. *Rhantus gutticollis* (9.8 to 13.0 mm) (pl. 27) is the most widespread species of predaceous diving beetle in California and is found throughout western North America. It prefers scavenging and hunting among vegetation and plant debris or on mineral deposits in quiet areas of small, clear streams. The head is black with a pale spot. The pronotum is yellow with two spots. The elytra are yellow with dark freckles that are sometimes expanded to form three longitudinal series of spots over lines of pits.

The Giant Green Water Beetle *(Dytiscus marginicollis)* (26.7 to 33.0 mm) (pl. 28) is the largest predaceous diving beetle in the state. Its smooth, dark green elytra and thorax are surrounded by a distinct yellow margin. The large and conspicuous larva, known as a water tiger, uses its sharp, slender mandibles to subdue small vertebrate and invertebrate prey. It inhabits a variety of semipermanent or permanent aquatic habitats, including hot springs and saline ponds surrounded by stands of scirpus *(Scirpus),* cattails *(Typha),* and rushes *(Juncus).* The Giant Green Water Beetle is distributed throughout western North America, from southwestern Canada to western Mexico. In California it is found in all but desert areas. Three other species of *Dytiscus* are recorded in the state.

The genus *Thermonectus* contains about 20 species, all occurring in temperate and tropical regions of the New World. Most *Thermonectus* prefer temporary ponds, grassy margins of ponds, and woodland pools. Of the eight species known from North America, two are recorded from California. The Sunburst Diving Beetle *(T. marmoratus)* (10.0 to 15.0 mm) (pl. 29), also called the Yellow-spotted Diving Beetle, is the most colorful species of predaceous diving beetle in the state and is often kept as a display animal in insect zoos. Broadly oval in outline and shiny black above, it is boldly marked with 10 or 11 bright yellow spots on each wing cover. The underside of its body is distinctly reddish. The Sunburst Diving Beetle is found in slow-moving seasonal streams, called arroyos, with rocky bottoms covered with little if any aquatic vegetation. The bright, contrasting color pattern is both cryptic and disruptive in function. The pattern renders the beetle inconspicuous as it rests and swims over sun-dappled, gravelly substrates. It also serves to break up its image as a beetle, shielding it from the attentions of predators. California records are primarily from canyon streams in the Peninsular Ranges.

Thermonectus intermedius (formerly known as *T. basillaris*) (9.0 to 11.0 mm) is found primarily in the northern and central portions of the state. It is distinguished from the Sunburst Diving Beetle by having black or reddish black elytra irregularly trimmed with a yellowish margin.

Eretes sticticus (12.7 to 17.0 mm) (pl. 30) is the only species of the genus occurring in North America. Its narrow and flat form, pale colors, fragmented and indistinct black band across the elytra, and wing covers tipped with short flat spines and golden setae are distinctive. It prefers warm, exposed ponds, especially temporary pools with cloudy water. A strong flyer, *E. sticticus* is often taken at lights in late summer and fall, far from breeding sites throughout the drier regions of central and southern California. This species, previously known for a time as *E. occidentalis,* also occurs in Arizona and Texas southward to Peru, as well as the Middle East and Africa.

Cybister is similar in appearance to *Dytiscus,* but the elytra reach their maximum width just behind the middle. Two species are known from California. *Cybister explanatus* (25.0 to 30.0 mm) occurs throughout California, while *C. ellipticus* (26.0 to 30.0 mm) is known from Imperial, Inyo, Orange, Riverside, San Bernardino, and San Diego Counties. In *C. explanatus,* the tip of the hind tibia is developed into a spine, while that of *C. ellipticus* is not.

COLLECTING METHODS: Even the most casual observer can see predaceous diving beetles in small pools as they move through open water and spaces between plants. A quick sweep of a kitchen sieve, aquarium net, or dip net will suffice to capture these beetles. Look under submerged objects and on aquatic vegetation for a variety of species. Sweeping a heavy-duty dip net through the vegetated shallows of ponds, streams, and lakes will produce other species. As you wade through dense emergent vegetation, sweep your net through the disturbed area several times to collect a variety of beetles. Practice this technique sparingly, as this form of collecting can be damaging to aquatic plants, especially in small pools and springs. Some predaceous diving beetles are readily attracted to lights or blacklight traps located near water. Shiny car surfaces or wet pavement often attract aquatic beetles in search of new habitats. Some collectors take advantage of this phenomenon by laying out sheets of Mylar or a thermal blanket to attract water beetles drawn to shiny surfaces. A useful technique for collecting predaceous diving beetles in the water is the bottle trap. Take a one-liter plastic bottle and cut the top off at the shoulders, where it begins to narrow down to the neck. Turn the funnel-like top around and insert it back into the bottle so that the cut edges are flush with one another, and seal them together with wire or silicone. Then bait the bottle trap with dog treats. Trapped beetles will drown in the trap unless there is an air bubble trapped in the container. Place the traps along the vegetated edges of ponds and slow-moving streams at night for the best results.

WATER SCAVENGER BEETLES   Hydrophilidae

The water scavenger beetles are one of the largest families of aquatic beetles in California, second only to the predaceous diving beetles (Dytiscidae). Most water scavenger beetles live in ponds, streams, and lakes with an abundance of plants or organic debris and are among the first arrivals in new aquatic habitats. They are particularly common along the vegetated edges of standing bodies of water, including rice fields and slow-moving streams or springs. A few are found in brackish water. Like other aquatic beetles, water scavengers are attracted to the shiny surfaces of cars and even blue tarps. They are sometimes confused with predaceous diving beetles but are easily distinguished by their humpbacked appearance, swimming style, and method for acquiring air bubbles. They also possess long, threadlike mouthparts that are easily mistaken for antennae.

The common name of this family is a bit of a misnomer. Water scavenger beetles live in a wide variety of habitats, including wet sand, rich organic soil, moist leaf litter, fresh mammal dung, and extremely decayed animal carcasses, as well as in water. The larvae are almost always predators, whereas the adults may be vegetarians, omnivores, or occasionally predators or scavengers. Predaceous species feed on a variety of animal foods, including snails and other small invertebrates, whereas omnivorous species add spores, algae, and decaying vegetation to their diets.

Aquatic water scavenger beetles capture and store a bubble of air along the ventral surface of their thorax and in the space under their elytra. They swim to the surface headfirst and break through the surface tension with their antennae, in contrast to the predaceous diving beetles, which break through the surface with the tip of their abdomens. A makeshift funnel is formed by the dense velvety setae on the water scavengers' antennae, mouthparts, and thorax, connecting the surface air with the thorax and subelytral space. Air is exchanged by the pumping action of the elytra and abdomen.

Although many aquatic water scavengers are good swimmers, some are slow and awkward. For better or worse, they propel themselves through the water by moving their legs in an alternate fashion, unlike the predaceous diving beetles, which move their legs in unison. The setae on the legs of both adults and larvae may or may not enhance their ability to swim. In at least one genus, *Berosus,* rows of hairs on the legs are repeatedly drawn through the air bubble in an apparent effort to increase aeration. This strategy might be especially useful for beetles living in stagnant bodies of water with low oxygen content. These hairs may also be part of a grooming system associated with special pores covering the abdomen.

Many aquatic species are capable of stridulating, producing sounds by rubbing their elytra and abdomen together. These sounds are emitted when the beetle is handled or under attack and is apparently a deterrent against predators. Sound production may be used by some species as a part of their courting behavior.

Up to 100 or more eggs are laid in a protective silken case that varies from a loose submerged pouch to a floatation device fitted with a sail-like breathing tube. The case is usually attached to the substrate or to vegetation and perhaps serves as a means of preventing the drowning of the eggs.

The larvae are carnivorous, feeding upon other invertebrates. The growth rate is dependent upon temperature and the availability of food. They molt three times before pupation. Mature larvae leave the water to construct a pupal chamber of mud near the shore. The chamber is either buried in the soil or tucked beneath a rock or other object. The pupae are neatly suspended within the cell by strategically placed setae. The newly emerged adults usually remain in the chamber until their body has darkened and hardened.

Fish, amphibians, reptiles, and water birds all prey upon aquatic water scavengers. At night, the beetles often take to the air in search of new habitats and are sometimes attracted to lights. During these nocturnal explorations they are subject to attack by still more birds and bats.

North American water scavenger beetles are of little economic importance, although larger species are reported to be pests in fish hatcheries. A few may be of some benefit as predators of mosquito larvae.

IDENTIFICATION: California water scavenger beetles are broadly oval, distinctly arched on top, and flattened or sunken underneath. They range in length up to 40 mm. The head is hypognathous and bears gradually clubbed antennae with six to 10 segments. The last three segments form a variable club that is usually nested in the preceding cuplike segment. The maxillary palps often exceed the length of the antennae, but in terrestrial and semiaquatic species they are usually equal in length or shorter. The pronotum is broader than the head and usually wider than long. The scutellum is visible. The elytra are widest at the middle and are broader at their base than the pronotum. The elytra are either smooth or rough and may be covered with rows of small pits and completely conceal the abdomen. The tarsal formula is 5-5-5, 5-4-4, or rarely 4-5-5. The claws are generally simple but sometimes modified in the male. The abdomen usually has five, or rarely six, segments visible from underneath. Water scavengers are usually black, black with brownish markings, or rarely greenish or with cream markings.

The gray or yellowish brown larvae are quite variable in shape and are somewhat long, flattened, cylindrical, and spindle or cone shaped. The body is sometimes marked with patterns of dark setae. The distinct and flattened head is prognathous and bears six or fewer pairs of simple eyes. The antennae are three- or four-segmented. The first thoracic segment usually has a distinctly armored plate on the back. The five-segmented, including a single claw, legs are present in most aquatic species. The eight-to 10-segmented abdomen may or may not have gills on the sides. Paired projections on the tip of the abdomen may have from one to three segments.

SIMILAR CALIFORNIA FAMILIES:
- burrowing water beetles (Noteridae)—antennae threadlike; mouthparts inconspicuous; body not flattened underneath
- predaceous diving beetles (Dytiscidae)—antennae threadlike; mouthparts inconspicuous; body not flattened underneath
- minute moss beetles (Hydraenidae)—abdomen with six or seven visible segments underneath
- minute flower beetles (Phalacridae)—small (1.0 to 3.0 mm); maxillary palps short
- riffle beetles (Elmidae)—small; legs long with large claws
- dung beetles (Scarabaeidae)—antennal club with flat segments

CALIFORNIA FAUNA: 71 species in 20 genera.

Species of the genus *Berosus* are extremely humpbacked. Their bodies are yellowish brown with dark spots. They are good swimmers and dive easily and may play dead when disturbed. *Berosus striatus* (4.0 to 6.5 mm) is known throughout the United States and southern Canada. In California it occurs throughout the state in a variety of shallow ponds, living among mats of algae and rooted vegetation. The tips of the elytra of both the male and the female are entire; those of the female have a small tooth at the suture near the tip of each elytron. The distinctly spotted *B. punctatissimus* (6.0 to 8.0 mm) (pl. 31) has bicolorous femora, the bases of which are black. The tip of each elytron has two spines separated by a distinct notch. It is found from Washington to Baja California and Arizona. These and other species of *Berosus* are frequently collected at lights. Species of *Berosus* produce chirping sounds while feeding, when handled, upon encountering other individuals, and during courtship.

The genus *Hydrochara* is distributed nearly worldwide. The striking pale yellowish green or dark bluish green *H. lineata* (13.5 to 17.0 mm) (pl. 32) is distinctive among California's aquatic beetles. It is widely distributed in California and throughout the southwest, including Baja California, and is sometimes found in mineralized water and hot springs.

The genus *Tropisternus* is very common throughout the warmer months, usually preferring shallow aquatic habitats with plenty of organic debris or algae. When disturbed, these beetles stridulate loudly. In some species the females stridulate to attract males. The upper side of their body is strongly arched. A sharp, backward-pointing spine or keel is evident below. *Tropisternus* occurs throughout the state at all elevations in ponds and streams of varying water quality, including brackish lagoons. The larvae may be important predators of mosquito larvae in rice fields. Eight species occur in California. *Tropisternus ellipticus* (8.0 to 12.0 mm) (pl. 33) is common throughout the western United States and is widespread in California. The adult overwinters in the soil or under rocks in ponds. In spring males and females migrate to other bodies of water, where they breed through summer. The female fastens its silken egg cases onto objects beneath the water surface. Pupation occurs in moist soil at the water's edge. The adult consumes algae, detritus, and animal remains. In fall the beetles migrate to other pools, where they overwinter. Another species, *T. californicus* (9.5 to 11.5 mm), is not as strongly humpbacked as *T. ellipticus* and has finer punctures on the pronotum. It appears to be coastal in its distribution and ranges from Oregon to Baja California.

The Giant Black Water Beetle *(Hydrophilus triangularis)* (33.0 to 40.0 mm) (pl. 34) is the largest beetle in the family and is found throughout the United States. In California it occurs east of the Sierra Nevada, along the Colorado River, and in the Great Central Valley. It was once more common throughout the state. Its shiny black elytra exhibit a greenish tinge. The eggs are laid in a distinctive brownish case that is either attached to emergent vegetation or set afloat. The adult is often attracted to lights at night and is frequently found in temporary desert rain pools. Like the previous species, the larva of the Giant Black Water Beetle may be an important predator of mosquito larvae in rice fields.

Accidentally introduced from Europe, the Spotted Dung Beetle *(Sphaeridium scarabaeoides)* (5.0 to 7.0 mm) (pl. 35) is now found throughout much of North America. It was first reported from California early in the twentieth century. By the early 1920s it was widely established in the northern and central parts of the state. Its round, shiny body is black with red and tan markings on the elytra. It prefers fresh, almost soupy, cow dung in which to feed and breed and is sometimes common wherever cattle are grazing during the spring and summer months. The larva feeds on both the dung and maggots developing in the dung.

COLLECTING METHODS: Sweeping an aquatic net along the shallow, vegetated margins of ponds, lakes, and streams is usually productive. Aquatic habitats with lots of submerged organic debris and algae are particularly rewarding, as is searching shallow margins at night with a flashlight. Another technique is to disturb aquatic substrates and vegetation to dislodge small, nonswimming species that become trapped on the surface film. Organic debris raked up on shore in an open sandy area or on a white shower curtain will reveal numerous beetles as they attempt to escape back into the water. Smaller amounts of debris can be placed in a white pan for sorting. Picking through wet leaf litter, moist cow dung, or very rotten carcasses is also productive for terrestrial species. A Berlese funnel is useful for removing individuals from leaf litter. Light traps placed near bodies of water will attract water scavengers in large numbers that might not be collected otherwise. Check well-lit storefronts and under streetlights at night for aquatic species.

CLOWN BEETLES   Histeridae

The clown beetle family is fascinating yet poorly understood in terms of its biology and distribution in California. Both adults and larvae are primarily carnivorous, preying on insects. Many species are found on dung, carrion, decaying plants, fungus, sapping wounds, or under bark, where they feed on the eggs and larvae of insects and mites. Others are specialists that live in ant nests and scavenge or prey on all stages of their hosts. A few ant-nest dwellers are so specialized in their form and behavior and have integrated themselves to such an extent into the colony that they are actually fed by the ants. Other species living with ants feed on other insects and surplus food collected by their hosts. Still other species live in reptile, bird, or mammal nests, especially those of pack rats *(Neotoma)*, ground squirrels *(Spermophilus)*, and kangaroo rats *(Dipodomys)*. A few (e.g., subgenus *Spilodiscus* of *Hister*, and *Hypocaccus*) are found in sand at the base of coastal or desert dune plants, where they probably feed on the larvae of scarabs (Scarabaeidae), weevils (Curculionidae), and flies. When threatened, adults pull their head inside their prothorax and tuck their legs tightly beneath their shiny, round, and compact bodies.

The larvae are unusual among the beetles, molting only twice, including the molt to the pupal stage. The thoracic legs are small and not useful for walking. Instead, they apparently move by waves of contractions of the abdomen. They feed on liquids, and before ingestion they must digest their food outside of the body using digestive fluids.

As predators, some clown beetles are moderately beneficial. Species living in dung are probably of greater importance because they prey on pestiferous flies infesting accumulations of animal waste in cattle pastures and poultry farms. Several dung-inhabiting *Hister* have been investigated as biological controls for the Horn Fly *(Haematobia irritans)*. A few cylindrical species live in trees, where they attack bark beetles (Curculionidae) inside their tunnels.

IDENTIFICATION: The majority of clown beetles are shiny black, but a few species have distinct red markings or may be reddish or metallic blue or green. The head is usually prognathous but may be hypognathous and typically bears prominent, sometimes very large mandibles. The 11-segmented antennae are elbowed and tipped with a compact three-segmented club that is often clothed in patches of sensory hairs. The elytra are distinctly grooved and short, appearing cut off, usually exposing the last two abdominal segments. The scutellum is visible. The legs are widely separated. The tarsal formula is 5-5-5 or 5-5-4, with claws equal in size and simple.

The mature larvae are generally long and cylindrical. The head, thoracic segments, and abdominal projections are thickly armored and dark, while the rest of the body is soft and membranous. The head is prognathous and partially withdrawn into the prothorax. The antennae are three-segmented, and there is a single pair of simple eyes. The thoracic segments have one or more plates on the back. The short legs are five-segmented, each with a single claw with two bristles. The abdomen is 10-segmented; the ninth segment usually bears a pair of two-segmented projections.

SIMILAR CALIFORNIA FAMILIES:

- false clown beetles (Sphaeritidae)—head not as withdrawn into prothorax; bases of front legs nearly touching at base; elytra loosely covering abdomen, exposing only one abdominal segment
- shining fungus beetles (Scaphidiinae, Staphylinidae)—

antennae weakly clubbed, long, and not tucked under body; abdomen pointed
- spotted dung beetles (*Sphaeridium*, Hydrophilidae) — red and tan markings on elytra; elytra completely covering abdomen; antennal club not as compact
- sap beetles (Nitidulidae) — antenna not elbowed; tarsi usually expanded and hairy beneath; fourth tarsal segment reduced
- short-winged flower beetles (Kateretidae) — antenna not elbowed

CALIFORNIA FAUNA: Approximately 140 species in 37 genera.

The shining black *Neopachylopus sulcifrons* (5.7 to 7.7 mm) is found under piles of seaweed on coastal beaches. It apparently feeds on the eggs and larvae of kelp flies and other insects.

Xerosaprinus (pl. 36) is probably the most commonly encountered genus of clown beetle in California, where it is always found in dung and carrion. At least 13 species are recorded in California.

Six species of *Saprinus* are known from California. The most common species, *S. lugens* (4.5 to 8.0 mm) (pl. 37), is a large clown beetle common in coastal areas and low to middle elevations throughout the state. It is found on carrion and dung and is sometimes attracted to foul-smelling flowers that depend on carrion-visiting insects for pollination.

Hololepta are all broad and flat. Most live under the bark of decaying hardwoods. *Hololepta populnea* (3.5 to 5.0 mm) is found under the bark of cottonwood *(Populus)* in southern California. *Hololepta vicina* (4.0 to 5.0 mm) is found in decaying cacti and other rotting plant materials. It is widespread along the Pacific coast of the United States. *Hololepta yucateca* (8.0 to 10.0 mm) is found in decaying flower stalks of yucca *(Yucca)* and the fruits of wild gourds in southern California, Arizona, New Mexico, and Texas. Two additional species of *Hololepta* live in the state.

Another flat species associated with rotting cacti in winter and spring is *Iliotona cacti* (4.5 to 7.5 mm) (pl. 38), which occurs in Texas, New Mexico, Arizona, and California. It is known from southern California along the coast near San Diego. It differs from *Hololepta* by the teeth of the middle and hind tibiae that are equally spaced rather than unequally spaced.

Atholus bimaculatus (3.0 to 5.3 mm) (pl. 39) is shiny black with large red or reddish brown spots on the elytra. It is commonly found in dung, where it feeds on fly larvae. This European species is widely distributed and was also introduced into Hawai'i to control the larvae of the Horn Fly *(Haematobia irritans)*.

Margarinotus sexstriatus (6.0 to 8.0 mm) is oval, black, and shiny, with three complete grooves on each elytron. It is found in the Coast, Transverse, and Peninsular Ranges from San Francisco Bay to Baja California. It is active in early spring under dung and is possibly associated with rodent burrows.

The large genus *Hister,* with 33 species in the United States, lives primarily in forested areas. Adults are often found in dung, carrion, and fungi, whereas some species live in mammal burrows. The Striated Hister Beetle *(H. abbreviatus)* (3.5 to 4.3 mm) is probably the most widely distributed clown beetle in North America. This northern California species is found throughout the United States and into parts of southern Canada and northern Mexico. It is easily distinguished from other members of the genus by the distinctly patterned elytra, each with four deep grooves. This species is found in cow and horse dung as well as carrion and fungus, where it feeds on dung-breeding flies. Another species, *H. sellatus* (4.1 to 6.6 mm), occurs primarily in California but is also known from Oregon and Washington. It prefers living in sandy habitats along coastal beaches and montane riparian habitats above 4,500 ft. Along the coast it is found year-round but is primarily active from February through June. Once widespread, this species has disappeared from much of its former range as a result of coastal urbanization and possibly from habitat degradation by nonnative plants such as ice plant *(Carpobrotus)* and European beach grass *(Ammophila arenaria)*.

COLLECTING METHODS: Examine dead animals, dung, rotting vegetation, tree wounds, and under the bark of recently killed trees, especially pines *(Pinus)*. Some species can be found by sifting sand from beneath coastal and desert plants. Others *(Euspilotus, Eremosaprinus, Geomysaprinus,* and *Aphelosternus)* are found in rodent burrows. Pitfall traps baited with carrion are also productive.

PRIMITIVE CARRION BEETLES   Agyrtidae

Once recognized as a subfamily of the carrion beetles (Silphidae), the primitive carrion beetles are now recognized as a full family. The little information known of the biology of these beetles indicates that both the larvae and adults are scavengers of dead or decaying organic material found in cool, wet habitats. They are often associated with beach and river drift, the margins of mountain streams, and high-elevation snowfields. They are primarily active during late fall, winter, and early spring. Nearly all of California's primitive carrion beetles are found in the mountains of the northern and central parts of the state, but one species, *Necrophilus hydrophiloides,* ranges as far south as coastal southern California.

IDENTIFICATION: Primitive carrion beetles are oval to elongate-oval, slightly flattened, and brownish. They range in size from 4.0 to 14.0 mm in length. The head is prognathous or pointed slightly downward. The 11-segmented antennae are threadlike or with a distinct or weak four- or five-segmented club. The pronotum is broader than the head and is distinctly margined at the sides. The scutellum is visible. The elytra always completely cover the abdomen. The surface of each elytron is grooved with nine or 10 rows of pits. The tarsal formula is 5-5-5, with claws equal in size and simple. The abdomen has five, or rarely six, segments visible from below.

The larvae are elongate, flattened, and more or less straight sided. Color usually ranges from dark brown to black. The distinct head is prognathous with six pairs of simple eyes. The antennae are long and three-segmented. The thoracic segments have one or more armored plates on the back. The long legs are five-segmented, including the claw. The abdomen is 10-segmented; the ninth segment has a pair of two-segmented projections.

SIMILAR CALIFORNIA FAMILIES:
- ground beetles (Carabidae)—antennae not clubbed
- carrion beetles (Silphidae)—elytra ribbed, rough, or smooth, never grooved

CALIFORNIA FAUNA: Seven species in four genera.

Necrophilus hydrophiloides (10.0 to 13.0 mm) (pl. 40) is a somewhat oval, brown beetle. Each elytron has nine distinct punctured grooves. It is found along the Pacific coast, from the Alaskan panhandle into southern California. Both the adult and larva are scavengers on decaying plant and animal tissues. The adult is primarily active November through May and is most often encountered at carrion, in garbage, or in rotting vegetable material.

COLLECTING METHODS: Primitive carrion beetles are rarely collected and are difficult to find. *Necrophilus* is attracted to carrion-baited traps during the cooler parts of the year. Sifting through river debris, fungi, or material under rotting bark may produce specimens of *Agyrtes* or *Ipelates*. *Apteroloma* is found among beach drift, amid gravel and moss on the banks of mountain streams, or under rocks at the edges of high-elevation snowfields.

CARRION BEETLES Silphidae

Most carrion beetles feed primarily on decaying animal material or rotting fungi, whereas a few species prefer to eat living plants and can become minor garden pests. The larvae of most carrion-feeding species appear to feed exclusively on dead animals, whereas some adults feed on both carrion and insects, especially maggots. Carrion beetles locate dead animals by smell, using their antennae to detect hydrogen sulfide and some cyclic carbon compounds that are released as the carcass decays. Phoretic mites are often found wandering about the bodies of burying beetles found on carcasses or at lights. Their presence may benefit the beetles because they feed on fly eggs that might otherwise hatch and compete with the beetles for the carcass.

Adult *Nicrophorus*, commonly referred to as burying beetles, stridulate by rubbing a pair of files located on their abdomen against the edge of their elytra, creating a scraping sound. Stridulation occurs during stress, mating, and confrontations with other beetles, and to communicate with the larvae.

Adult pairs of burying beetles demonstrate the most advanced behaviors of parental care known in beetles. Small animal carcasses are buried to reduce competition with flies, ants, and other insects in an effort to provide food for their larvae. Either the male or female initiates the carrion burying process, during which time the arrival of a mate is likely. The carcass of a mouse, bird, or other animal is carefully buried and meticulously prepared in a chamber beneath the ground by the adults. Mating occurs only after the carrion has been secured and the chamber formed. Feathers, hair, and skin are removed, and the carcass is kneaded into a pear-shaped ball. The female then lays her eggs in the walls of the burial chamber. The male may then leave the chamber and seek new mating opportunities. The female stridulates to indicate the location of the food ball to the newly hatched larvae. A broad depression, made by the female in the upper surface of the carcass, is where the young larvae gather. The newly hatched young are fed droplets of regurgitated tissue from one or both parents. The larvae receive care from the mother or both parents throughout their development, and only when the larvae burrow into the soil to pupate will the female voluntarily leave the brood chamber.

IDENTIFICATION: California carrion beetles are slightly to strongly flattened and black, sometimes with orange or reddish orange markings on the elytra. They range in size from 7.0 to 22.0 mm in length. The head is prognathous. The 11-segmented antennae are gradually or abruptly clubbed, and the clubs are velvety in appearance. The pronotum is broader than the head and has strong margins. The scutellum is visible. Sometimes the tips of the elytra are truncate, appearing as though they have been squarely cut off. In such cases, one or more segments of the abdomen are exposed. The surface of the elytra is rough or smooth, sometimes with three raised ribs, but never grooved. The tarsal formula is 5-5-5, with claws equal in size and simple. The abdomen has six, or rarely seven, segments visible from below.

The mature larvae are elongate, flattened, and straight sided. Color usually ranges from dark brown to black. The distinct head is prognathous. The antennae are long and three-segmented. There are six pairs *(Aclypea, Heterosilpha,* and *Thanatophilus)* or one pair *(Nicrophorus)* of simple eyes. The thorax and abdomen each have one or more armored plates across the back. The long legs are five-segmented, including the claw. The 10 abdominal segments are equal in length; the ninth bears a pair of one- or two-segmented projections.

SIMILAR CALIFORNIA FAMILIES:

- primitive carrion beetles (Agyrtidae) — elytra with distinct grooves

CALIFORNIA FAUNA: 10 species in four genera.

The Garden Carrion Beetle *(Heterosilpha ramosa)* (11.0 to 17.0 mm) (pl. 41) is dull black. Each wing cover has three shiny, branched ribs running lengthwise. The adult is active March through October and is commonly found in damp lawns, fields, and mountain meadows. The adult overwinters and becomes active the following spring. Eggs are laid in the soil around a carcass or rotting vegetable matter and take approximately five days to hatch. The larva resembles a sowbug, with dorsal plates extending laterally to form a shield that obscures the head and legs from view. The larval stage lasts approximately two to three weeks. The pupal stage lasts 8 to 9 days. Garden Carrion Beetles may produce two generations per year. Both the adult and larva are general feeders, consuming plant matter and living or dead insects that feed on decaying organic matter. It ranges as far north as southern Canada and is found west of Lake Superior to British Columbia, west of a line from northeastern Minnesota to south-central New Mexico, and as far south as Sonora and northern Baja California. The adult has been found feeding on dead Devasting Grasshoppers *(Melanoplus devastator)* and Brown Garden Snails *(Helix aspersa)*. The only other species in the genus is *Heterosilpha aenescens* (10.0 to 15.0 mm), which occurs primarily along the Pacific coast, from southern Oregon to northern Baja California. The elytra in some specimens have a shiny, metallic appearance. The adult has been found year-round. The male of this species is distinguished from *H. ramosa* by having normal, not expanded, front and middle tarsi. Also, the apex of the female's elytron is gradually rounded rather than prolonged as in *H. ramosa*.

The Satin Carrion Beetle *(Thanatophilus lapponicus)* (8.0 to 15.0 mm) ranges across northern Europe and Asia. In North America it is found in most of southern Canada and Alaska and coast to coast in the northern United States. It also occurs along the Rocky Mountains to New Mexico and Arizona, and southward along the Pacific states to Baja California. It is similar in appearance to *Heterosilpha ramosa* but is smaller and bears small bumps between the elytral ribs. This species occurs in a wide range of elevations in less-disturbed habitats. In more northern localities or at higher elevations it seems to prefer forested or open habitats. The adult has been collected from March to October.

Four species of the genus *Nicrophorus* are known from California. The Red and Black Burying Beetle *(N. marginatus)* (13.9 to 22.0 mm) (pl. 42) occurs in southern Canada, most of the United States, and into northern Mexico. This shiny black beetle has orange red antennal clubs and two broad, orange red marks on each wing cover, as well as a similarly colored spot on the head. The adult feeds extensively on fly larvae at carrion. It is active in summer and seems to prefer open fields, montane meadows, grasslands, and desert woodlands. Another red and black species, *N. defodiens* (12.0 to 18 mm), is found along the Pacific coast, the Rocky Mountain states, northern and central United States, and northeastern North America southward through the Appalachian Mountains. It occurs in the northern and central parts of California, as well as along the southern coastal region. It is distinguished from *N. marginatus* by having a completely black antennal club. The Black Burying Beetle *(N. nigrita)* (13.0 to 18.0 mm) is similar in appearance but is entirely black and has a brownish patch of setae on the underside of its thorax. It lives along the Pacific coast, from British Columbia to Baja California. The adult is nocturnal and active from February through November in coastal forests and open habitats. *Nicrophorus guttula* (14.0 to 20.0 mm) (pl. 43) inhabits dry forests, prairies, and deserts throughout western North America. The adult is active during the day from spring through early fall and has been collected at human dung as well as at carrion. The antennal club is orange, or with the basal segment black and the remaining segments orange. The thorax is clothed below in a bright yellow patch of setae. The elytra are black, variably marked with orange, or solid black with a small, inconspicuous spot on each side at the shoulders.

COLLECTING METHODS: Search through dead animal remains or decaying vegetable matter. *Heterosilpha* can be collected during the day walking on well-watered lawns. Look beneath naturally occurring animal carcasses for *Nicrophorus,* or deliberately set out whole animal carcasses to check regularly for beetles. Pigs, rabbits, chickens, or juvenile turkeys can be obtained cheaply from farms and butchers. If raccoons, skunks, or coyotes are present in the trapping area you might want to consider securing the carcass with chicken wire anchored by rocks, then covering it with plywood and weighing everything down with another rock. This will not only keep out the predators but will prevent rain from diluting the attractiveness of the bait. Traps baited with different types of carrion (squid, fish, chicken legs, or chicken wings) placed in more or less open situations are effective for both *Heterosilpha* and *Nicrophorus.* Species of *Nicrophorus* are sometimes attracted to lights at night.

ROVE BEETLES Staphylinidae

Although rove beetles comprise the largest family of beetles in California, most species are relatively small and secretive in their habits. The majority of species have short elytra, usually exposing five abdominal segments. They occur in every kind of habitat from arctic tundra and alpine timberlines to tropical forests. Most species live in forest leaf litter and other decaying plant debris. Some species prefer to live in wet habitats, such as lakeshores and beaches. Dung and carrion are also favored habitats for rove beetles. Other species live under bark or on trees and shrubs, where they prey upon aphids and bark beetles (Curculionidae) or feed on pollen or fungi. Some species are specialists, living in the nests of mammals and birds, among ants and termites, or as parasites of insect pupae. A few of these specialists gain access to ants and termites by mimicking the behavior and the chemical communication systems of their hosts. Some of these species even look like their hosts to avoid being attacked by them.

When excited, rove beetles often curl their abdomens over their backs, appearing as if they might sting, even though they are harmless. Adults and some larvae produce noxious defensive secretions. Adult *Paederus* have blistering chemicals in their blood. Predators are exposed to the blistering agent when they puncture the rove beetle's body. These beetles are unable to produce the chemicals themselves. Instead, they rely on symbiotic bacteria to make the defensive concoction. Mothers pass along the bacteria to their offspring via the eggs.

Little is known about rove beetle larvae. Most species molt three times before becoming a pupa. They usually occupy the same habitats as the adults but are more secretive, preferring moist and hidden habitats. Many have feeding habits similar to those of the adults. The larvae of *Aleochara* are parasites on the pupae of flies, whereas other species are apparently fungus feeders.

IDENTIFICATION: Most rove beetles are black or brown and are usually small (1.0 to 10.0 mm), but a few are large (30 mm or more) and brightly colored. The head is prognathous. The antennae are usually threadlike, beadlike, or thickened at the tip, but some species have a distinct club with one to four segments. The pronotum is usually broader than the head, with the borders often margined. The scutellum is usually visible. The elytra are usually short and usually expose five (sometimes fewer) abdominal segments. The tarsal formula is usually 5-5-5, but may be 4-5-5, 4-4-5, 4-4-4, 5-4-4, 3-3-3, or 2-2-2. The abdomen has six or seven segments visible from below.

Mature larvae are elongate and often flattened. The distinct head has a Y-shaped suture and is usually prognathous. The antennae are three- or four-segmented. One to six pairs of simple eyes are usually present (occasionally none). Each thoracic segment has a pair of armored plates on the back. The legs are five-segmented, including the claw. The abdomen is 10-segmented, with segments one through six equal in size, and seven through nine becoming gradually narrower. The ninth segment usually bears a pair of articulating, one- or two-segmented (very rarely three-segmented) projections with conspicuous bristles at the tips.

SIMILAR CALIFORNIA FAMILIES:
- sap beetles (Nitidulidae)—antennal club abrupt; only five abdominal segments visible from below
- short-winged flower beetles (Kateretidae)—antennal club abrupt; only five abdominal segments visible from below
- skiff beetles (Hydroscaphidae)—small (1 to 2 mm); aquatic; last antennal segment is long

CALIFORNIA FAUNA: Approximately 1,200 species.

Although the majority of California's rove beetles are small and inconspicuous in their color and habits, the following species are exceptional. The Hairy Rove Beetle *(Creophilus maxillosus)* (11.0 to 23.0 mm) (pl. 44) lives along the strand line of lakeshores and rivers or on dead animals, where both the larva and adult feed on the maggots of flies. It is widely distributed throughout the state in the lower and middle elevations.

The Devil's Coach Horse *(Ocypus olens)* (17.0 to 33.0 mm) (pl. 45) is black and is a native of Europe. It was first recorded in California in 1931 and has since become established in the San Francisco Bay region and coastal southern California. When disturbed, this beetle opens its powerful jaws and raises its abdomen up over the head. Expanding glands located at the tip of the abdomen turn inside out to emit a foul smelling yellowish fluid, a habit that gives this species its scientific name *olens,* or "stinking." The predatory adult is active in fall, whereas the voracious larva is out in spring, searching for slugs and snails. It has been considered as a biological control agent for another unwelcome introduction from Europe, the Brown Garden Snail *(Helix aspersa).*

The Pictured Rove Beetle *(Thinopinus pictus)* (12.0 to 22.0 mm) (pl. 46) is a flightless, pale yellowish brown rove beetle with striking black markings. The coloration varies to match the coloration of the sand; darker beetles are found on dark beaches, and lighter individuals are found on paler sand. It may be found in great numbers throughout the year along the length of the Pacific coast of North America, from Alaska to California. Both the adult and larva are nocturnal, hunting for and ambushing small invertebrates, particularly crustaceans such as sand fleas. They spend their days in temporary burrows at the upper part of the beach. The larva resembles the adult, except its mandibles lack teeth, the elytra are absent, and much of the thorax is black.

COLLECTING METHODS: Rove beetles require moist places, such as along streams, lakeshores, and beaches and in decaying organic material. Searching for them in humid habitats is the most productive, but also look for them in moist spots in dry places such as rotting cactus and temporary flood debris in deserts. Many species are found on or near animal dung and carrion, whereas others are found beneath bark, on fungi, under objects on the ground, and along the shores of freshwater and marine habitats. Sweeping flowers and foliage will produce some specimens. Sifting leaf litter and other organic debris is also worthwhile. Some species, especially those living near water, are attracted to lights around dusk and early evening, sometimes in tremendous numbers.

STAG BEETLES   Lucanidae

The common name "stag beetle" comes from the enormous antlerlike development of the mandibles in some males. The size of the mandibles is sometimes directly proportionate to the size of the body. These huge mouthparts are used in combat with other males of the same species to win the right to mate with nearby females. In most California species, both males and females have relatively inconspicuous mandibles.

Stag beetles are more or less restricted to wet habitats, where dead, decomposing wood is plentiful. They breed in stumps and roots of both coniferous and deciduous trees, laying their eggs in crevices. The grubs feed on the decaying wood of logs, stumps, and roots but do not damage homes or other buildings. The grubs are C shaped, resembling those of scarab beetles (Scarabaeidae). Although some adult stag beetles are known to feed at sap flows on the trunks of trees or on flowers, adults of California species are not known to feed at all.

IDENTIFICATION: California stag beetles are black, reddish brown, dull bronze, or deep, shiny blue beetles that range in length from 9.0 to 19.0 mm. Their bodies are elongate and either cylindrical or somewhat flattened. The head is prognathous and lacks a horn, except for the males of *Sinodendron*. The 10-segmented antennae are usually elbowed, except in *Ceruchus* and *Sinodendron*. The antennal club is asymmetrical, with three or four thick, velvety segments. Unlike scarab beetles, stag beetles cannot close these antennal segments into a compact club. Their mandibles are not particularly prominent. The pronotum is narrower at the base than the adjacent elytra and lacks ridges, grooves, horns, or other projections, except in *Sinodendron*. The elytra are long, almost or completely concealing the abdomen from above. The surface of the elytra is smooth or rough, with or without fine longitudinal grooves or punctures. The tarsal formula is 5-5-5, with the tarsal claws equal in size and simple. The abdomen has five segments visible from below, occasionally with a portion of a sixth segment visible as well.

Stag beetle larvae are creamy white or yellowish and have long, cylindrical, and C-shaped bodies. The end of the body is translucent and may be darkened by waste materials inside the digestive tract. The distinct head is darker and harder than the rest of the body. The antennae are three- or four-segmented. Simple eyes are absent in all genera except *Platycerus*. The middle and hind legs bear a sound-producing apparatus. All legs are five-segmented, including the claw. The abdomen is 10-segmented, with the last two segments short, and is without projections.

SIMILAR CALIFORNIA FAMILIES:

- ground beetles (Carabidae)—antennae threadlike, not clubbed
- scarab beetles (Scarabaeidae)—antennal club segments fold tightly together
- some bark-gnawing beetles (*Temnoscheila*, Trogossitidae)—antennal club symmetrical
- some darkling beetles (Tenebrionidae)—antennae beadlike
- bostrichid beetles (Bostrichidae)—antennal club not elongated; head hypognathous; pronotum usually with bumps or small horns

CALIFORNIA FAUNA: Nine species in five genera.

Three species of *Platyceroides* are known from California. The Oak Stag Beetle *(P. agassizi)* (9.0 to 10.3 mm) (pl 47) is dark reddish brown overall with a mildly bronzed sheen. It is found along the coast of northern California and Oregon. The adult is found among redwood groves, perched during the late morning hours on vegetation and split rail fences lining trails in late spring and early summer. It breeds in dead Pacific madrone *(Arbutus menziesii)*, coast live oak *(Quercus agrifolia)*, and tanbark-oak *(Lithocarpus densiflorus)*.

The bodies of *Platycerus* are somewhat flattened. Two species are found in California. The Oregon Stag Beetle *(P. oregonensis)* (8.3 to 10.3 mm) (pl. 48) is found in the northern and central Coast Ranges and on the western slope of the Sierra Nevada northward to British Columbia. The head of the adult is black, while the pronotum and elytra are a deep, shiny blue. The elytra are smooth with rows of shallow pits. The adult is active in summer and is seen flying in the afternoon or resting on vegetation. Both the adult and the larva are found beneath the bark of toyon *(Heteromeles arbutifolia)*, mountain maple *(Acer glabrum)*, bigleaf maple *(A. macrophyllum)*, California laurel *(Umbellularia californica)*, Pacific madrone, coast live oak, California black oak *(Quercus kelloggii)*, blue gum *(Eucalyptus globulus)*, and red alder *(Alnus rubra)*.

The Rugose Stag Beetle *(Sinodendron rugosum)* (11.0 to 18.0 mm) (pl. 49) is cylindrical in form. It lives in isolated wet canyons of the Transverse and Peninsular Ranges in southern California and throughout the wooded areas of northern and central California northward to British Columbia. The mandibles of both the male and female are not strongly developed and do not project much beyond the head. The male has a single, well-developed horn on the head and a tubercle on the pronotum. The female has these structures in reduced form. The adult is occasionally found flying on warm summer afternoons. The larva, pupa, and adult can be common beneath the bark of wet, rotten logs of alder, ash *(Fraxinus)*, maple, California laurel, willow *(Salix)*, cherry *(Prunus)*, white poplar *(Populus alba)*, Fremont cottonwood *(P. fremontii)*, water birch *(Betula occidentalis)*, and oak. Hundreds of beetles may emerge from a single section of log only two or three feet in length.

Both California species of *Ceruchus* are shining black with relatively prominent mandibles in the male. The head is as broad as the pronotum in the male, but narrower in the female. Unlike most other California stag beetles, the antennae of *Ceruchus* are not elbowed. Although other stag beetles tend to breed in hardwoods, *Ceruchus* prefers conifers such as firs and pines. The adult flies at dusk or in the evening and is occasionally attracted to lights. The elytra of the Striated Stag Beetle *(C. striatus)* (13.0 to 19.0 mm) are distinctly grooved. It ranges from the northern Coast Ranges to southwestern British Columbia and has been taken from rotten logs of mountain hemlock *(Tsuga mertensiana)* and coast redwood *(Sequoia sempervirens)*. The Punctate Stag Beetle *(C. punctatus)* (8.0 to 16.0 mm) has elytra that are densely pitted and only feebly grooved. It is distributed along the entire length of the Sierra Nevada and Klamath Mountains northward to Idaho and British Columbia. It is found beneath the bark of white fir *(Abies concolor)* and ponderosa pine *(Pinus ponderosa)*.

COLLECTING METHODS: Look for stag beetles in wet, wooded habitats of coastal and montane California, particularly in northern and central parts of the state. In late spring through midsummer, adult stag beetles are encountered flying about in the afternoon or evening, walking on downed logs, or resting on vegetation. Adults and larvae may be found beneath bark of rotten logs or stumps of conifers and hardwoods throughout the year, especially along wet canyon bottoms or sandy coastlines. Partially submerged rotten logs may be very productive. Adults are collected by hand, netted on the wing, or caught by beating and sweeping vegetation. Some species are attracted to lights at night.

HIDE BEETLES Trogidae

These grayish, often mud- and mite-covered beetles are widely distributed throughout the world. The family name is derived from the Greek word *trog,* meaning "to gnaw or chew." Larvae and adults are scavengers upon dried animal carcasses, where they feed on keratin-rich hair, hooves, and feathers. The feeding activities of hide beetles represent the final stage of carrion decomposition, as they consume the remains left behind by other scavengers. Some species prefer the concentrations of feathers or hair that accumulate in bird nests or rodent burrows. A few species are attracted to the hair-laden feces of carnivores.

When alarmed, adult beetles pull in their legs and feign death. Encrusted with debris, they look very much like a pebble, clump of earth, or some other inanimate object to avoid the attention of predators. They can also stridulate, or make a chirping sound, by rubbing a patch of ridges on their next to last abdominal segment against the inner margin of the elytra. Many species are commonly attracted to lights at night. The larvae occur in shallow burrows in the soil directly beneath where the adults are feeding. Hide beetles were, until recently, considered a subfamily of the scarab beetles (Scarabaeidae).

IDENTIFICATION: Hide beetles are easily recognized by their overall warty, brown, gray, or black, dirt-encrusted appearance, velvety three-segmented lamellate antennal clubs, and flat, five-segmented abdomen completely concealed by the elytra. They are convex beetles, with sides slightly rounded or parallel. They are usually shades of dark brown or gray and are often covered in a thick gray or brown crust that is sometimes organized into a pattern. They range in size from 5.0 to 14.0 mm in length. The head is hypognathous, lacks horns, and bears a pair of 10-segmented antennae tipped with fanlike segments that fold tightly into a lopsided three-segmented club. The pronotum is squarish or rectangular with sharp side margins. The scutellum is visible and is either hatchet shaped *(Omorgus)* or oval *(Trox)*. The elytra are strongly ridged or covered with rows of small raised bumps and completely conceal the abdomen. The tarsal formula is 5-5-5, with all claws equal in size and not toothed. The abdomen has five segments visible from below.

The white larvae are cylindrical and C shaped. The head is distinct, dark in color, and hypognathous. The antennae are three-segmented, and the simple eyes are absent. The legs are five-segmented, including the claw. The 10-segmented abdomen does not have any projections.

SIMILAR CALIFORNIA FAMILIES:

- stag beetles (Lucanidae)—body long; lamellate antennal segments do not form a compact club
- 5.1 mm); color pale brown
- horned fungus beetles (Tenebrionidae)—beneath bark; antennae not abruptly clubbed; male pronotum horned

CALIFORNIA FAUNA: Eight species in two genera.

The scutellum of *Omorgus* is hatchet shaped. *Omorgus suberosus* (9.0 to 14.0 mm) (pl. 50) is found throughout the United States, Mexico, and Central and South America and was accidentally introduced into Africa, Australia, and Europe. Its large body and relatively smooth elytra are distinctive among the genus in California. In Argentina this species is credited with the destruction of locust egg pods by eating the protective coating, exposing the eggs to desiccation and fungal growths. Specimens are found under carrion and dry cow manure, and at lights throughout the valleys and desert regions of the state. Clean specimens have an almost checkered appearance. The only other California species is *O. punctatus* (10.5 to 15.0 mm), which is taken at lights in the Providence Mountains in the eastern Mojave Desert and is widely distributed elsewhere in the western United States. It has distinct bumps on the elytra.

The genus *Trox* has a long and oval scutellum. California has six species. *Trox gemmulatus* (9.0 to 12.0 mm) (pl. 51) is restricted to coastal chaparral communities of Baja California and southern California. It becomes active at the onset of the fall rains in October or November and remains so until the following spring. This species is commonly found feeding on coyote scats or dog dung in Los Angeles, Orange, Riverside, and San Diego Counties. *Trox fascifer* (5.0 to 7.0 mm) is a small, somewhat shiny brown species widely distributed throughout the Pacific Northwest from southwestern British Columbia and western Washington and Oregon, southward to the San Francisco Bay region. It is usually taken at lights.

COLLECTING METHODS: Hide beetles are most commonly collected on carcasses, in animal nests and burrows, or at lights. *Trox gemmulatus* is found on coyote scats packed with hair. The soil beneath carcasses and scats should be carefully checked for larval burrows and buried adults.

RAIN BEETLES   Pleocomidae

The rain beetle family contains the single genus *Pleocoma,* whose species are distributed from southern Washington to Baja California. Previously published records of rain beetles from Alaska and Utah have proved to be erroneous. California's rain beetles occur throughout the mountainous regions of the state, except in the deserts. Small, isolated populations are also known in the Sacramento Valley and the coastal plain of San Diego County. The known modern distribution of this apparently ancient lineage of beetles is restricted by the flightless females and is more or less correlated to areas of land that have never been subjected to glaciation or inundation by inland seas during the last two or three million years. However, areas around San Francisco Bay, the Santa Monica Mountains, inland valleys, and some coastal areas subject to water inundation support rain beetle populations, suggesting that they have migrated into these areas.

Rain beetles are large, robust, and shiny beetles. The thick layer of hair covering their undersides is remarkably ineffective as insulation, especially for flying or rapidly crawling males who must maintain high body temperatures in cold damp weather. Males can attain an internal temperature of 95 degrees F by shivering, or vibrating, their thoracic muscles. The thick pile probably functions to protect both sexes from abrasion as they burrow through the soil. Males and females dig with powerful, rakelike legs. Digging males also benefit from the V-shaped scoop on the front of the head. Males are fully winged and capable of flight, whereas the hind wings of the flightless females are reduced to small flaps of tissue. Females are usually considerably larger and more heavy bodied than males.

Lacking functional mouthparts, adult rain beetles cannot chew and are unable to feed. They must instead rely on fat stored in their bodies while they were root-feeding grubs. Because of their limited energy stores, adults are active for only a short time. On average, males of some rain beetle species have only enough energy stored as fat to give them about two hours of air time and live only a few days. The more sedentary females require less energy and may live for months after fall and winter storms.

In most species of rain beetles, male activity is triggered by weather conditions in fall, winter, or early spring. They require sufficient amounts of rainfall or snowmelt before becoming active. Depending upon circumstances, males may take to the air at dawn or at dusk, or they may fly during evening showers. Others are encountered flying late in the morning on sunny days following recent heavy rains or snowmelt.

Males fly low over the ground, searching for females releasing pheromones from the entrances of their burrows. Amorous males are capable of tracking "calling" females over considerable distances, often through dense vegetation. Dozens of males may descend upon a single female, clambering over each other as they jockey for position to mate. Mating takes place on the ground or in the female's burrow. After mating, the males leave and die a short time later. It is unknown whether males or females mate multiple times. During their nuptial flights males are frequently attracted to lights or shiny pools of water. Females crawl back down their burrows and may wait up to several months for their eggs to mature. They eventually lay 40 to 50 eggs in a spiral pattern at the end of the burrow as much as 10 ft below the surface. The eggs hatch in about two months.

Upon hatching, the small grubs use their powerful legs and jaws to tunnel deep in hard and compact soils to follow the root systems of their host plant. They feed upon roots of grasses, shrubs, and trees. In Oregon, the larvae of some rain beetles are considered pests when they attack the roots of strawberries, pears *(Pyrus)*, apples *(Malus)*, and cherries *(Prunus)*. Unlike most scarab beetles (Scarabaeidae), which have three instars, *Pleocoma* larvae molt seven or more times and may take up to 13 years before reaching maturity. Pupation occurs in a simple, elongate chamber.

Rain beetles have numerous enemies, below and above ground. Predatory robber fly maggots attack both the larvae and pupae. There is also a record of an adult female infested internally by nematode worms. Coyotes *(Canis latrans)*, foxes *(Vulpes)*, skunks *(Mephitis)*, Raccoons *(Procyon lotor)*, and owls feast upon the often slow-moving adults, plucking the males out of the air. At the height of the flight season it is not unusual to find the droppings or pellets of these predators filled with the indigestible bits of rain beetle legs and wing covers.

IDENTIFICATION: Adult California rain beetles are among the largest beetles in North America, resembling large June bugs (Scarabaeidae) in overall body shape. The males are usually reddish brown to black, whereas females are generally reddish brown. Their bodies are robust and broadly oval. The upper surface is shiny, while the underside is densely clothed in hairlike setae. The winged males (16.5 to 29.0 mm) are easily distinguished from the larger, flightless females (19.5 to 44.5 mm). The head is armed with a horn and bears antennae consisting of 11 segments tipped with four to eight fanlike segments that can be folded tightly into a club. The club is larger in males than females. Adults lack functional chewing mouthparts. The prothorax is broad and unique among other scarabaeoid beetle families in that the cavities into which the front legs are inserted are open toward the back. The scutellum is visible. The elytra are smooth, shining, or distinctly grooved and almost completely conceal the abdomen. The powerful forelegs are equipped with rakelike teeth. The tarsal formula is 5-5-5, with tarsal claws equal in size and simple. The abdomen has six segments visible from below.

The larvae are creamy white, cylindrical, and C-shaped grubs. The tip of the abdomen may be darkened by waste material inside the body. The distinct head is shiny yellowish or reddish brown and hypognathous. The antennae are three-segmented, and simple eyes are absent. The middle and hind legs have sound-producing patches, and all legs are five-segmented, including the claw. The 10-segmented abdomen does not have any projections.

SIMILAR CALIFORNIA FAMILIES: Their large size, hairy undersides, horned heads of the males, 11-segmented antennae, and their fall and winter activity periods easily distinguish rain beetles from other California scarablike beetles and their relatives.

CALIFORNIA FAUNA: Approximately 20 species in one genus.

Two species of rain beetles are common in southern California. The Black Rain Beetle *(Pleocoma puncticollis)* (males 26.0 to 31.0 mm) is known from the Santa Monica Mountains, Del Mar on the coastal plain of San Diego County, and the Peninsular Ranges into Baja California. The shiny black male is clothed in black or rusty black setae and flies after the second or third soaking rains in late November and December, usually at dawn or in the evening during showers. Both the adult and the larva are associated with the roots of California-lilac *(Ceanothus)*. The Southern Rain Beetle *(P. australis)* (males 24.0 to 28.0 mm) (pl. 52) is known from portions of the Transverse and Peninsular Ranges and flies during the first fall rains in October and November. Although males have been observed flying on bright sunny mornings following showers, they are most abundant during light drizzle at dusk or just after dark. The male has a dark reddish brown head and pronotum, black elytra, and reddish brown setae. The larva feeds on the roots of canyon live oak *(Quercus chrysolepis)*.

Most the state's rain beetles live in central and northern California. Behren's Rain Beetle *(P. behrensi)* (males 21.0 to 27.0 mm) is common in the Oakland Hills in the San Francisco Bay region. The male is black with a slightly brownish pronotum and is clothed with yellowish setae. The larval host plant is chaparral broom *(Baccharis pilularis)*. The Fimbriate Rain Beetle *(P. fimbriata)* (males 27.0 to 34.0 mm) is known from the northern foothills of the Sierra Nevada. The male is black with pale yellowish setae. The larva feeds on the roots of buckbrush *(Ceanothus cuneatus)*. The Santa Cruz Rain Beetle *(P. conjugens)* (20.0 to 23.5 mm), of the Santa Cruz Mountains, was proposed for endangered species status in 1994 but was later withdrawn. Surveys suggested that this species is more widespread than previously thought. However, all of its known range is located within an area that is rapidly being swallowed up by urban development, so concern for the species is still warranted.

COLLECTING METHODS: When collecting rain beetles, expect conditions to be cold, wet, and uncomfortable. And you still may come up empty-handed, even when conditions seem right. Although males are sometimes quite numerous when they emerge, their colonies can be very localized and are easily overlooked. Although large numbers of males can be easily collected from several well-known sites throughout California, it is more rewarding scientifically to explore areas between recognized populations where few or no beetles have been observed. Search for flying males just after or during the second or third soaking rains in fall, especially in the evening or at dawn. Males flying during the day can be netted, while species active at dawn or in the evening are readily attracted to black lights. Light traps with timers placed out in remote mountain locales during the fall and winter wet season may produce new records as well as new species. Enlisting the assistance of interested townspeople in California's mountain communities is also effective, since the males are very conspicuous when they are attracted to streetlights and lit house windows. Also check puddles and swimming pools for struggling or drowned specimens. Flightless females must be dug from their burrows, which may be located by following the males. Caged females will release pheromones and can be used to attract males.

EARTH-BORING SCARAB BEETLES   Geotrupidae

Earth-boring scarab beetles are related to the scarab beetles (Scarabaeidae) and were previously considered a subfamily within the scarabs. The habits of California species are poorly known. Adults feed on fungi and leaf litter and, outside of California, occasionally dung. All earth-boring scarabs are diggers, burrowing down into the soil, sometimes to depths of 6 ft or more. The burrows of earth-boring scarabs are frequently marked by "push ups," piles of soil or sand dug up from the burrow below. At the end of the burrows they construct one or more chambers and provision them with leaf litter or fungi. These materials are pushed down into the burrow, where they are formed into a plug on which the grubs feed. The adults do not stay in the burrow to care for the young. All adult earth-boring beetles are capable of producing sound by stridulating, the purpose of which is unknown. Adults are nocturnal and are frequently attracted to lights. Earth-boring scarabs are of no economic importance, although the burrowing activities of *Ceratophyus* have occasionally caused damage to lawns and golf courses.

IDENTIFICATION: California earth-boring scarab beetles are similar to other scarabs by virtue of their lamellate antennae. The 11-segmented antennae, tipped with a compact, velvety, three-segmented club and exposed mandibles, will distinguish them from most other scarabs. Adults are usually oval or round and are yellowish or reddish brown, brown, or black. They range in length up to 23.0 mm. The head usually has a distinct horn, tubercle, or ridge and is prognathous, with the mandibles clearly visible from above. The last two segments of the antennal club are partially contained within a cup-shaped segment. The pronotum is convex, and its base is wider than or nearly equal in width to the base of the elytra. The pronotum may have tubercles, ridges, horns, or deep grooves. The scutellum is visible. The elytra are convex, smooth, or distinctly grooved and completely conceal the abdomen. The tarsal formula is 5-5-5, with claws equal in size and simple. The abdomen has six segments visible from below.

The larvae (those of *Bolbelasmus* and *Bolbocerastes* have not been described) are C shaped and creamy white or yellow, except at the tip of the abdomen, which may be darkened by accumulated feces. The head is brown to dark brown. The antennae are three-segmented, and simple eyes are absent. The legs are two-segmented *(Bolboceras)* or three-segmented, with the last pair of legs reduced in size *(Ceratophyus)*. The 10-segmented abdomen does not have any projections.

SIMILAR CALIFORNIA FAMILIES:

- sand-loving scarab beetles (Ochodaeidae)—small (3.0 to 10.0 mm); antennae nine- or 10-segmented; eyes bulging; spurs on ends of middle tibiae appear feathery or finely notched

CALIFORNIA FAUNA: Five species in four genera.

The biology of the light reddish brown *Bolbelasmus hornii* (up to 14.0 mm) is unknown. It is widely distributed throughout the Coast Ranges, Sierra Nevada, and Transverse and Peninsular Ranges and is frequently attracted to lights from January through May. It appears to be most active in April. *Bolbelasmus* is distinguished from *Bolboceras,* the only other montane earth-boring beetle in California, by having its compound eyes completely divided into two halves and by the light reddish brown color.

The dark reddish brown to black *Bolboceras obesus* (6.5 to 12.0 mm) (pl. 53) is found throughout western North America. In California it is found in the northern Coast Ranges, Sierra Nevada, and Transverse and Peninsular Ranges. The adult is usually encountered at lights throughout the year, but this species reaches its peak activity in spring and early summer. The head of the male bears a slender curved horn that may extend backward over the pronotum to the scutellum. Both the larva and pupa have been dug from sandy soil at a depth of 3 to 6 in. among a stand of manzanitas *(Arctostaphylos)* in Lassen County. This beetle's pushups have also been found in wooded areas.

The genus *Bolbocerastes* consists of robust, hemispherical, reddish brown beetles. Both sexes have a scooplike head backed up by a bulldozerlike pronotum. The front of the head is extended into a short, broad horn that appears to have been cut off, whereas the pronotum bulges at the middle and is flanked by blade-shaped flanges. The male of *B. imperialis* (11.0 to 19.0 mm) has a narrow horn that appears almost cylindrical in cross section. This species is found in Arizona, Colorado, New Mexico, Utah, Texas, and Baja California. In California it is most abundant in the Colorado Desert. It is active May through September and is collected at lights or dug from burrows whose entrances are marked by pushups. The pushups have been found in moist sand at or near the bases of creosote bush *(Larrea tridentata)* and honey mesquite *(Prosopis glandulosa* var. *torreyana).* The male *B. regalis* (10.0 to 21.0 mm) (pl. 54) has broad horns that are flat in cross section. This species is distributed in Arizona, Nevada, and northwestern Sonora. In California it is found in the Mojave and Colorado Deserts and also on San Clemente Island. *B. regalis* is active in March through August, when it is also found at lights and dug from burrows. Both species sometimes occur in the same locality.

The large and distinct Gopher Beetle *(Ceratophyus gopherinus)* (15.0 to 23.0 mm) (pl. 55) is seldom seen. The male has a sharp, upturned horn on the front of its head and an equally sharp horn on the pronotum extending forward. Both of these horns are less developed in the female. This large, black, shiny species was first discovered in 1962 at Vandenburg Village, north of Lompoc in Santa Barbara County, where the burrowing activities of adults were reported to damage lawns and golf courses. Later it was discovered in the scrub oak thickets on Vandenburg Air Force Base, where the pale pushups stood in stark contrast to the dark layer of plant debris covering the ground. The nocturnal adult is active January through April, digging its winding tunnels up to a depth of several feet. A horizontal chamber is constructed at the end of the burrow and provisioned with dry surface sand, leaf litter, and twigs of chamise *(Adenostoma fasciculatum)*, manzanita, and California-lilac *(Ceanothus)*. *Ceratophyus* is primarily an Old World genus of beetles. It was originally suggested that the Gopher Beetle was transported to California from Europe in ship's ballast. However, recent research indicates that *C. gopherinus* is a native relict of a genus that once had a much wider distribution, an idea that is bolstered by the fact that this conspicuous species has yet to be discovered in Europe or Asia.

COLLECTING METHODS: In the more humid eastern United States, some earth-boring scarab beetles are attracted to fermenting malt and molasses, but these baits are seldom effective in the drier climates of the west. *Bolbelasmus, Bolboceras,* and *Bolbocerastes* are attracted to lights, sometimes in numbers. Beetles attracted to lights on buildings often bury themselves in the soft earth or litter at the base of walls during the day. Check for pushups with fresh or moist soil along mountain trails and roads or at the base of desert shrubs in moist sand, especially in spring. Old pushups may indicate the presence of larvae below. *Ceratophyus* is extremely localized and spends much of its time buried in soil.

BUMBLEBEE SCARAB BEETLES — Glaphyridae

This small family of odd beetles contains eight genera distributed in Europe, Asia, and North America. Of these only *Lichnanthe* occurs in North America, with eight species restricted to the far eastern or western states. One of the two species recorded from the eastern United States, the Cranberry Grub *(L. vulpina)*, is a pest of cranberry bogs in the northeastern states, where the larva attacks the roots.

Bumblebee scarabs are so named because the adults are often hairy and brightly colored. Fast and agile fliers, they are easily mistaken for bees. Active from midmorning to midafternoon, their peak activity is between 10:00 A.M. and 1:00 P.M. They are often observed hovering near flowers and foliage or flying over sandy areas. *Lichnanthe* are sometimes found on flowers, but it is not clear whether they actually feed on pollen. At the end of the daily flight period both males and females burrow into the soil.

The life histories of most species of *Lichnanthe* are poorly documented. The larvae feed on layers of decaying leaf litter and other plant debris buried in coastal dunes or sandy areas along streams and rivers.

IDENTIFICATION: Bumblebee scarabs are long and hairy beetles, ranging in length up to 17.5 mm. The color is quite variable, ranging from pale yellowish to black, often with metallic reflections. Two or more color varieties based on hair color may be present within populations, with white, yellow, orange, brown, or black forms in a single species. The head is hypognathous, with 10-segmented antennae tipped with three fanlike segments that fold tightly into a velvety club. The pronotum is bulging, slightly wider than long, and usually densely pitted and hairy. The scutellum is visible and U shaped. The elytra are smooth and transparent, their tips diverging to expose the abdomen. The tarsal formula is 5-5-5, with claws equal in size and toothed. The abdomen has six segments visible from underneath.

The larvae are bluish white, cylindrical, C-shaped grubs that turn yellow just before pupation. The tip of the abdomen may be darkened by waste material inside the body. The reddish brown head has a conspicuous depression on the middle of the forehead. The antennae are four-segmented, and simple eyes are absent. The well-developed legs are five-segmented, including the claw. The 10-segmented abdomen does not have any projections.

SIMILAR CALIFORNIA FAMILIES:

- scarab beetles (Scarabaeidae)—elytra meet along their entire length at the suture; abdomen not exposed from above

CALIFORNIA FAUNA: Six species in one genus.

The six species of California *Lichnanthe* are patchily distributed along the coast or live along inland rivers and streams. Two species are restricted to coastal dunes and are considered to be sensitive species. The White Sand Bear Scarab *(L. albipilosa)* (13.5 to 17.5 mm) is covered with thick, white hair and is found only at the coastal sand dunes at Oso Flaco Lake in San Luis Obispo County, on the central coast. It is active April through May. *Lichnanthe ursina* (12.9 to 17.2 mm) is restricted to coastal dunes in northern California, from Sonoma County to San Mateo County. Of the two distinct color phases, the light form is the most common. This species is most active during the months of May and June. Males are often encountered flying close to the surface of the sand as they search for mates. The Bee Scarab *(L. apina)* (9.7 to 13.5) (pl. 56) lives along streams and rivers in the Coast, Transverse, and Peninsular Ranges, as well as the Great Central Valley. The male is found in June and July flying down sandy paths winding through thickets of willow *(Salix)* or bamboo along rivers of the southern coastal plain. Mating occurs on the ground or on vegetation. With its bright metallic green or blue green prothorax, the Bee Scarab resembles a sweat bee in flight. The male has been observed resting on yarrow *(Achillea millefolium)* along the upper reaches of the Santa Ana River in the San Bernardino Mountains. The orange yellow form is much more common than the black and white variety. Rathvon's Bumblebee Scarab *(L. rathvoni)* (10.6 to 16.6 mm) is widespread in western North America from British Columbia to California and to Idaho, Utah, and Nevada. It is most common in July. In California, Rathvon's Bumblebee Scarab is known throughout the state and appears to be most abundant in sandy, streamside areas in the north. The orange color form is the most common, followed by the yellow and black varieties. This species has also been observed on yarrow at Hallelujah Junction in Lassen County.

COLLECTING METHODS: Bumblebee scarabs are collected on the wing or while resting on the ground or vegetation. Males are sometimes found on flowers. They are usually active during the late morning and early afternoon hours on sunny or slightly overcast days during spring and summer.

SCARAB BEETLES Scarabaeidae

The scarabs comprise one of the largest and most diverse families of beetles in California. They have long been popular with collectors and naturalists because of their large size, beautiful colors, and interesting behaviors. The life histories of the family are incredibly diverse. Eggs are laid in or near a suitable substrate, including dung, compost, or leaf litter beneath the adult's food plant. The distinctive, C-shaped larvae, called white grubs, molt twice before constructing a chamber and transforming into a pupa. Some larvae (e.g., *Cotinis, Cremastocheilus,* and *Euphoria*) construct pupal chambers from their own fecal material. Most species overwinter as larvae or adults. Depending upon moisture and temperature conditions, adults emerge the following spring or summer and begin feeding on dung, compost, detritus, or roots. Some species are specialists, preferring to breed in the nests of ants, termites, birds, or rodents. Dung scarabs may provide food for their young, but most other species simply lay their eggs and move on. The biology and behavior of most California scarabs remain unknown.

Some species are diurnal or crepuscular, but most are nocturnal. In California, several genera (e.g., *Serica, Diplotaxis, Amblonoxia, Polyphylla,* and *Cyclocephala*) are readily familiar to homeowners in residential areas and mountain communities because they are attracted to porch lights at night, sometimes in large numbers. The robust May beetles and June beetles, or "June bugs," of the genus *Phyllophaga,* a common sight swarming about lights in the Midwest and the eastern and southern United States, are seldom encountered in California.

A few species are of minor concern as pests in California. The feeding activities of some larval *Aphodius* and *Ataenius* damage lawns, especially in parks and golf courses. The larvae of masked chafers *(Cyclocephala)* are also turf pests. Adult *Serica, Dichelonyx, Diplotaxis, Polyphylla,* and *Hoplia* may defoliate deciduous garden shrubs and orchard trees, as well as some conifers. Their larvae may be particularly destructive in forest nurseries, where they feed on the tender roots of seedling trees.

IDENTIFICATION: California scarab beetles vary considerably in shape and are oval, long, square, cylindrical, or slightly flattened and range in length up to 30.0 mm or more. They are black, brown, yellowish brown, and occasionally green, metallic, or scaled with blotched or striped patterns. The head is weakly hypognathous. The eight- to 10-segmented antennae are tipped with three to seven fanlike segments that fold tightly into a flat, lopsided club. The antennal club appears velvety in the dung scarabs but is nearly bare in all other species. The pronotum is variable, with or without horns and tubercles, but the sides are always distinctly margined. The scutellum is hidden in the dung beetle genus *Onthophagus,* but visible in all others. The elytra are slightly rounded or parallel sided, with the surface smooth, distinctly pitted, grooved, or covered with scales, and completely conceal the abdomen only in some dung scarabs. The tarsal formula is 5-5-5, with the claws equal in size or not, and simple or toothed. The abdomen has six segments visible from below.

The whitish, yellowish, or creamy white larvae, or grubs, are C shaped. In some dung scarabs (e.g., *Canthon, Liatongus,* and *Onthophagus*) the larvae appear humpbacked due to the enlargement of the thoracic segments. The head is distinct and darker than the rest of the body and has a pair of four-segmented antennae. Simple eyes are present, faintly indicated by pigmented spots, or absent. The legs are two-segmented in some dung scarabs (e.g., *Canthon, Liatongus,* and *Onthophagus*) and four-segmented in all other groups. The 10-segmented abdomen lacks projections.

SIMILAR CALIFORNIA FAMILIES:
- stag beetles (Lucanidae)—antennae usually elbowed, with club segments unable to form a compressed club
- false stag beetles (Diphyllostomatidae)—antennae with club segments unable to form a compressed club; mandibles usually protruding; abdomen with seven segments visible from below
- enigmatic scarab beetles (Glaresidae)—abdomen with five segments visible from below; first segment of antennal club cup-shaped to receive remaining segments
- hide beetles (Trogidae)—abdomen with five segments visible from below
- rain beetles (Pleocomidae)—antennae 11-segmented; mandibles absent
- earth-boring scarab beetles (Geotrupidae)—antennae 11-segmented
- sand-loving scarab beetles (Ochodaeidae)—mandibles exposed; spurs on ends of middle tibiae appear feathery or finely notched
- scavenger scarab beetles (Hybosoridae)—mandibles exposed; first segment of antennal club cup-shaped to receive remaining segments
- bumblebee scarab beetles (Glaphyridae)—elytra short, with distinctly separated tips exposing abdominal segments above

CALIFORNIA FAUNA: 291 species in 47 genera.

The genus *Aphodius* has nearly 2,000 species worldwide, 46 of which are recorded from California. Species indigenous to North America are often associated with deer dung, live inside rodent nests, or are associated with decaying plant material buried in sand dunes. However, the most common and conspicuous species in California are those accidentally introduced from Europe. Long established in North America, these species are usually found in cow and horse manure. The European Dung Beetle *(A. fimetarius)* (6.0 to 9.0 mm) (pl. 57) is black with distinctive brick red elytra and spots on the pronotum. It was first reported from northern California in 1923. It is sometimes found in older cow dung. Another European import, *A. fossor* (8.0 to 11.0 mm), is completely black and has three small tubercles across its head. First recorded in North America from British Columbia in 1952, *A. fossor* is now known from Alaska, across southern Canada and the northern United States to the Pacific Northwest. It is locally abundant in northern California. Another large bicolored species, *A. hamatus* (5.5 to 8.3 mm), has pale brownish, whitish, or yellowish elytra. It is found across the northern United States. In California it is abundant in several localities along the east side of the Sierra in cattle dung. The larva is a root feeder. *Aphodius granarius* (3.0 to 5.0 mm) is a common, smaller, black species found in cow dung. The larva prefers to develop among the roots of grass. Both the adult and larva of the cosmopolitan *A. lividus* (3.0 to 6.0 mm) (pl. 58) are found in cattle dung. This beetle is distinctively marked with a broad, pale band running the length of each elytron and is very similar in appearance to another introduced species, *A. pseudolividus*. *Aphodius lividus* has thick, shovellike spurs at the tips of its hind tibiae, whereas *A. pseudolividus* has spurs that are more tapered. The larva of *A. pardalis* (4.0 to 6.0 mm) is a grass-root feeder, damaging lawns, golf courses, and bowling greens throughout the Pacific Northwest. The small, speckled adults are common at lights near grassy areas. Both larval and adult stages of *A. vittatus* (3.0 to 5.0 mm) are found in cow manure. The shiny black elytra of the adult is distinctly marked with red.

Small and cylindrical, species of *Ataenius* are similar in appearance to *Aphodius* but differ in having a distinct ridge crossing the middle of their hind tibiae. The adult is active at night and, unlike its larvae, does not feed on the roots of grasses. The Shining Black Turfgrass Ataenius *(Ataenius spretulus)* (3.6 to 5.5 mm) has parallel grooves on the elytra and is a turf pest throughout much of the United States. The larva feeds underground and causes damage by pruning the roots just below the thatch, resulting in patches of thinning turf that eventually turn brown. Damaged sod is easily rolled back, revealing the C-shaped white grubs. The adults are commonly attracted to lights. *Ataenius platensis* (3.6 to 4.9 mm) (pl. 59) occurs primarily throughout southern United States but is also established in the coastal plain of southern California. It is sometimes attracted to lights. Eleven other species in this genus occur in the state.

Dung-rolling scarabs are represented by one species in California. *Canthon simplex* (5.0 to 9.0 mm) (pl. 60) is broadly distributed from British Columbia and Alberta through Washington, Idaho, Montana, and Oregon. In California it occurs in various habitats throughout the Sierra Nevada and Peninsular Ranges, from coastal chaparral to pine forest. Using its legs to carve out a piece of cow or horse dung, the female fashions a dung ball, places a single egg inside, rolls it a short distance away, and buries it. The dung provides both food and shelter for the developing grub. It uses cattle dung or rolls the pellets of the Beechey Ground Squirrel *(Spermophilus beecheyi)* and the Yellow-bellied Marmot *(Marmota flaviventris).* A form of this species, distinguished by reddish spots on its shoulders, is distributed along the foothills in the western and southern slopes of the Sierra Nevada, up to 9,000 ft.

Euoniticellus intermedius (7.0 to 9.0 mm) is light brown with distinctive symmetrical markings on the pronotum. The male has a single, blunt, curved horn on the head. A native of sub-Saharan Africa, *E. intermedius* searches for fresh dung pads during the day. An adult lives one or two months, during which time the female can lay more than 100 eggs. The nest structure varies depending on the availability of moisture. During the wetter parts of the year the nest of *E. intermedius* may consist of a main burrow with several branched galleries, each provided with a pear-shaped ball of dung. Under drier conditions the brood balls are placed immediately below the pad to take advantage of its available moisture.

Liatongus californicus (11.0 to 12.0 mm) is native to North America and lives in extreme southern Oregon and in northern and central California in the foothills of the Sierra Nevada. Both male and female are black, but the male is easily distinguished by an erect horn on its head. It is found in summer on cattle dung.

Onitis alexis (12.0 to 20.0 mm) is an oblong, robust species native to Africa. It also occurs in southern Europe and has been successfully introduced into Australia in an effort to control dung-breeding flies. *Onitis* is easily distinguished from the other dung beetles in California by its large size, dark brown or coppery green pronotum with two distinct pits centrally located on the rear margin, and distinctly grooved, light brown elytra. Both the male and female lack horns. A single hooklike spine located on the inside of each hind femur distinguishes the male. The adult is active at dusk and prefers dung that is two days old or less. *Onitis* is established along the coastal plain of southern California.

The Brown Dung Beetle *(Onthophagus gazella)* (8.0 to 13.0 mm) (pl. 61) is an African-Asian species that was purposely introduced into North America via Texas in 1972. Since then it has spread throughout much of southern United States. It was also released at various sites in central and southern California and is now well established in the coastal plains of Ventura County and southward, the Coachella Valley, and the coastal chaparral regions of Orange and San Diego Counties. Much of the body is greenish or coppery black, with a narrow, pale band along the sides and base of the pronotum, and brownish yellow elytra and portions of the underside. The male has a pair of short, straight horns on its head, and the female lacks horns. It flies to fresh pads of horse and cow dung from dusk to dawn and is readily attracted to lights. *Onthophagus taurus* (6.0 to 11.5 mm) is a small species native to southern Europe and Asia Minor. It is nearly all black, sometimes with a dark greenish sheen. It is widely established in California, especially in urban areas, where it is often attracted to dog feces. Major males have a long pair of curved horns sweeping back over the pronotum, but smaller males have short, straight horns or none at all. Both exotic *Onthophagus* species bury their dung balls in chambers dug next to the dung pad.

The adults of the genus *Hoplia* are small, robust beetles often covered in brightly colored scales. They are found feeding on foliage and flowers of various wild and garden plants in spring and summer. *Hoplia dispar* (6.0 to 13.0 mm) is variably hued with light solid yellow, pale green, orange to light brown, or the same colors superimposed with a black pattern. The underside and the exposed upper portion of the abdomen are always covered with pearly or silvery scales. It is distributed in the Cascade Range and Sierra Nevada and is commonly found on the flowers of California-lilac *(Ceanothus),* especially from May through July.

The Grapevine Hoplia *(H. callipyge)* (5.0 to 11.0 mm) (pl. 62) is brownish and widely distributed throughout California, especially in the coastal plains, valleys, and foothills. The adult feeds on the leaves and flowers of a variety of plants, including grapes *(Vitis)*, roses *(Rosa)*, ceanothus *(Ceanothus)*, lupines *(Lupinus)*, willows *(Salix)*, owl's clovers *(Castilleja)*, California buckeye *(Aesculus californica)*, and yerba santa *(Eriodictyon)*. The scales on the elytra vary from oval to long. Several other populations of *Hoplia* are found in the coastal plain and desert regions of southern California and require further study before they can be reliably assigned to species. Four species are currently recognized in the state.

Over 40 species of *Serica* are known from the state, including several undescribed species. These small (6.0 to 10.0 mm), round beetles are brown, reddish brown, or black, with shallow grooves on the elytra. They often have a trace of iridescence under strong light. Positive identification to species is usually possible only by dissection and examination of the male genitalia. The nocturnal adults feed on the leaves and flowers of various plants and are commonly attracted to lights. They sometimes appear in great numbers and do serious damage by defoliating fruit trees. The reddish brown *S. perigonia* (7.0 to 9.0 mm) (pl. 63) is common throughout the valleys and foothills of southern California, where it is encountered at lights and on the flowers of California buckwheat *(Eriogonum fasciculatum)*.

Species of the genus *Dinacoma* lack distinct stripes on the elytra and have antennal clubs that are twice as long as the antennal pedicel in the males. *Dinacoma marginata* (15.1 to 21.1 mm) (pl. 64) is known primarily during the month of June from the vicinities of Hemet, Riverside County, and Del Mar, San Diego County. Additional populations may be found in the few remaining areas of desertlike habitat scattered along some of the major river drainages flowing out of the San Bernardino and San Gabriel Mountains. The scales near the elytral suture are white, but the rest are golden brown. The elytral scales of *D. caseyi* (14.3 to 17.8 mm) are uniformly white. This species is active in April and May in the vicinity of Palm Canyon and Palm Springs, Riverside County. Because of the destruction of its habitat due to development, this species is of special concern.

The Dusty June Beetle *(Amblonoxia palpalis)* (16.0 to 26.0 mm) (pl. 65) is so named because of the grayish appearance of its scaled body. It is commonly attracted to lights during summer in the coastal flood plains of southern California. *Amblonoxia harfordi* (16.0 to 25.0 mm) is similar in appearance and is found in the San Francisco Bay region, the adjacent Coast Ranges, and northern Great Central Valley. The females of these and other species of *Amblonoxia* are seldom attracted to light and generally have fewer scales. A similar species, *Plectrodes pubescens* (15.2 to 22.0 mm), is reddish and clothed with yellowish hairs. It is common at lights in the San Joaquin Valley. The males of *Plectrodes* are distinguished from those of *Amblonoxia* by having distinct grooves separating the abdominal segments.

Fourteen species of *Polyphylla* are found throughout the state of California. The Ten-lined June Beetle *(P. decemlineata)* (18.0 to 31.0 mm) (pl. 66) is widely distributed throughout western North America. It is found in all areas of California except the Mojave and Colorado Deserts. The larva feeds on the roots of various plants, whereas the adult has been recorded to feed on the needles of ponderosa pine *(Pinus ponderosa)*. The adult flies at dusk and in the evening from late spring through summer and is commonly attracted to lights. *Polyphylla nigra* (21.0 to 30.2 mm) is known throughout northern California, especially along the coast, and is also common in the higher elevations of the Transverse and Peninsular Ranges. Individuals have also been collected from the Channel Islands. It is distinguished from *P. decemlineata* by having scattered hairs on the pronotum and stripes with a slightly rougher edge. The White-lined June Beetle *(P. crinita)* is very similar to *P. nigra* but is usually 30 mm or greater in length. The male lacks hairs on the pronotum and the base color under the scales of both sexes is olive green. It is best known from the Coast Ranges of northern and central California but is occasionally found in the higher elevations of the Transverse and Peninsular Ranges. This species is sometimes attacked and eaten by robber flies. The Mount Hermon June Beetle, *P. barbata* (20.0 to 22.0 mm), is known only from the Zayante sand hills of the Mount Hermon, Ben Lomond, and Scots Valley areas of Santa Cruz County. It was listed as a federally endangered species in 1997. This beetle is similar to other beetles of the genus *Polyphylla,* except the elytra have scattered erect hairs and its stripes are broken up into scattered clumps. The larva lives underground, where it feeds on roots. The Mount Hermon June Beetle may complete its life cycle in one year. The adult male emerges in summer and takes flight in search of females at the entrances of their burrows by homing in on their pheromones. The adult male is commonly attracted to lights. Sand mining, urban development, fire suppression, and agriculture threaten the Zayante sand hills habitat.

Although most of the 237 species of *Diplotaxis* are from Mexico, 18 species occur in California. *Diplotaxis sierrae* (11.0 to 14.5 mm) (pl. 67) is a large, dark red to black species found throughout the Sierra Nevada, Panamint Mountains, and Transverse and Peninsular Ranges. The reddish brown to black *D. moerens moerens* (8.0 to 12.0 mm) is slightly smaller and is found in Arizona, Nevada, Utah, southern California, and the northern half of the Baja California Peninsula. A broad, raised area across the forehead easily distinguishes the smaller *D. subangulata* (6.0 to 9.0 mm). This species is common at lights in the deserts and surrounding foothills during late spring and summer.

The genus *Coenonycha* is distributed throughout desert, chaparral, and Great Basin sage habitats in the western United States and Baja California. These beetles become active in late December or early January along the coast and in the Colorado Desert of southern California. Foothill and Mojave Desert populations reach their peak in March and April, while high-elevation populations along the slopes of the eastern Sierra and in the Great Basin become active in May and June. *Coenonycha* are found after dark in large numbers on the tips of branches of various plants, especially on Great Basin sagebrush *(Artemisia tridentata),* California buckwheat *(Eriogonum fasciculatum),* and chamise *(Adenostoma fasciculatum),* where they remain motionless, often with their forelegs extended. Many species are attracted to light. Several species have reduced flight wings and are incapable of flight. The yellowish brown *C. testacea* (7.0 to 11.0 mm) (pl. 68) is known from the coastal plain and foothill areas in San Luis Obispo, Santa Barbara, Ventura, and Los Angeles Counties. Several similar, yet undescribed species are found in other regions of California. The dark reddish brown *C. fusca* (8.8 to 10.8 mm) is widely distributed in the foothills of the Coast Ranges and the western Sierra Nevada, where it is found perching and feeding on chamise. The large *C. ampla* (10.0 to 13.0 mm) is light to dark reddish brown. It is found in the southern Coast Ranges and interior slopes of the Transverse Ranges, where it feeds on California juniper *(Juniperus californica).* Twenty-six described species are known from the state, and several more await description.

Dichelonyx is widely distributed in the United States and Canada, with both diurnal and nocturnal species. Several species are commonly attracted to lights at night. In California 15 species are found in chaparral and conifer forests, where they feed primarily on chamise, oak *(Quercus),* pine *(Pinus),* and fir *(Abies).* Although not considered to be of economic importance, the adults of some species are reported to defoliate conifers and shade trees. They are similar in appearance to *Coenonycha,* but nearly all species have at least a metallic luster. *Dichelonyx truncata* (5.6 to 7.8 mm) is grayish black or yellowish with a pale elytral border and tinged overall with a metallic bronze sheen. It is found in the southern Sierra Nevada and Coast, Tranverse, and Peninsular Ranges, and in some isolated desert mountain ranges. It feeds on a wide variety of plants, including ceanothus, manzanita *(Arctostaphylos),* willow, mountain-mahogany *(Cercocarpus),* rose, and California flannelbush *(Fremontodendron californicum).* *Dichelonyx pusilla* (7.0 to 9.0 mm) is a dark brown species with faintly to distinctly white-striped elytra. It is common on chamise and other plants in coastal chaparral communities in the coastal plains of southern California and the coastal slopes of the Transverse and Peninsular Ranges. *Dichelonyx valida vicina* (8.5 to 14.5 mm) (pl. 69) has bright metallic green elytra and is widely distributed in the mountainous regions of the state, where it feeds on pine and fir. It is sometimes collected with another bright metallic green species, *D. backi* (formerly known as *D. fulgida)* (6.5 to 12.2 mm). This species is distinguished from *D. valida vicina* by not having a distinct groove running down the middle of the pronotum.

Species of the genus *Phobetus* are hairy, black, brown, tan, or pale yellowish beetles with a thick pile of hair underneath their bodies. Day-active species are often encountered flying low over plants and sandy washes in the morning or afternoon from January through April. The male of the dark yellowish brown Hairy June Beetle *(P. comatus)* (12.0 to 17.0 mm) (pl. 70) flies at dusk in chaparral and oak woodlands in foothill regions throughout the state in March through May and is occasionally attracted to lights, while the female remains in its burrow. The larva feeds on the roots of canyon live oak *(Quercus chrysolepis),* Nutall's scrub oak *(Q. dumosa),* and California sagebrush *(Artemisia californica).* *Phobetus mojavus* (14.0 to 17.0 mm) is pale yellow with long, white hair and is found at lights or perching on desert plants at night in the Mojave Desert. *Phobetus saylori* (11.0 to 12.0 mm) is black with pale, translucent patches on the elytra and long white hair. It is found on the eastern slopes of the southern Sierra Nevada, desert slopes of the Transverse Ranges, and juniper-pinyon woodlands of the Santa Rosa Mountains in the Peninsular Ranges. Ten species are known in California, plus several undescribed species from the Channel Islands and elsewhere in the state.

During the months of April and May, visitors to the Colorado River may come across *Cotalpa flavida* (22.0 to 26.0 mm) (pl. 71), which sometimes feeds in numbers during the day on the leaves of willows or Fremont cottonwood *(Populus fremontii)* or is attracted to lights. This large, yellow scarab has reddish legs with black feet. It also ranges up the Colorado River and its tributaries, in Nevada, Arizona, and Utah.

Four species of *Paracotalpa* occur in California. The Little Bear *(P. ursina)* (10.0 to 23.0 mm) (pl. 72) is a robust, round beetle with a black, steel blue, or metallic green head and pronotum and black or reddish elytra. Adults appear in large numbers during the day in spring, flying low over grasslands and chaparral. It is common along the ranges of the southern Great Central Valley and is often associated with chamise in the chaparral communities of southern California. *Paracotalpa puncticollis* (17.5 to 22.0 mm) (pl. 73) has bright yellow elytra with thin, black stripes and is restricted to the juniper woodlands in the mountains of the eastern Mojave Desert, the southeastern slopes of the San Bernardino Mountains, and the inland slopes of the Peninsular Ranges. *Paracotalpa granicollis* (13.0 to 18.0 mm) is similar in appearance to *P. ursina* but is distinguished by the coarsely pitted and metallic green head and pronotum. It occupies the juniper woodlands of the northeastern and central Great Basin Ranges and ranges into similar habitats in Idaho, Nevada, Oregon, Washington, and Utah. The black *P. deserta* (14.5 to 19.0 mm) is restricted in California to the southwestern reaches of the Colorado Desert of Imperial and San Diego Counties. In February and March it is found during the afternoons emerging from the bases of creosote bush *(Larrea tridentata)* and burr sage *(Franseria)*, or feeding on the leaves and blossoms of desert annuals such as *Abronia villosa* and *Camissonia claviformis*. It is sometimes found impaled on the spines of burr sage, possibly put there by Loggerhead Shrikes *(Lanius ludovicianus)*.

Masked chafers of the genus *Cyclocephala* are restricted to the New World, with 250 species known primarily from the tropics. The name "masked chafer" is derived from the fact that some species have a dark band connecting the eyes. Five species in California are distributed throughout the coastal plains, valleys, foothills, and deserts in summer. *Cyclocephala hirta* (10.5 to 14.0 mm) is yellowish brown with short hairs scattered over its body and is the only California species with a "bead," or lip, along the posterior margin of the pronotum. It is often encountered in large numbers on summer nights, swarming over grassy areas at dusk or attracted to lights at night. A slightly darker, less hairy species, *C. pasadenae* (12.0 to 14.0 mm) (pl. 74) is also widespread and common in residential areas. *Cyclocephala longula* (11.0 to 14.3 mm) is a bit more slender than the previous species, and its head and pronotum are slightly darker than the elytra. This species is more common in the deserts and surrounding foothills. Another species, distinctly bicolored, *C. melanocephala* (10.6 to 15.0 mm), is occasionally found at lights but is more often found deep inside the flowers of Jimson weed *(Datura)*. The elytra are pale yellowish brown, but the head and pronotum are dark reddish brown to black. It is distributed from the southwestern United States to Argentina.

The Western Carrot Beetle *(Tomarus* [formerly known as *Ligyrus] gibbosus obsoletus)* (14.0 to 16.0 mm) (pl. 75), also known as the Muck Beetle, is a robust, shiny reddish brown beetle. It is common throughout the state and is frequently found at lights in late winter, spring, and summer. The larva feeds on the roots of grasses and other plants, sometimes attacking potted plants in nurseries, while the adult works above and below the ground attacking beets, carrots, celery, corn, cotton, dahlias, elm, oak, parsnip, potatoes, and sunflowers.

Sleeper's Elephant Beetle *(Megasoma sleeperi)* (24.8 to 30.5 mm) (pl. 76) is known from the Colorado Desert. Both the male and female fly at dusk during the late summer and are occasionally attracted to lights and traps baited with malt. This species is sometimes found emerging from deep sand at the Algodones Dunes in Imperial County. Adults emerging from the sand are occasionally attacked and dismembered by rodents, probably kangaroo rats *(Dipodomys)*. The female can be observed clinging to the branches of blue palo verde *(Cercidium floridum)* in the early morning hours. The larva most likely feeds on dead or dying palo verde and honey mesquite *(Prosopis glandulosa* var. *torreyana)*.

Hemiphileurus illatus (18.0 to 22.0 mm) (pl. 77) is a shiny dark brown and somewhat elongate beetle. The elytra are distinctly grooved and the front tibiae have four strong teeth. It occurs in California in both the Colorado and Mojave Deserts. In the region east of the Algodones Sand Dunes of Imperial County it is occasionally attracted to lights placed in sandy washes lined with blue palo verdes and ironwood *(Olneya tesota).* This species is also found in Arizona, the Baja California Peninsula, and western Mexico. It is active in summer.

The Green Fig Beetle *(Cotinis mutabilis,* often misspelled as "*Cotinus*") (20.0 to 30.0 mm) (pl. 78), also called the Peach Beetle, is erroneously referred to as the Chinese Beetle or Japanese Beetle. It is common in the coastal plains of southern California, portions of the Antelope Valley, and the length of the Great Central Valley, where it breeds in compost and dung heaps. It also occurs in Arizona and Mexico. The adult produces a loud buzzing noise as it clumsily flies through yards, orchards, and parks. The large grub (up to 50 mm) has well-developed legs but prefers to walk on its back and is sometimes called "back crawler," a habit characteristic among scarabs known as fruit chafers. The adult attacks peaches, figs, nectarines, and other fruits damaged by birds or that have fallen on the ground. Adults are active from late June through November, with peak emergence occurring in July and August. This species was also known as *C. texanus.*

The brightly colored *Euphoria fascifera trapezium* (12.8 to 13.1 mm) (pl. 79) is sometimes encountered flying during the day after heavy summer rains in the mountains and desert flats of the Colorado and Mojave Deserts. It breeds in the accumulations of plant material and dung inside the nests of pack rats *(Neotoma).* On the wing this black and yellow orange beetle resembles a bumblebee in flight. It is occasionally attracted to malt traps.

The ant-loving scarab beetles of the genus *Cremastocheilus* become active on the first warm days of spring or after infrequent desert showers in late summer. They are flattened, black beetles that often have faint white markings. Males are seen flying in a zigzag pattern as they track pheromone trails of the females. Eggs and larvae develop inside ant nests or rodent burrows, where the larvae feed on decaying plant matter. The larvae lack any sort of appeasement glands, but they may be able to absorb nest odors directly into their exoskeleton. The larvae pupate in earthen chambers. Beetles land near the entrance of the ant's colony (primarily those of the genera *Formica, Pogonomyrmex,* and *Camponotus*) and enter the nest on their own or are dragged in by the ants. Once inside the nest, *Cremastocheilus* may be persecuted or left unmolested to feed on the eggs and larvae of the ants. Twelve species and one subspecies of ant-loving scarabs are recorded in California. *Cremastocheilus angularis* (10.0 to 13.0 mm) is known only from California, where it ranges throughout much of the state. *Cremastocheilus armatus* (10.0 to 13.0 mm) occurs throughout the Pacific Northwest, Idaho, Montana, and Nevada. In California it appears to be most common in the northern and central parts of the state. The bases of the elytra of *C. armatus* have yellowish patches, while those of *C. angularis* do not. Both species are associated with mound-building ants *(Formica)*. *Cremastocheilus westwoodi* (7.0 to 9.0 mm) (pl. 80) is found along the eastern side of the Coast Ranges, along the northern edge of the Transverse Ranges, and in Owens Valley. It is most common in spring and is thought to be associated with black harvester ants of the genus *Messor*. This species is very similar to *C. schaumii*, but the segments of the hind tarsi are loosely joined together. The hind tarsus is also about two-thirds the length of the hind femora.

Valgus californicus (8.3 to 9.0 mm) (pl. 81) breeds beneath the bark of pine stumps and dead logs infested with Pacific Dampwood Termites *(Zootermopsis angusticollis)*. The larva is found in moist termite frass, while the pupa is found in small chambers within hollows. The adult is found beneath bark or in the blind ends of galleries in dry wood. This beetle is rarely encountered in flight. Eastern species of *Valgus* are frequently encountered on flowers.

COLLECTING METHODS: California scarabs are generally quite common and easy to collect. The most effective way to find nocturnal species is with light traps, especially those with ultraviolet or mercury vapor lights. Sweeping, beating, and searching vegetation at night is useful for locating *Serica, Coenonycha,* and *Diplotaxis*. Once located, carefully inspect the plants with a headlamp or flashlight to locate mating pairs or feeding individuals to establish host plant records. Carefully digging through the soil under the leaf litter of chamise, California buckwheat, mountain-mahogany, oaks, and other adult host plants will also produce the larvae and pupae of these species. Look for dung beetles in and beneath relatively fresh cow dung; old, dried out chips are seldom productive. Dog dung is particularly attractive to *Onthophagus taurus* in the suburbs, whereas cattle dung lures *O. gazella* in parts of rural southern California. Pitfall traps baited with dung may be productive in some areas, especially for introduced species. Native species of *Aphodius* are associated with rodent burrows and deer dung. Traps baited with malt and molasses are attractive to some scarabs in desert regions, such as *Euphoria* and *Megasoma*. Rotting fruit will attract *Cotinis*. Spring flowers in the deserts, especially members of the sunflower family, are particularly attractive to several genera of smaller scarabs, including *Oncerus, Chnaunanthus,* and *Gymnopyge*. Compost, dung heaps, and other moist accumulations of plant material are especially attractive to the larva and adult female of *Cotinis mutabilis*.

SOFT-BODIED PLANT BEETLES   Dascillidae

Very little is known about the habits of soft-bodied plant beetles. The larvae of *Dascillus* are found in moist sandy soil or under rocks and apparently feed on the roots of plants. They resemble white grubs (Scarabaeidae) but are not as strongly C shaped and are larger toward the head, tapering behind and slightly flattened. Adults are often found on grass stems in spring and feeding on fruit trees. Male *Anorus* are commonly attracted to lights in the drier regions of southern California, but the females have greatly reduced wings and are rarely seen.

IDENTIFICATION: California soft-bodied plant beetles are of two distinct types. *Dascillus* species are black to mottled gray brownish bullet-shaped beetles with hypognathous heads. They range in length from 8.0 to 20.0 mm. The head is prognathous in *Anorus* species. These are soft, pale to brown beetles that range from 7.0 to 13.0 mm. The antennae in both genera are long, up to half the body length, and consist of 11 segments. The pronotum is broader than the head and equal in width or slightly narrower than the base of the elytra. The scutellum is visible. The elytra are covered with fine hairs and rows of tiny pits and completely conceal the abdomen, except in *Anorus* females. The tarsal formula is 5-5-5, with the claws equal in size and simple. The abdomen has five segments visible from below.

Only the larvae of *Dascillus* are known. Mature larvae are yellowish, elongate, slightly curved, with a somewhat large, flattened prominent head. The antennae are three-segmented, and simple eyes are absent. The well-developed legs are five-segmented, including the claw. The 10-segmented abdomen does not have any projections, although the European *D. cervinus* does have two small projections on the ninth segment.

SIMILAR CALIFORNIA FAMILIES:

- click beetles (Elateridae)—prothorax is loosely hinged to rest of the body, with hind angles extended backward
- some false click beetles (*Perithopus,* Eucnemidae)—prothorax with hind angles extending backward

CALIFORNIA FAUNA: Four species in two genera.

Two species of genus *Dascillus* (males 8.0 to 13.5 mm; females to 20.0 mm) occur in northern California. Davidson's Beetle *(D. davidsoni)* (pl. 82) is dull brown or blackish and covered with minute, gray hairs, except for two irregular, bare crossbands. The larva has been found on the roots of apple *(Malus),* cherry *(Prunus),* acacia *(Acacia),* snowberry *(Symphoricarpos),* and other native plants. The adult appears in spring and feeds on various fruit trees. It has also been beaten from toyon *(Heteromeles arbutifolia). Dascillus plumbeus* is black and covered with gray pubescence, giving it an overall slate color. The adult of this species also feeds on fruit trees and is found in the western foothills of the northern Sierra Nevada.

Anorus is represented by three described species in North America, one each in California and Arizona, and the third shared between both states. The female *Anorus* has reduced wings and has been found at the entrance to its burrow in the soil, in a house near baseboards recently treated with insecticides for termites, and in a dead catclaw *(Acacia greggii)* root in association with the subterranean dry-rot termite *Paraneotermes simplicicornis. Anorus piceus* (7.0 to 10.0 mm) (pl. 83) is an elongate, parallel-sided beetle that is completely brown and clothed throughout with fine, light-colored hairs. This beetle has large, curved mandibles nearly as long as the head, with one or two prominent teeth. This species flies at dusk and is commonly attracted to lights in southern California. A second species, *A. parvicollis,* also occurs in the deserts of southern California. The sides of the pronotum are angulate at the middle, while those of *A. piceus* are evenly rounded.

COLLECTING METHODS: Inspect fruit trees in early spring for *Dascillus* in the northern and central parts of the state. *Anorus* is attracted to lights in late spring and early summer in the deserts and adjacent foothills of southern California.

CEDAR BEETLES or
CICADA PARASITE BEETLES Rhipiceridae

The cedar beetle or cicada parasite beetle family contains five genera and about 50 species worldwide, six of which occur in the United States. Little is known about these beetles' biology, although some information has been published on the observations of *Sandalus niger* from eastern United States. This species gathers in mating aggregations in late summer on the trunks of American elm *(Ulmus americanus)*, shingle oak *(Quercus imbricarius)*, and other trees. Males and females are found crawling or copulating on the bark, resting on nearby grass, or flying nearby. Females lay their eggs in the cracks and crevices of the bark with an ovipositor that is nearly as long as their body.

The larvae feed externally on the subterranean larvae of cicadas. Females lay large numbers of eggs in the holes and cracks of bark, preferably in areas where there are plenty of cicadas. The larvae develop by hypermetamorphosis. The eggs hatch into triungulins, highly active, silverfishlike larvae with well-developed legs. They make their way into the soil and locate young cicada larvae. The intermediate stages of development are unknown. A single pupa with a shed larval exoskeleton of *S. niger* was found inside the hollowed-out husk of a larval cicada. The remains of the larval exoskeleton revealed a more sedentary, grublike larva. Label data accompanying some California *Sandalus* also indicate a larval relationship with cicadas.

IDENTIFICATION: California cedar beetles are long, convex, and usually black. The head is weakly hypognathous. The 11-segmented antennnae are distinctly fan shaped in the males but are more saw-toothed in females. The pronotum is narrowed at the head and wider behind, but not nearly as wide as the base of the elytra. The scutellum is visible. The elytra are long and completely conceal the abdomen. The surface of the elytra is vaguely ribbed and coarsely pitted. The tarsal formula is 5-5-5, with each of the first four tarsomeres distinctly heart shaped and padded. The claws are equal in size and simple. The abdomen has five segments visible from below.

The larvae of the California cedar beetles, *S. californicus* and *S. cribricollis,* are unknown. The first-stage, silverfish-shaped triungulin larva of the eastern *S. niger* is long, white, and flattened in appearance. The antennae are three-segmented and the simple eyes are absent. The five-segmented legs are long and tipped with a long, spinelike claw that is half the length of the rest of the leg. The 10-segmented abdomen does not end in any projections. The mature larva of *S. niger* is known only from a shed exoskeleton. The remains indicate a white and more grublike, spindle-shaped body with an enlarged abdomen. The head lacks simple eyes. Each leg is short and three-segmented, including a hooked claw. The abdomen is 10-segmented, and the ninth segment bears a pair of small, upturned projections.

SIMILAR CALIFORNIA FAMILIES: The fan-shaped antennae, elongate body form, and distinctly lobed tarsi of *Sandalus* are distinctive.

CALIFORNIA FAUNA: Two species in one genus.

Twenty-five species of *Sandalus* occur in North and South America, Africa, India, Southeast Asia, China, and Japan. Only two species are found in California. The Black California Cedar Beetle *(S. californicus)* (16.0 to 17.0 mm) occurs in Idaho, Nevada, Oregon, and California. It is active during August and September in the Coast Ranges south of San Francisco Bay and the Great Central Valley, Cascade Ranges, Sierra Nevada, Great Basin Desert, and coastal plains of southern California. Individuals have been found "on grass at base of elm tree" and crawling up the branches of Great Basin sagebrush *(Artemisia tridentata),* whereas others were collected at lights or light traps. The second species, *S. cribricollis* (19.0 mm) (pl. 84), was described from Loyalton in Sierra County and is also found in the southern part of the state, including Santa Catalina Island. *Sandalus cribricollis* is distinguished from *S. californicus* by having very coarse pits on the head and prothorax, whereas the latter is densely, yet finely pitted. Specimens of both species range from dark reddish brown to black.

COLLECTING METHODS: Although large and conspicuous, cedar beetles are seldom collected in California. Look for *Sandalus* on the trunks of trees, on or at the bases of shrubs, or in flight in late summer. They are sometimes attracted to lights.

SCHIZOPODID BEETLES or
FALSE JEWEL BEETLES Schizopodidae

Schizopodid beetles or false jewel beetles are similar to metallic wood-boring beetles or jewel beetles (Buprestidae) but are distinguished by the deeply bilobed fourth tarsal segment of the adults. A small family with only three genera and seven species, schizopodid beetles are restricted to southwestern Arizona, California, southern Nevada, and Baja California.

Eggs are probably laid in the soil. The immature stages are poorly known and are thought to be external root feeders.

IDENTIFICATION: Schizopodid beetles are somewhat stout and very convex beetles ranging up to 19.5 mm in length. They vary from light brown, blue, purple, coppery green overall, or green to blue with orange elytra. The head is hypognathous and has somewhat serrate 11- or 12-segmented antennae. The prothorax is wider than the head, but narrower than the base of the elytra. The scutellum is visible. The elytra are smooth or coarsely pitted, sometimes with faint traces of ribs, and completely cover the abdomen. The tarsal formula is 5-5-5, with each tarsal claw equal in size and notched. The females' abdomen has five segments visible from below, whereas the males' abdomen has six.

The long, cylindrical, legless larvae are parallel sided. The head has three well-developed yet unequal simple eyes on either side. Abdominal segments one through eight have a pair of well-developed prolegs ending in hoof-shaped structures. Segments one through seven have a pair of glands with an extended duct underneath.

SIMILAR CALIFORNIA FAMILIES:
- metallic wood-boring beetles or jewel beetles (Buprestidae)—fourth tarsal segment of adult not deeply notched or distinctly bilobed

CALIFORNIA FAUNA: Seven species and two subspecies in three genera.

Species of *Schizopus* have 11-segmented antennae. *Schizopus laetus* (9.9 to 18.0 mm) (pl. 85) is found in the Colorado Desert in spring, where it feeds on the flowers of desert-sunflower *(Geraea canescens)*. It has also been collected feeding or resting on brittlebush *(Encelia farinosa)*, spiny senna *(Senna armata)*, ringstem *(Anulocaulis annulata)*, desert tea *(Ephedra californica)*, pebble pincushion *(Chaenactis carphoclinia)*, *Psorothamnus schottii*, and poppies *(Eschscholzia)*. The surface texture of the elytra is coarse. The male is green to blue with yellow orange or red orange elytra, while the female is entirely green, blue, or purple. It is also found in southwestern Arizona, southern Nevada, and Baja California from March through June. Another species, *S. sallei sallei* (9.0 to 19.5 mm), is found on the eastern slope of the Great Central Valley from mid-May to early June. Adults are collected on dry grass and the stems of brodiaea *(Brodiaea)*. Both the male and female are brownish yellow with a darker head, pronotum, and broad elytral stripe down each wing cover. The subspecies *S. sallaei nigricans* is only known from localities in San Benito and San Luis Obispo Counties on the western slopes bordering the Great Central Valley. The male of this subspecies is similar to its eastern counterpart, but the female is black. It is found from early May to June clinging to dry grasses.

The genera *Dystaxia* and *Glyptoscelimorpha* both have 12-segmented antennae. *Dystaxia elegans* (10.5 to 14.5 mm) is active from late May through August and is distributed in the Coast Ranges from Alameda County southward to the Transverse and Peninsular Ranges. The iridescent green adult is found on the leaves of coast live oak *(Quercus agrifolia)* and scrub oak *(Q. berberidifolia)*. Another species, *D. murrayi* (9.5 to 17.0 mm), is also widely distributed, ranging from the northern Coast Ranges to the Transverse and Peninsular Ranges. It also occurs in the lower elevations of the southern Sierra, especially in the vicinity of Lake Isabella, where it is commonly found on rubber rabbitbrush *(Chrysothamnus nauseosus* subsp. *mohavensis)*. In other areas it is found on scrub oak. It is somewhat larger than *D. elegans,* and its antennal segments are elongate and less triangular.

The genus *Glyptoscelimorpha* has three species and one subspecies, all of which are found in California. *Glyptoscelimorpha viridis* (6.2 to 8.7 mm) (pl. 86) is iridescent green to coppery, with yellowish or whitish flattened hairs or setae. It is found in the juniper belts growing along the northern slopes of the Transverse Ranges and in and around the Peninsular Ranges. It is collected on California juniper *(Juniperus californica)* from mid-June through mid-August. A similar species, *G. juniperae juniperae* (8.1 to 11.7 mm), is found primarily on junipers growing along the northern and western edges of the Colorado Desert. The subspecies *G. juniperae viridiceps* lives on Utah juniper *(J. osteospermae)* in Inyo County. The distribution of the brassy bronze

G. marmorata (6.9 to 9.5 mm) is similar to that of *D. murrayi*.
COLLECTING METHODS: Adults are collected during the day on desert flowers, beaten from oaks, or hand picked or swept from herbs and dry grasses during spring and summer.

METALLIC WOOD-BORING BEETLES or JEWEL BEETLES   Buprestidae

Metallic wood-boring beetles or jewel beetles are named for their beautiful and iridescent colors. Many species are brightly marked with orange, yellow, or red bands and spots. Their streamlined, almost bullet-shaped bodies resemble those of click beetles (Elateridae); however, their rigid bodies, saw-toothed antennae, and metallic colors underneath easily distinguish them. Most species are somewhat flattened, as indicated by the typical elliptical or oval emergence holes left behind in trunks and branches. Although metallic wood-boring beetles are among the most destructive of borers in managed timber regions, they are an important link in the recycling of dead trees and downed wood.

The adults feed on foliage, pollen, and nectar. Active on hot, sunny days, they readily fly when disturbed. Most species are strong flyers, and some make a loud buzzing noise when airborne. They are often seen running rapidly over trees, probing the bark and wood with their ovipositors extended in preparation for laying eggs. Females lay their eggs directly on the trunks and branches or in crevices in the bark or wood.

Although a few species attack healthy trees, most metallic wood-boring beetles prefer to breed in trees or shrubs weakened by drought, fire, injury, or infestations by other insects. Recently killed trees, especially those felled by logging operations, are particularly attractive.

The legless larvae are often shaped like horseshoe nails by virtue of their broad, flat thoracic segments, which suggested the misapplied name "flatheaded" borers. The larvae of many species mine the sapwood of branches, trunks, and roots, whereas others bore extensively into the heartwood, and some species work both. Their tunneling activities can hasten the death of already weakened trees. The larvae of some *Agrilus* species produce knotty swellings of living plant tissues known as galls. Girdlers construct spiral galleries around small stems, killing the terminal end of the branch. Still other species *(Brachys* and *Taphrocerus)* are stem and leaf miners of herbs and woody plants. A few species attack the cones of conifers *(Chrysobothris cupressicona* and *Chrysophana conicola)* or seasoned wood *(Buprestis)*.

The larval galleries are broad, flat, and form long, linear or meandering tracts beneath the bark or in the heartwood. The galleries are usually tightly packed with dust and frass. This material is often arranged in finely reticulate ridges that resemble fingerprints. Most species overwinter as larvae, but a few species pupate in fall and overwinter as adults. The development from egg to adult may take one or more years. Individuals of some species may take longer under extraordinary circumstances.

IDENTIFICATION: California metallic wood-boring beetles are elongate, flattened, or cylindrical beetles ranging in length up to 31 mm. They are usually metallic or black with orange, red, or yellow markings. The head is tightly tucked inside the slightly broader prothorax. The head is hypognathous, and the sawtoothed antennae consist of 11 segments. The scutellum may or may not be visible. The elytra are smooth, ribbed, or sculptured and usually almost completely conceal the abdomen. The tarsal formula is 5-5-5, with each tarsal claw equal in size and simple, lobed, or notched. The abdomen has five segments visible from below, the first two of which are distinctly fused together.

The mature, legless larvae are creamy white, white, or yellowish and are long, slender, and legless. Their small eyeless head bears three-segmented antennae and can be retracted within the usually much broader thoracic segments. The thoracic segments have a distinct, V-shaped groove on the back. The slender, 10-segmented abdomen is sharply narrowed behind the thorax. The ninth segment has two fleshy lobes or a pair of short, thick projections, or toothed forks.

SIMILAR CALIFORNIA FAMILIES:

- schizopodid beetles or false jewel beetles (Schizopodidae)—fourth tarsal segment of adult is heart shaped, deeply notched with two distinct lobes
- false click beetles (Eucnemidae)—never metallic; body distinctly flexible between prothorax and elytra
- click beetles (Elateridae)—rarely metallic; body distinctly flexible between prothorax and elytra
- lizard beetles (Languriinae, Erotylidae)—antennae clubbed

CALIFORNIA FAUNA: Approximately 264 species in 33 genera.

Eight species of *Polycesta* are found in California. Adults are sometimes found overwintering in the wood of their host trees. *Polycesta californica* (9.0 to 19.5 mm) (pl. 87) is widespread throughout the state. The larva attacks a variety of trees, including oaks *(Quercus)*, silk tassel bush *(Garrya)*, Pacific madrone *(Arbutus menziesii)*, maple *(Acer)*, California-lilac *(Ceanothus)*, mountain-mahogany *(Cercocarpus)*, willow *(Salix)*, and even the introduced blue gum *(Eucalyptus globulus)*.

Approximately 70 species of *Acmaeodera* occur in California. They are usually black with bright markings on their elytra. The adults frequent flowers, where they feed on pollen. At first glance they appear bee- or wasplike, hovering over flowers with their contrastingly marked elytra closed and slightly lifted over their backs with flight wings fully extended. They are most often found in the deserts and chaparral plant communities but also occur in the mountains. Their bodies are clothed in soft hairs, suggesting that some species may be important pollinators. The larva of the Spotted Flower Buprestid *(A. connexa)* (7.2 to 13.0 mm) (pl. 88) mines injured oaks in the Sierra Nevada and Coast, Transverse, and Peninsular Ranges. The adult is commonly found on a wide variety of flowers. *Acmaeodera cribricollis* (8.0 to 10.0 mm) occurs widely in the deserts of southern California and has been collected on the flowers of honey mesquite *(Prosopis glandulosa* var. *torreyana)* and creosote bush *(Larrea tridentata)*. *Acmaeodera gibbula* (10.0 to 12.0 mm) (pl. 89) occurs in the desert regions of southern California, Arizona, and New Mexico and the length of Baja California. Its larva develops in several desert thorn trees including mesquite *(Prosopis)*, catclaw *(Acacia greggii)*, and willow.

The species of *Squamodera* are similar to those of *Acmaeodera* but have thicker antennae and are clothed beneath with dense, white scales. In California, *S. barri* (5.3 to 9.4 mm) is restricted to the Colorado Desert. The adult rests on various plants from June through July, but the larva is only known from indigo bush *(Psorothamnus arborescens)* and *P. emoryi*, and Spanish needle *(Palafoxia arida)*. Another species, *S. vanduzee* (5.2 to 8.3 mm), occurs in the Mojave and Colorado Desert regions, reaching as far north as Big Pine in Inyo County. The adult is found resting on a variety of desert shrubs from June through August. Larval hosts include big saltbush *(Atriplex lentiformis)*, a palo verde *(Cercidium microphyllum)*, desert trumpet *(Eriogonum inflatum)*, and globemallow *(Sphaeralcea)*.

Adult *Acmaeoderopsis* resemble those of *Acmaeodera*, but the females have abdominal segments three to five clothed with long, curved, hairlike setae that are different from the setae on the rest of the body. *Acmaeoderopsis hualpaiana* (5.5 to 6.5 mm) is found in Arizona and the Colorado Desert of southern California during the summer months. It is commonly encountered during the late morning and early afternoon on honey mesquite growing in isolated sandy areas. Other species include *A. jaguarina* (4.0 to 6.0 mm), which can be found on flowers of screw-bean *(Prosopis pubescens)* in Imperial County, and the chaparral species *A. guttifera* (4.0 to 7.0 mm), which is mostly associated with dead branches and stems of scrub oak *(Q. berberidifolia)*.

Paratyndaris olneyae (6.2 to 9.5 mm) is blackish and covered with white hairs. The elytra have a purplish tinge and four orange spots near the base. In California the adult is found on a variety of plants, including saltbush, palo verde, condalia *(Condalia)*, arrow weed *(Pluchea sericea)*, and screw-bean. The larva bores into catclaw, blue palo verde *(Cercidium floridum)*, ironwood *(Olneya tesota)*, and mesquite. This species is found in both the Mojave and Colorado Deserts and also occurs in Arizona, Nevada, New Mexico, Utah, and the length of the Baja California peninsula. *Paratyndaris knulli* (4.0 to 5.5 mm) is shiny black with brassy or purplish tints and covered with white hairs. The elytra have many irregular yellow spots. It is also found in the deserts of southern California, as well as Arizona, Nevada, Baja California, and Sonora. The larvae have been reared from the dead branches of mesquite. Adults are taken on catclaw, palo verde, snakeweed *(Gutierrezia)*, ironwood, and mesquite.

The Sculptured Pine Borer *(Chalcophora angulicollis)* (22.0 to 31.0) (pl. 90) is California's largest jewel beetle. It is a shining black beetle with a bronze luster. The upper surface of the pronotum and elytra is sculptured with irregular ribbing, which makes this beetle especially difficult to see sitting on the trunk of a burned tree. The large white larva develops under the bark of pine *(Pinus)*, fir *(Abies)*, and Douglas-fir *(Pseudotsuga menziesii)*. The slow-flying adult makes a loud buzzing noise when airborne and is common in logging areas with recently felled trees.

Gyascutus dianae (8.4 to 20.4 mm) (pl. 91) is known from the Colorado and Mojave Deserts. The adult is found May through September on a variety of plants, especially Mormon tea *(Ephedra)* and desert buckwheat *(Eriogonum deserticola)*. The males are sometimes found on the flowers while the females perch on stems below. The larva has been reared on iodine bush *(Allenrolfea occidentalis)*, smoke tree *(Psorothamnus spinosus)*, and a Jimson weed *(Datura)*. This species also occurs in southern Arizona, Baja California, and Sonora.

Species of *Nanularia* are robust, nearly cylindrical beetles usually less than 15 mm long. They are dressed in erect, white setae, giving them a whitish or pale grayish look. The larva develops in the woody stems and roots of buckwheat *(Eriogonum)*. Six species occur in California. *Nanularia brunneata* (6.8 to 12.0 mm) is active May through September. It is the most widespread species in the genus and is found in Arizona, Idaho, Nevada, Utah, and the Colorado and Mojave Deserts of southern California. *Nanularia monoensis* (8.0 to 12.5 mm) lives in the Great Basin Desert near Mono Lake. Another species, *N. obrienorum* (7.0 to 15.0 mm), occurs in the region bordered by the southern Great Central Valley and western Antelope Valley. The adult is found during the months of July and August on buckwheats *(E. elongatum* and *E. nudum)*.

Adult *Dicerca* overwinter under bark or in the wood of the larval host tree. The Poplar Dicerca *(D. tenebrica)* (14.5 to 26.0 mm) is a robust, metallic brassy to black beetle with small, inconspicuous raised areas on the elytra. The adult is active March through November. It lays its eggs on the sun-drenched trunks of cottonwoods and aspens *(Populus)*. It ranges across the United States and Canada. In California it is known primarily from the northeastern part of the state. *Dicerca tenebrosa* (17.0 to 19.0 mm) is dark bronze or gray with numerous raised, short ridges that are shiny black. It feeds on pine, fir, spruce, and Douglas-fir and is attracted to freshly cut logs. The widespread and common *D. hornii* (14.0 to 22.0 mm) (pl. 92) is a dark bronze color with a coppery luster underneath. The longitudinally ridged elytra bear scattered, small, dark dots between the ridges. The dorsum of the abdomen is a brilliant, metallic green. The larva attacks a wide variety of injured trees, including white alder *(Alnus rhombifolia)* snags and stumps, and is sometimes a destructive pest of fruit trees. This species is distributed in the Cascades, Sierra Nevada, and Transverse and Peninsular Ranges.

Poecilonota salicis (13.8 to 25.5 mm) is polished gray above and bronze underneath. The body is oval in cross section. The saw-toothed antennae are not notched on the tips. The maxillary palps are widened at the tips. The pronotum has a fine ridge down the middle, and the scutellum is broader than long. It is found along the inland foothills of the Transverse and Peninsular Ranges, as well as on both sides of the southern Sierra. The adult has been reared from willows.

The adult of *Spectralia purpurascens* (8.0 to 12.0 mm) is purplish bronze and is coarsely and evenly punctured. The body is oval in cross section. The saw-toothed antennae are not notched on the tips. The maxillary palps are slender at the tips. The tips of the elytra each have two small teeth. This species is found in the Colorado Desert during the months of June and July clinging to the stems of its larval host plant, beloperone or chuparosa *(Justicia californica)*.

The elytra of *Buprestis* are not ribbed, as in *Cypriacis*, but are grooved. Most species breed in conifers. *Buprestis subornata* (17.0 to 20.0 mm) lacks any markings and is green to blue, violet, or coppery green to dark copper. Its hosts include pine, fir, and Douglas-fir. It is found at the height of summer basking on logs and feeding on young pine needles. However, *B. viridisuturalis* (11.0 to 22.0 mm) feeds only on deciduous trees. Its head and pronotum are green to blue, but the elytra are bright yellow or yellowish brown, usually with a broad, wavy band of blue or green along the elytral suture. This species occurs in both California and Oregon, wherever cottonwoods and white alder *(Alnus rhombifolia)* grow. This species is especially common in woodlands along streams and rivers in the Great Central Valley and Mojave Desert.

The Golden Buprestid *(Cypriacis* [formerly *Buprestis*] *aurulenta)* (12.0 to 22.0 mm) (pl. 93) is one of California's most spectacular beetles. The distinctly ribbed elytra are mostly iridescent green or blue green, while the elytral suture and margins are shiny copper. Eggs are laid only in and around fire or lightning scars and wounds on conifers, especially ponderosa pine *(Pinus ponderosa)* and Douglas-fir, but this beetle will also use fir and spruce *(Picea)*. The larva is often associated with pitchy scars on living trees but will also develop in stumps, exposed roots, and downed logs. Emerging adults cause considerable damage to wooden storage tanks built with infested lumber. Building timbers and boards with finished surfaces are also damaged in this manner. Surrounding forests, lumberyards, and exposed wooden structures may act as reservoirs for future beetle infestations. The

Golden Buprestid is most common in the pine forests of the Sierra Nevada and the Transverse and Peninsular Ranges. Under natural forest conditions the life cycle may take two to four years, sometimes longer. Although it uses only unseasoned, unhealthy, or injured wood as egg-laying sites, infested wood is sometimes processed as lumber for use in the construction of buildings. Larvae feeding in lumber may take up to 30 years or more to develop and emerge as adults. There is one record of a beetle emerging after 51 years!

Juniperella mirabilis (21.0 mm) (pl. 94) is another spectacular, though rarely encountered, metallic wood-boring beetle. It lives among the juniper belts growing along the foothills of the Transverse and Peninsular Ranges. The larva mines the thick trunks of California juniper *(Juniperus californica)* and leaves a distinctive elliptical emergence hole near the ground. The adult rests on the foliage in summer. When disturbed, this beetle takes to the air with a loud buzzing noise.

The genus *Trachykele* contains six species in the United States, four of which occur in California. The scutellum is not visible in the adults of this genus. The larva of the Western Cedar Borer *(T. blondeli)* (11.0 to 17.0 mm) attacks western red cedar *(Thuja plicata)* but also develops in cypress *(Cupressus),* juniper, and possibly incense-cedar *(Calodecrus decurrens).* The mining activities of the larvae sometimes degrade the value of trees cut for poles, shingles, and other products. The bright emerald green adult has a slight golden sheen. It occurs throughout the Pacific Northwest. A similar species, *T. opulenta* (12.0 mm), attacks both young suppressed trees and the fire scars of mature trees. Larval food plants include giant sequoia *(Sequoiadendron giganteum),* incense-cedar, and western red cedar. It occurs in California and Oregon. The larva of *T. nimbosa* (14.0 to 18.0 mm) (pl. 95) attacks mountain hemlock *(Tsuga mertensiana)* and firs. This species extends from California to British Columbia and Idaho. The bronze-colored *T. hartmanni* (16.0 to 22.0 mm) is found in central California, where its larva mines the wood of living Sargent's cypress *(Cupressus sargentii).*

The Charcoal Beetle *(Melanophila consputa)* (8.0 to 13.0 mm), also called Firebug, is commonly attracted to smoke generated from forest fires and sawmills, as well as the acrid fumes produced by smelters and plumes of dust released by cement plants. This black beetle has 12 yellow spots on the elytra and is occasionally reported to nibble on firefighters. It lays its eggs on burned or smoldering wood. The black *M. occidentalis* (6.0 to 12.0 mm) occurs widely in western North America and attacks oaks and other broadleaf trees and shrubs. Species of *Melanophila* are occasionally taken at lights.

The California Flatheaded Borer *(Phaenops californica)* (7.0 to 9.0 mm) (pl. 96) attacks several species of pines. It prefers old, stressed, or diseased trees growing on rocky slopes and other dry situations. The California Flatheaded Borer is greenish bronze above and brassy green below. Many individuals have up to three yellow spots on each elytron. The adult feeds on pine needles, and the female must eat before it can lay eggs. For a few months up to four years, the larva feeds in the sapwood without harming the tree. If the tree is not killed, the larva soon dies; but if the tree is killed, the larva will complete its development quickly. Often associated with this species is the Flatheaded Pine Borer *(P. gentilis)*. This bright bluish green beetle is similar to the Calfornia Flatheaded Borer but lacks yellow spots.

Thirty-six species of the genus *Anthaxia* (pl. 97) occur in North America north of Mexico, most of them in the west. The larvae feed on the branches of coniferous and broadleaf trees and shrubs. These small beetles (4.0 to 8.0 mm) are bronzed brown or black, or metallic green and are extremely difficult to distinguish from one another. Adults are usually found on yellow or white flowers throughout the mountainous regions of the state, especially wallflowers *(Erysimum)*. *Anthaxia inornata* (formerly known as *A. expansa*), *A. aenescens*, and *A. prasina* are the most commonly encountered species.

Approximately 50 species of *Chrysobothris* occur throughout California. *Chrysobothris biramosa calida* (7.0 to 11.0 mm) is found on saltbush in the Colorado Desert, especially in the vicinity of the Salton Sea. *Chrysobothris caurina* (9.5 to 13.0 mm) (formerly known as *C. beeri* and sometimes confused with *C. leechi* from Nevada) is restricted to the Pacific coast, Nevada, and British Columbia. The adult Flatheaded Appletree Borer *(C. femorata)* (7.0 to 16.0 mm) is metallic and indistinctly marked with gray spots and irregular bands. It appears from March through November but is most common in May. It attacks a wide variety of deciduous trees, including oak, maple, apple *(Malus),* and poplar, including sick trees that have been recently transplanted or injured by insects and other animals. The adult is found resting on tree trunks, where the female is often observed probing the bark with its ovipositor in search of egg-laying sites. The larva feeds just beneath the bark before burrowing into the sapwood to pupate. The Pacific Flatheaded Borer *(C. mali)* (6.0 to 11.0 mm) also feeds on a variety of broadleaf trees, including sycamore *(Platanus),* fruit, and nut trees. It is dark bronze to reddish copper and has distinct copper spots on the elytra. This beetle is most common during the spring and summer months. The feeding tunnels of the larvae are shallow and wind through the inner bark and outer sapwood. They sometimes girdle branches and the trunks of small trees, causing disfigurement or death of the tree. *Chrysobothris monticola* (10.2 to 16.5 mm) (pl. 98) is widely distributed in western North America. Its larva develops in pine. The larva of the Flatheaded Cedar Borer *(C. nixa)* (9.0 to 14.5 mm) mines living Sargent's cypress in central California and tecate cypress *(Cupressus forbesii)* in the southern part of the state. It also attacks ornamental junipers and arborvitae *(C. macrocarpa)* in nurseries. *Chrysobothris octocala* (10.2 to 17.0 mm) (pl. 99) is brassy or coppery brown with shiny green or golden spots on the elytra. It is active in fall and is common throughout the Mojave, Colorado, and Sonoran Deserts. It was accidentally introduced into Hawai'i. The adult is found on catclaw, big saltbush, and arrow weed and is occasionally taken at lights. The larva mines mesquite and blue palo verde *(Cercidium floridum)* and has been recorded from dooryard peach *(Prunus persica).*

With nearly 2,800 known species, and many more yet to be named, the genus *Agrilus* is certainly among the largest genera of organisms. Twenty species are known in California. Adults are slender, cylindrical, metallic beetles. The larvae have only slightly broadened thoracic segments and two spinelike projections on the tip of their abdomen. They develop in broadleaf trees, where they bore long, winding, shallow galleries between the bark and the sapwood of trunks and roots. Some species occasionally girdle and kill branches with their spiral galleries. The dark coppery Pacific Oak Twig Girdler *(A. angelicus)* (5.0 to 7.0 mm) is a serious pest of live oaks in southern California, attacking trees already weakened by drought or fire. The Common Willow Agrilus *(A. politus)* (5.0 to 8.5 mm) ranges throughout the United States and Canada, wherever maples and willows grow. The adult emerges from May through August and is variably colored, ranging through brassy, coppery, purplish, greenish, or bluish. The girdling activities of the larva produces galls and damages the outer bark, eventually killing the affected branch. Pupation occurs in May and June in chambers deep in the heartwood of branches and trunks. *Agrilus walsinghami* (9.0 to 13.0 mm) (pl. 100) is a relatively large and distinctive fall species with bronzy green or blue elytra distinctly marked with white spots. In California it visits the flowers of rubber rabbitbrush *(Chrysothamnus nauseosus)* in late summer throughout the western Mojave Desert and Great Basin. This species is also known from British Columbia, Idaho, New Mexico, Nevada, and Oregon.

COLLECTING METHODS: Metallic wood-boring beetles are most active during the hottest parts of the day and are hand collected or netted on the flowers and branches of their larval hosts and other plants. Look for woodland species resting on dead or dying tree trunks and limbs. Many species are attracted to freshly cut wood. Flower-visitors are commonly attracted to the yellow blooms of composites and mustards. Some of these species are also attracted to yellow pan traps. However, the most effective method of collecting metallic wood-boring beetles is by beating and sweeping vegetation during the early morning hours. Using an aspirator will allow you to suck up specimens quickly and easily. As the heat of the day increases, dislodged beetles will hit the beating sheet and quickly fly away before they can be captured. Rearing beetles from infested wood can be especially productive.

RIFFLE BEETLES   Elmidae

Commonly known as riffle beetles because of their preference for living in shallow rapids, or riffles, of streams and rivers, 27 genera and nearly 100 species of this family are known in North America. Although they often occur in large numbers, their small size (3 mm or less) and bottom-dwelling lifestyle make them inconspicuous. The larvae and nearly all of the adults are truly aquatic and are often found together, crawling over, under, and in between submerged substrates. Some species are associated with aquatic mosses.

Unlike other aquatic beetles (e.g., predaceous diving beetles [Dytiscidae] and water scavenger beetles [Hydrophilidae]), the legs of adult riffle beetles are not adapted for swimming, and they do not need to return to the surface periodically to replenish their oxygen supply. Instead, they are covered with a dense layer of short hairs, enveloping them in a silvery blanket of air called a plastron. Oxygen from the surrounding water diffuses into the plastron, while carbon dioxide diffuses out. This form of respiration is sufficient in shallow, well-oxygenated water. Once submerged, they may never have to surface or leave the water again.

Little is known of their biology. Males and females mate underwater. Females glue their eggs singly or in small batches under rocks and other submerged objects. The larvae are equipped with retractile tracheal gills located on the tip of their abdomen. They move about slowly, feeding on waterlogged wood, algae, plants, and other detritus. The larvae undergo from six to eight instars and may require several years before pupation. Mature larvae crawl out of the water to pupate in small cells in moist sand beneath rocks, under loose bark, in wet moss, or in other protected sites near the water's edge.

Adults typically emerge from their pupae at dusk on warm summer nights. Fully winged species take to the air only once to disperse and are commonly attracted to lights, although some species apparently fly only during daylight hours. In captivity adult beetles may live two years or more. Long-lived beetles in nature may be covered in mineral and organic deposits and small aquatic organisms.

Riffle beetles are found in all kinds of streams but are seldom found in those with seasonal flow, heavy sediment load, muddy or sandy bottoms, or sluggish current. Because of their habitat requirements, they are gaining increasing recognition as indicators of water quality in streams.

IDENTIFICATION: California riffle beetle adults are black, gray, or brown dorsally, while the ventral surface is clothed in a thick pile that appears as a silvery gray. Some species have faint reddish markings on the elytra. They are long to oval and range in length from 2.0 to 3.0 mm. The head is hypognathous and is often hidden from above. Their eight- *(Zaitzevia)*, 10-, or 11-segmented antennae are threadlike or gradually clubbed. The nearly rectangular pronotum is broader than the head and somewhat pointed in front, and the margins are often notched or slightly saw-toothed, with ridges on top. The scutellum is visible. The elytra are pitted, rough, and sometimes ribbed and completely conceal the abdomen. The legs are generally long, and the tarsal formula is 5-5-5, with large claws equal in size and not toothed. The legs are not modified for swimming. The abdomen has five segments visible from below.

Mature larvae are light brown or nearly black and are long, cylindrical, slightly flattened, or triangular in cross section. The prognathous head is distinctly narrower than the thorax and bears one pair of simple eyes and a pair of three-segmented antennae. The long legs are five-segmented, including the claw. The abdomen is nine-segmented. The tip of the abdomen is variously modified but lacks any projections; the retractile gills and a pair of hooks are located under a plate underneath.

SIMILAR CALIFORNIA FAMILIES:

- minute mud-loving beetles (Georissinae, Hydrophilidae)—legs short, claws small
- minute moss beetles (Hydraenidae)—mouthparts (maxillary palps) are long
- long-toed water beetles (Dryopidae)—antennae short, most segments wider than long
- minute marsh-loving beetles (Limnichidae)—body covered with short, scalelike hairs
- water penny beetles (Psephenidae)—strongly oval in form; antennae feathery or saw-toothed

CALIFORNIA FAUNA: 23 species in 14 genera.

Dubiraphia has 10- or 11-segmented antennae, fringed front tibiae, a smooth pronotum without ridges on the side, and four-segmented maxillary palpi. The adult and larva of *D. giulianii* (2.1 to 2.3 mm) are usually found on roots of streamside vegetation and on submerged plants. A similar species, *D. brunnescens* (1.8 to 2.5 mm), is restricted to Clear Lake, Lake County, where it occurs on submerged willow *(Salix)* roots. It may prove to be synonymous with *D. guilianii*.

Heterlimnius koebelei (2.0 to 2.5 mm) (pl. 101) usually has 11 antennal segments, although some specimens have 10 on one side and 11 on the other. The larva and adult are found in gravel in colder, high-elevation streams.

Optioservus has a broad pronotum with finely saw-toothed margins. Adults and larvae are most abundant on rocks and in gravel of riffles of cool, swift, clear streams. Little is known of their life cycles, although the larvae have been described. Five species are known in California. *Optioservus divergens* (2.2 to 2.7 mm) is reddish to shining black and is widely distributed throughout the mountains of western North America. It is known from the northern Coast and Transverse Ranges. *Optioservus seriatus* (1.8 to 2.3 mm) is black with red elytral shoulders and tips and is found in northern California, Oregon, Washington, and British Columbia, with scattered populations in Colorado, Utah, Wyoming, and Idaho. A similar species from the Sierra Nevada and elsewhere in northern California, *O. quadrimaculatus* (1.8 to 2.5 mm) (pl. 102), can be reliably distinguished from *O. seriatus* only by examination of the male genitalia.

Two species of *Zaitzevia* occur in California. *Zaitzevia parvula* (2.0 to 2.5 mm) (pl. 103) is one of the most widespread and commonly encountered of our riffle beetles. The adult is easily recognized by its parallel-sided body, lateral ridges on the elytra, and eight antennal segments. Another species, *Z. posthonia* (2.2 to 2.5 mm), occurs in the north-central part of the state.

Adults of *Lara* are distinctive because of their relatively large size. They are not truly aquatic but are instead found among accumulations of wood in cold mountain streams, on trash at the water's edge, or on the underside of undercut stream banks. Two species occur in the state. *Lara avara* (6.0 to 7.0 mm) (pl. 104) is associated with decaying wood in mountain and foothill streams in the western portion of Canada and the United States. It is black and more active than most riffle beetles and will quickly hide when exposed, even crawling over the surface of the water to reach shore. The larva chews through soft, submerged wood. Pupation takes place under moss on the upper surface of partly submerged logs.

COLLECTING METHODS: Riffle beetles are most readily collected by removing and examining vegetation and debris from the riffle areas of streams with gravelly or rocky bottoms. Another method is to stir up the bottom of riffles or submerged plant debris and let the current wash the dislodged beetles into a net or screen placed just downstream. Dump the net's contents into a shallow, light-colored pan, where the small, dark-colored larvae and adults will stand out against the light background. Beetles can be extracted from submerged plant debris by using a Berlese funnel. Adults are sometimes attracted in large numbers to lights placed near streams.

LONG-TOED WATER BEETLES   Dryopidae

Commonly known as the long-toed water beetles, these beetles have unusually long claws, or "toes," on their feet. Of the 14 species known in North America, five occur in California. Unlike most aquatic beetles, the adults are fully aquatic, whereas the larvae are terrestrial. The larvae of *Helichus* are found in moist sand several feet from the edge of streams, where they probably feed on roots or decaying vegetation. Adult California long-toed water beetles are found in the shallow rapids and riffles of streams, and in slow-moving water.

Like riffle beetles (Elmidae), adult long-toed water beetles are covered with a dense layer of short hairs that traps a silvery blanket of air on their bodies as they enter the water. This layer of air, known as a plastron, surrounds the body and functions as a gill. As the oxygen in the plastron is used by the beetle, it is replaced by a fresh supply that diffuses in from the surrounding water. This method of respiration restricts long-toed water beetles to relatively shallow, fast-flowing, cool waters with plenty of dissolved oxygen.

As the females age, their flight muscles become reduced or disappear altogether. Most females have well-developed egg-laying tubes equipped with blades, enabling them to deposit their eggs in the soil or in plant tissues. Larval development requires two to three years for *Helichus suturalis*, and four to five years for *Postelichus productus*. Pupation takes place on land.

The snapping edges of the abdominal segments in some beetle pupae, known as gin traps, are thought to defend the pupae from predators. In *Helichus* and *Postelichus*, however, the gin traps probably serve instead to anchor the pupae within the cast skin of the last larval instar to ease the emergence of the adult. Newly emerged adults fly or crawl to running water and are sometimes attracted to nearby lights in large numbers.

IDENTIFICATION: California long-toed water beetle adults are dull dark gray to dark brown or nearly black. They are long or oval beetles, ranging up to 8.0 mm in length. Their hard body is often encrusted with minerals, obscuring the upper surfaces. The head is hypognathous and is distinctly retracted into the prothorax. The antennae are short and 11-segmented. The second segment is greatly expanded into an earlike process that covers most of the remaining segments. Segments four through 11 are expanded sideways, forming a loose club. The pronotum is larger than the head, and the front margin is broadly curved backward. The elytra are sometimes covered with short, upright, golden hairs and completely conceal the abdomen. The legs are not modified for swimming. The tarsal formula is 5-5-5, with unusually long and simple claws. The abdomen has five segments visible from below.

The mature larvae resemble small wireworms (Elateridae) with rounded ends. The head is prognathous and is partially retracted within the thorax. The antennae are three-segmented, and six pairs of simple eyes are present or completely absent. The legs are short and five-segmented, including the claw. The nine-segmented abdomen does not have any projections.

SIMILAR CALIFORNIA FAMILIES:

- minute mud-loving beetles (Georissinae, Hydrophilidae) — legs short, claws small
- minute moss beetles (Hydraenidae) — maxillary palps elongate
- riffle beetles (Elmidae) — antennae long, with most segments longer than wide
- water penny beetles (Psephenidae) — strongly oval in form; feathery or saw-toothed antennae
- minute marsh-loving beetles (Limnichidae) — body covered with short, scalelike hairs

CALIFORNIA FAUNA: Five species in three genera.

Adults of *Helichus* have pubescence on the last abdominal segment that differs noticeably from that of the preceding segments, which appear almost bare. Adults are found in a variety of stream microhabitats, but they are especially common among submerged roots and debris piles of sticks and leaves. They can be the most abundant insects in some southwestern streams and rivers. The larval stage of *H. suturalis* (3.6 to 5.3 mm) takes two to three years to develop. Found in moist sand above the streamline, the larva presumably feeds on roots or other plant debris. The adult is reddish brown to black.

Postelichus species live in the same habitat as *Helichus* but are much larger (up to 8.0 mm). The abdominal segments of *Postelichus* (5.5 to 7.0 mm) are densely and uniformly pubescent. Both *P. immsi* (5.9 to 8.0 mm) (pl. 105) and *P. productus* (6.0 to 8.0 mm) are black and can be distinguished only by the relative shape of the male genitalia. Both species live in the warmer streams of the valleys and foothills of the southern Coast Ranges and southern California.

COLLECTING METHODS: Search for active adults in stream pools at night with a light. *Helichus* and *Postelichus* are both readily collected by removing and examining vegetation, including algae, and debris from stream bottoms. Another method is to stir up the bottom of riffles or submerged plant debris and let the current wash the dislodged beetles into an aquatic net or screen placed just downstream. Dump the net's contents into a shallow, light-colored pan, where the dark-colored adults will stand out against the light background. Beetles can be extracted from submerged plant debris with a Berlese funnel. Adults of recently emerged long-toed water beetles are adept flyers and are attracted to lights placed near streams.

VARIEGATED MUD-LOVING BEETLES   Heteroceridae

Variegated mud-loving beetles, so named because of the zigzag pattern on the elytra of many species, are distinctly shaped, small, flat beetles. Both adults and larvae are riparian, living in galleries dug in sand or mudflats along the shores of ponds, lakes, and streams. The rakelike forelegs of the adults are adapted for digging. The sides of each tibia are equipped with rows of heavy spines. As the beetle ploughs forward through the sediment, the legs clear the sediment away from the head and rake it to the sides and rear. Although they sometimes live in naturally occurring cracks and crevices, they prefer to dig their own tunnels.

Their tunnels do not form permanent shelters but are instead dug as beetles feed on algae, zooplankton, or other organic debris washed ashore. The adults are fast runners and fly readily, but only for short distances. They are attracted to lights at night, often in large numbers on warm evenings.

Females lay their eggs in small clutches inside their tunnels. The young larvae use the tunnels dug by the adults to feed until they are large enough to dig their own galleries. The life cycle is rapid. The larvae are fully developed in about a week, and the pupal stage takes only three to six days. Pupation takes place within a cocoon.

The lives of variegated mud-loving beetles are closely tied to the rapidly changing shore environment. Shifting water levels may dry out or inundate the immature stages, but the highly mobile adults are seldom affected.

IDENTIFICATION: California variegated mud-loving beetle adults are long, robust, somewhat flattened, covered with dense silky hair, and measure up to 8 mm long. Most species are dark colored with contrasting pale, zigzag markings on the elytra. The head is prognathous and has a set of prominent, flattened mandibles, especially in the males. The antennae are short and nine- to 11-segmented, with the last six segments forming a somewhat oblong and saw-toothed club. The pronotum is broader than the head, and the scutellum is visible. The elytra completely cover the abdomen and are clothed in dense, silky hair that often forms a zigzag pattern. The legs, particularly the front pair, are rakelike for digging. The tarsal formula is 4-4-4, and the claws are large, slender, and simple. Five abdominal segments are visible from below.

The mature larvae are long, nearly cylindrical, and widest at the thorax. They bear conspicuous hairs arranged in scattered rows. The distinct head is prognathous. The antennae are two-segmented, and there are five pairs of simple eyes. The large and well-developed legs are five-segmented, including the claw. The 10-segmented abdomen lacks projections, although the last segment may be modified into a fleshy, leglike lobe.

SIMILAR CALIFORNIA FAMILIES: The protruding mandibles, flattened body, zigzag patches of hair on the elytra, and spiny, rakelike legs are distinctive.

CALIFORNIA FAUNA: 10 species in seven genera.

Lanternarius is the largest genus in California, with four species. *Lanternarius brunneus* (3.0 to 5.0 mm) (pl. 106) is black to brown with various elytral markings. It is tolerant of poor water quality and has been reported from acid mine drainage with elevated sulfate and heavy metal concentrations. Other frequently encountered species in this genus are *L. parrotus* (4.0 to 5.0 mm), with indistinct elytral markings, and *L. sinuosus* (3.8 to 4.7 mm), with distinctive dark brown markings and slightly pale legs.

Neoheterocerus gnatho (4.0 to 7.0 mm) (pl. 107) is the largest species in the state and is often attracted to lights in large numbers in the Colorado Desert during summer and fall. Some males have relatively large mandibles in relation to the rest of their body. The elytra are black with wavy and broken bands of yellow; some individuals may have an entire yellow border.

The genus *Tropicus* has two species in the United States, one of which is found in California. They are distinguished from other members of the family by their small size and nine-segmented antennae. *Tropicus pusillus* (2.0 to 3.0 mm) is a small species widespread in the United States. It is usually uniformly pale yellow, but some specimens have a broad, dark, medial stripe extending along the suture of the elytra. This species is often attracted to lights in large numbers.

COLLECTING METHODS: Variegated mud-loving beetles are sometimes found in large numbers along the banks of streams, ponds, and lakes during the warmer months. Splashing water onto muddy banks will drive the beetles from their underground galleries into plain view. Some species are attracted to lights at night.

WATER PENNY BEETLES Psephenidae

The larvae of water pennies are distinctly oval and flat, resembling small trilobites. They live on or under submerged stones and wood in slow to moderately rapid streams and rivers with good water quality. The streamlined bodies of water penny larvae allows them to squeeze in between and under rocks, where they scrape off algae from submerged objects to feed. Their low profile also allows them to cling to exposed surfaces in fast-moving water, protected within the boundary layer, the thin region of "no flow" close to the substrate. Larval water pennies can flatten themselves even further to reduce drag as water current velocities increase, reducing their chances of being swept downstream. Most water penny larvae leave the water to construct their pupal chambers on land in moist streamside soils, but a few species pupate under water in air-filled chambers.

Adult water penny beetles are relatively short lived, brownish or blackish terrestrial beetles found on streamside rocks or vegetation. Females lay their eggs on rock surfaces beneath the water. While submerged, the females breathe air from a bubble that envelops their hairy body.

IDENTIFICATION: Adult water pennies are small (up to 5.0 mm), oval shaped, slightly flattened beetles. The head is either hypognathous and partly hidden by the pronotum, or prognathous and visible from above. The conspicuous 11-segmented antennae are saw-toothed or feathery. The scutellum is visible. The elytra appear soft and leathery and completely cover the abdomen. The tarsal formula is 5-5-5, with claws equal in size and simple, toothed, or notched at the tip. The abdomen has five to seven segments visible from underneath.

Mature larvae are coppery in color, distinctly flattened, and armored like a "roly-poly" or pillbug (a crustacean), with thoracic and abdominal plates on their back that completely conceal their head and legs when viewed from above. Plates extend to the sides beyond the body and are fringed with dense hairs. The head is hypognathous, with three-segmented antennae. Simple eyes are clumped to form a single eye on each side of the head. The legs are five-segmented, including the claw. The abdomen has nine segments visible from above and does not end in any projections. Abdominal gills are visible from below or are concealed under a plate.

SIMILAR CALIFORNIA FAMILIES:

- marsh beetles (Scirtidae)—tarsi with fourth segment deeply lobed
- ptilodactylid beetles (Ptilodactylidae)—larger body; antennae inserted below eyes

CALIFORNIA FAUNA: Three species in three genera.

The most common water penny in California is *Eubrianax edwardsii* (4.0 to 5.0 mm) (pl. 108). The larva lives and forages for food on the stream bottom and is most common during summer. The larva has four pairs of filamentous gill tufts located ventrally on the abdomen and pupates outside of the water within its larval skin. The adult is sometimes common and is active for a short time in spring or early summer. It is most frequently found resting on rocks and vegetation near streams. Both the male and female have a black head and prothorax. The male has light brown to yellowish elytra and feathery antennae, while the female has dark brown elytra and saw-toothed antennae. This species also occurs in Nevada and Oregon.

The elytra of the Western Water Penny Beetle *(Psephenus falli)* (3.3 to 5.1 mm) are usually dark brown, while the rest of the body is black. The legs and second antennal segment are sometimes lighter. This species is widespread in nondesert mountain streams throughout the state but appears to be more secretive than *E. edwardsii*. The larva has five pairs of filamentous abdominal gill tufts. At least one population of this species is known to pupate underwater. The adult is found during summer on wet parts of emergent boulders in fast-flowing streams. The male and female often engage in frenzied chasing and bumping behaviors before mating. The female lays up to 500 yellow eggs in a flat mass on the undersurfaces of rocks in riffles or those splashed by water and projecting above the water line. The time from egg to adult is about 15 to 18 months. This species is also known from Idaho, Oregon, Washington, and Baja California.

COLLECTING METHODS: Picking up and examining stones found in flowing waters usually produces larval water pennies, but they are also found on wood. Dislodging and rubbing rocks upstream of a net is also a good way to collect these larvae. Adults are found by searching streamside rocks and plants or by sweeping streamside vegetation in late spring and summer.

FALSE CLICK BEETLES   Eucnemidae

Eighty-two species of false click beetles are known from America north of Mexico. Although details of their biology remain largely unknown, it is suspected that they play an important role in the interactions between trees, fungi, and forest regeneration. False click beetles could be used as important indicators of forest diversity. Two species, *Hemiopsida robusta* and *Palaeoxenus dohrni*, are restricted to southern California and may be of interest to conservation biologists.

The misconception that they differ in their clicking ability from click beetles (Elateridae) has resulted in their popular name "false click beetles." In fact, most North American false click beetles click just as well as click beetles. A few species have enough flexibility between the prothorax and elytra that they are capable of flipping themselves into the air with an audible click. Their clicking mechanism is the same as that of click beetles (see below) and likewise is thought to startle predators.

The larvae of some false click beetles resemble those of the metallic wood-boring beetles (Buprestidae), though most are distinctly flattened and the cutting edges of their mandible are located on the outer surfaces. Most larvae develop in wood infected with fungi that usually causes white-rot. They apparently ingest the fungus after extraoral digestion. Although some species prefer coniferous trees, most false click beetles prefer to attack decaying hardwoods. The larvae of some species are sometimes called cross-cut borers because they typically mine across the grain, while others feed in soft and fungusy wood.

IDENTIFICATION: California false click beetles resemble click beetles but lack an externally visible labrum and have all five abdominal segments fused together. They are long, not quite cylindrical, brownish or black beetles (except *P. dohrni*) that range up to 19.0 mm in length. The head is hypognathous and partially retracted within the prothorax. The 11-segmented antennae are beadlike, threadlike, or saw-toothed or feathery from the third or fourth segment on. The scutellum is visible. The prothorax is broader than the head. The elytra are parallel sided and completely cover the abdomen. The tarsal formula is 5-5-5, with claws equal, simple, toothed, or comblike. The abdomen has five segments visible from below.

The mature larvae are whitish, yellowish, or brownish. They are long and nearly cylindrical or somewhat flattened. The flattened head is prognathous and bears two-segmented antennae. The mouthparts are visible from above *(Melasis)* or not *(Palaeoxenus)*. Simple eyes are absent. Thoracic segments are sometimes twice as wide as the abdomen *(Melasis)*, resembling the larvae of flatheaded borers (Buprestidae). In *Palaeoxenus* the thorax is about the same width as the abdomen. Legs are absent or appear as unsegmented knobs. The nine-segmented abdomen is blunt at the end, without projections, except in *Palaeoxenus*.

SIMILAR CALIFORNIA FAMILIES:

- metallic wood-boring beetles or jewel beetles (Buprestidae)—most shiny or metallic underneath
- click beetles (Elateridae)—labrum visible; abdomen with three, four, or five fused segments
- soft-bodied plant beetles (Dascillidae)—hind angles of prothorax not forming sharp points
- throscid beetles (Throscidae)—antennae usually end in three-segment club
- Texas beetles (Brachyspectridae)—antennae saw-toothed from fifth segment on
- rare click beetles (Cerophytidae)—hind trochanters very long
- lizard beetles (Languriinae, Erotylidae)—antennae clubbed

CALIFORNIA FAUNA: 17 species in 12 genera.

Two species of *Anelastes* occur in California. Their protibiae have two well-developed apical spurs and the tarsi have simple claws. The tips of the antennae are excavated, or scooped out. The elytra have nine distinct rows of punctures and are sparsely clothed in short, fine setae. *Anelastes druryi* (7.5 to 13.0 mm) (pl. 109) is a dark brown species with an incomplete ridge on the prothorax. This species is often seen on the wing in coniferous forests throughout the state and is the most commonly collected false click beetle in western North America. The larva is unknown but is thought to develop in rotten wood in the soil.

Anelastes californicus (5.7 to 10.2) is usually pale brown and has a complete ridge on the side of the prothorax.

Dohrn's Elegant Euchemid Beetle *(Palaeoxenus dohrni)* (13.0 to 19.0 mm) (pl. 110) is the most striking California false click beetle. The head, first antennal segment, pronotum, elytral humeri, and tip of the elytra are bright blood red, while the rest of the body is black. Both the adult and the larva are found in the lower portions of incense-cedar *(Calocedrus decurrens)* stumps and under the bark of old trunks of bigcone Douglas-fir *(Pseudotsuga macrocarpa)* during spring and summer. It has also been reported, perhaps mistakenly, from dead sugar pine *(Pinus lambertiana)* trunks, and it has been observed flying at dusk. This beetle is apparently restricted to the Tranverse and Peninsular Ranges of southern California.

COLLECTING METHODS: False click beetles are seldom collected as commonly as click beetles. Some species are beaten or swept from vegetation, whereas others are found beneath the bark of their host trees. A few species can be netted on the wing at dusk or are attracted to lights. Flight-intercept traps, Malaise traps, and Lindgren funnel traps are the best methods for passive trapping of false click beetles.

CLICK BEETLES   Elateridae

Click beetles are found throughout California and reach their greatest diversity in the wetter montane forests and meadows. Click beetles are commonly found on vegetation or under bark and are often attracted to lights, particularly in shrub lands and desert habitats. Although some are active during the day, most species are nocturnal. Many feed on rotting fruit, flowers, nectar, pollen, fungi, and sapping wounds on shrubs and trees. Predatory species attack small invertebrates, especially wood-boring insects and plant hoppers.

The name "click beetle" is derived from the distinctly audible sound made by a special mechanism on the underside of the body between the pro- and mesothorax. The clicking action and sound is probably intended to startle predators. Stranded on their backs, click beetles also use this mechanism to right themselves by flipping up into the air, although several attempts may be required before landing on their feet. Smaller species can propel themselves up to 10 inches into the air, while larger species (e.g., *Alaus* and *Chalcolepidius*) flip only a few inches.

The tough, wiry larvae, or wireworms, resemble slender mealworms sold in pet stores and bait shops. Wireworms are found in soil, rich humus, or decaying plant materials, especially rotten wood. The larvae of wood-dwelling species prey upon small invertebrates, whereas other species scavenge fungi. Soil-dwellers feed on invertebrates and/or roots. Pest species damage seeds and roots of a variety of crops and garden plants. The larvae undergo three to five molts and may take up to three years to mature, depending upon the availability of food and moisture. Both adults and larvae overwinter in the ground, under bark, or in rotten wood.

IDENTIFICATION: Click beetles are easily distinguished from other beetles by their long, somewhat flattened bodies with large and loosely joined forebodies that usually have backward-pointing projections on the back corners. They are brown or black, but a few species bear distinct markings or have a slightly metallic upper surface *(Chalcolepidius)*. They range in length up to 45.0 mm. The head is slightly hypognathous, with the mandibles exposed (e.g., *Aplastus, Euthysanius,* and *Octinodes*) or not. The 11- or, rarely, 12-segmented antennae are usually saw-toothed or feathery (e.g., *Aplastus, Euthysanius,* and *Octinodes*) and attached close to the eyes. The scutellum is visible. The elytra are smooth, ribbed, occasionally hairy or scaly, and nearly always completely conceal the abdomen (except in female *Euthysanius*). The tarsal formula is 5-5-5, with the claws equal in size and simple, toothed, or comblike. The abdomen has five segments visible from below.

Mature larvae are elongate, slender, and cylindrical, with most of the abdominal segments similar in length. The color varies from pale white to dark yellowish brown. The head is wedge shaped and prognathous. The antennae are three-segmented. Simple eyes range from zero to six on either side of the head. The thoracic segments are equal. The legs are well developed and five-segmented, including the claw. The abdomen is 10-segmented but appears nine-segmented from above. The tip of abdominal segment nine is rounded or pointed, thin and flat, notched, or with two small bumps or projections curved upward.

SIMILAR CALIFORNIA FAMILIES:

- soft-bodied plant beetles (Dascillidae)—forebody not moveable
- metallic wood-boring beetles or jewel beetles (Buprestidae)—usually shiny or metallic underneath

- false click beetles (Eucnemidae)—labrum not visible; abdomen with five fused segments
- throscid beetles (Throscidae)—antennae clubbed, rarely saw-toothed
- lizard beetles (Languriinae, Erotylidae)—antennae clubbed
- some false darkling beetles (Melandryidae)—tarsi 5-5-4

CALIFORNIA FAUNA: Approximately 300 species.

At least five species of *Euthysanius* are found in California. The male of *E. lautus* (15.0 to 19.0 mm) (pl. 111) is reddish brown with grooved elytra and feathery, 12-segmented antennae. This beetle is found under the bark of pines *(Pinus)* and is attracted to lights throughout southern California. The adult female (up to 35.0 mm) (pl. 112) has very short elytra and lacks flight wings, exposing nearly the entire abdomen. It is found crawling over the ground.

The genus *Aplastus* has 10 species widely distributed in California and is similar to *Euthysanius* but has sawlike antennal segments. *Aplastus speratus* (15.0 mm) is distributed in the southern part of the state, whereas *A. molestus* (12.0 to 13.0 mm) and *A. optatus* (14.0 to 16.0 mm) are found in the San Francisco Bay region.

Two species of *Elater* occur in California. *Elater lecontei* (21.0 to 31.5 mm) (pl. 113) is smooth and ranges from dark reddish brown to black, lighter underneath. It is found on trees during the day and is often collected at lights in spring and summer throughout southern California and southern Arizona. Some individuals may carry up to 30 or more phoretic mites *(Uropoda)* that attach themselves to the tips of the elytra or the legs. *Elater pinguis* (19.0 to 30.0 mm) is a reddish black species found in northern California and Oregon.

Megapenthes tartareus (9.0 to 13.0 mm) (pl. 114) ranges from southern British Columbia to Utah, Arizona, and southern California. The body is black, and the mouthparts, antennae, and legs are reddish brown.

Ampedus contains numerous species in western North America. The slender, tan larvae feed under bark or in decaying wood. The adults are usually black or brown and often marked with red, orange, or yellow. *Ampedus cordifer* (8.0 to 11.0 mm) (pl. 115) has bright orange elytra with black patches on the tips of the elytra forming a heart-shaped pattern that reaches the elytral margins only at the tips. It is found in western Oregon and California. *Ampedus occidentalis* (8.0 to 10.0 mm) (formerly known as *A. bimaculatus*) ranges from British Columbia through Washington and Idaho to California. This species is similar in appearance to *A. cordifer* except that the black patches on the elytral tips are limited to an oblong spot on each elytron, and the setae on the pronotum are grayish brown rather than black.

Danosoma brevicornis (10.0 to 18.0 mm) (pl. 116) is brown with brown and yellow scales creating a distinctive pattern. The lighter scales dominate and may be more gray than yellow. The adult feeds beneath the bark of dead conifers, especially Jeffrey and ponderosa pines *(Pinus jeffreyi* and *P. ponderosa)* and Douglas-fir *(Pseudotsuga menziesii)*. It is found in the Sierra Nevada and Transverse and Peninsular Ranges in summer and early fall.

The black *Lacon sparsus* (12.0 to 15.0 mm) (pl. 117) is clothed with flattened, black, scalelike hairs intermixed with a few white scales. The prothorax is fitted with grooves underneath to receive both the antennae and the protarsi. Both the larva and adult are often found under the bark of dead pine and fir *(Abies)* logs and stumps. The larva is believed to be a predator. This species also inhabits southeastern British Columbia, eastern Washington, and Oregon. *Lacon rorulenta* (11.0 to 16.0 mm) (pl. 118) is reddish and covered with gold and brown scales with no definite pattern. It occurs in southern British Columbia and through Washington, Idaho, and Oregon. Both species occur in the coniferous forests of California.

Twenty-seven species of *Conoderus* are known in the United States, most of which occur in the eastern half of the country. The Sugar Cane Wireworm *(Conoderus exsul)* (9.5 to 11.0 mm) (pl. 119), also called the Pasture Wireworm, was first described from New Zealand and was introduced to Hawai'i in 1916. It was first reported from California in 1937 and has since spread throughout the Great Central Valley and the coastal regions of southern California. It is active in summer and is commonly attracted to lights. The body is dark brown and covered with short yellow hairs. The elytra are distinctly grooved, and the legs are yellow.

The Western Eyed Click Beetle *(Alaus melanops)* (20.0 to 35.0 mm) (pl. 120) is one of the largest click beetles in California and is easily recognized by the two oval, black eye spots on the pronotum, each surrounded by a narrow line of white scales. The nearly black elytra are sprinkled with white scales. This species is broadly distributed in British Columbia, Washington, Idaho, Oregon, Utah, Arizona, and New Mexico. In California it is found in the Sierra Nevada and the Cascade, Transverse, and Peninsular Ranges. Both the larva and adult are found beneath the bark of stumps and standing or fallen logs of Jeffrey pine, ponderosa pine, lodgepole pine *(Pinus contorta),* Douglas-fir, and oak *(Quercus)* throughout the year. Adults reach their peak of activity from May through July and are sometimes attracted to lights. The larva is shiny yellowish brown with the head and thoracic segments dark brown. It reaches 35.0 mm in length when fully grown. It is found under bark and is reported to attack various wood-boring insect larvae and pupae.

The Black and White Click Beetle *(Chalcolepidius webbi)* (25.0 to 38.0 mm) (pl. 121) is a dark, steely blue to black beetle with two broad, whitish bands trailing along the margins of the pronotum, each becoming narrower on the elytra. This species occurs in riparian gallery forests and oak woodlands throughout southern Arizona. The larva bores into ponderosa pine and other trees. In California this large, beautiful beetle is bluer than black and is encountered during July and August feeding on sapping willows *(Salix)* growing along the Colorado River and nearby irrigation ditches.

The known larvae of the genus *Athous* are found in forest duff or in rotten logs. Adults of some species are active during the day and can be found on vegetation. *Athous axillaris* (9.4 to 12.9 mm) (pl. 122) is dark brown to black with the head and pronotum a muddy orange. Some forms also have a dark orange, narrow, lateral stripe on the elytra and may also have inconspicuous dark orange spots on the shoulders of the elytra. This species is known only from California and occurs from Monterey County and Yosemite National Park southward. Seventeen species of *Athous* are found in California.

Of the 56 species of *Limonius* in North America, 20 occur in California. Adults are slender, dark brown beetles, sometimes with partially or wholly reddish brown elytra. The wireworms of certain lowland species are sometimes pests of garden plants, attacking the roots and seeds of numerous agricultural crops, including bell peppers, pumpkins, radishes, spinach, tomatoes, sweet corn, garlic, horseradish, potatoes, and sweet potatoes. The Sugar-beet Wireworm *(L. californicus)* (8.0 to 12.0 mm) (pl. 123) is generally black with a brassy head and pronotum, but the elytra are sometimes brown with distinct punctures. The yellow brown larva (up to 25.0 mm) is a typical wireworm with a hard, slender, and cylindrical body. It tunnels in or feeds on seeds, roots, seedlings, and tubers and is a pest in potatoes, beans, sugar beets, corn, cereals, and red clover. This species occurs throughout western North America and is widespread in California. The Pacific Coast Wireworm *(L. canus)* (9.5 to 11.0 mm), also known as the Grape Click Beetle, is often found on grapevines during the day, but its feeding activities seldom affect grape production. The male is black with elytra and tibiae reddish brown; the female is reddish with a dark spot on the pronotum and head. Both of these species are covered with wooly, white or yellowish hair.

The large, shining black *Pityobius murrayi* (21.0 to 30.0 mm) (pl. 124) occurs in the Sierra Nevada, where it has been found beneath the bark of rotten ponderosa pine logs. The pronotum is strongly convex, and its surface is coarse and divided lengthwise by a prominent groove that is deepest toward the rear and with two small depressions on either side. The elytra are deeply grooved. The male has feathery antennae, whereas those of the female are saw-toothed.

Until recently, species of *Ctenicera* (6.0 to 26.0 mm) were thought to occur in both Europe and North America. However, studies indicate that all but one species from the northern states and Canada belong in other genera. The larvae of a few of these New World species are considered pests, attacking the roots of grain and vegetable crops, as well as bulbs and flowers. Most are entirely black, sometimes with red markings on the elytra. *Ctenicera conjugens* (formerly *C. lecontei*), from the Sierra Nevada, has tan elytra with black zigzag markings.

COLLECTING METHODS: Sweeping, beating vegetation, and black lighting are the best methods for collecting adult click beetles. Many species are found beneath the bark of dead trees, downed logs, or stumps. Others are found beneath stones, boards, or debris. Most nocturnal species are beaten from vegetation or attracted to lights.

NET-WINGED BEETLES   Lycidae

Seventy-six species of net-winged beetles are known from America north of Mexico. They are called "net-winged" beetles because of the unique elytral sculpturing of most species, consisting of a series of distinct lengthwise ridges connected by numerous cross-ridges. The head is concealed above by a hoodlike pronotum. Their soft bodies are usually boldly marked with bright patches of red or orange.

The biology of California net-winged beetles is poorly known. The larvae are usually found in or under rotten fallen logs, as well as in leaf litter and under bark. Elsewhere there are reports that some larvae are predatory, but most probably feed on fungi, slime molds, and/or fluids associated with rotten wood. Adults are found on leaves and flowers and eat nectar and possibly honeydew. Many net-winged beetles are brightly colored as both adults and larvae, probably to warn potential predators that they are distasteful. Adults are short-lived and sometimes form conspicuous mating aggregations on plants.

IDENTIFICATION: California net-winged beetles are long, soft-bodied, flattened, black beetles with red, orange, or blue markings that range up to 12.5 mm in length. The head is mostly hidden by the pronotum and hypognathous. Antennae are usually 11-segmented and saw-toothed. The pronotum is flat, broader than the head, and sharply margined on each side. The scutellum is visible. The elytra are parallel sided or slightly expanded at the tips, have ribbed sculpturing (except *Calochromus*), and cover the abdomen completely. The tarsal formula is 5-5-5, with all claws equal in size and simple. The abdomen has seven (female) or eight (male) segments visible from below.

The mature larvae are somewhat long, flat, straight sided, or spindle shaped. The small head is prognathous and bears short, thick, two-segmented antennae. A single pair of simple eyes is present or absent. The well-developed legs are five-segmented, including the claw. The sides of the thoracic and abdominal segments are often distinctly margined or flanged. The 10-segmented abdomen appears nine-segmented from above, and it lacks projections.

SIMILAR CALIFORNIA FAMILIES:

- reticulated beetles (Cupedidae)—head exposed
- adult male glowworms (Phengodidae)—head exposed; elytra without ridges; abdomen slightly exposed; sicklelike mandibles distinct; antennae feathery
- false soldier beetles (Omethidae)—elytra without netlike ridges
- soldier beetles (Cantharidae)—head exposed

- fireflies and glowworms (Lampyridae)—elytra never with cross veins

CALIFORNIA FAUNA: Six species in five genera.

Dictyoptera simplicipes (6.5 to 12.5 mm) (pl. 125) is black with bright red pronotum, elytra, femora, and tibiae. The pronotum has a diamond-shaped black spot in the middle. In California it is known from the Sierra Nevada and the Transverse and Peninsular Ranges, where it is collected in spring and summer. It is also distributed from Alaska to Washington and to Idaho, Wyoming, and Arizona.

The body of *Plateros lictor* (3.5 to 8.3 mm) (pl. 126) (previously known as *P. californicus*) is black, with a reddish pronotum marked by a long black spot; the elytral shoulders are rarely reddish. This species is widespread across southern Canada and the northern and eastern United States. In California it has been found in Shasta and Trinity Counties, where adults have been observed in small aggregations in early summer. Mature larvae and pupae have been found beneath the bark of ponderosa pine *(Pinus ponderosa)* stumps.

COLLECTING METHODS: Net-winged beetles are more common in the wetter eastern and southern portions of the United States. In California look for them in moist, dense woods of montane forests during the day resting on vegetation or under bark in spring and early summer. Beating the vegetation in these habitats may produce additional specimens.

GLOWWORMS   Phengodidae

Glowworms are a small New World family of soft-bodied beetles similar in appearance to fireflies (Lampyridae) and soldier beetles (Cantharidae). The family is mostly tropical in distribution, with six genera occurring in America north of Mexico. Glowworms are particularly interesting because of the larviform condition of adult females and the brilliant bioluminescence of the eggs, larvae, and adult females. The upper surface of both thoracic and abdominal segments has luminous bands. The intensity of these lights may vary, but they do not flash on and off as in some fireflies. It has been suggested that the males are first attracted from a distance to the pheromones released by a calling female. Once in the vicinity the males can zero in on her lights.

Most glowworm larvae are bioluminescent, with spots of green, orange, or rarely, red light. The larvae of the South American genus *Phrixothrix* are unusual in producing two colors of light. The head glows a fiery red and is followed by a series of pale greenish yellow lights on the abdomen. California species produce a greenish glow.

The larvae live in leaf litter under forest trees or hide beneath bark or boards on the ground. They prey upon millipedes, insects, and possibly other invertebrates, while adult males and larviform females are not known to feed at all. The larvae of both fireflies and glowworms have sickle-shaped mandibles that are channeled, an unusual feature among the beetles. The channels serve to direct digestive enzymes into the body of the victim.

IDENTIFICATION: The protruding mandibles, distinct form and color, and feathery, double-branched antennal segments of the males distinguish glowworms from other California beetle families. Adult males are soft-bodied, elongate, and flattened in overall body shape. They range in length up to 23.0 mm. The head is prognathous and fitted with a pair of short, sickle-shaped mandibles. The 12-segmented antennae are feathery. The pronotum is flat, rectangular, and sharply keeled on each side. The scutellum is visible. The elytra of the male are soft and do not conceal the last three or more segments of the abdomen. The tarsal formula is 5-5-5, with all claws equal in size and simple. The abdomen has seven segments visible from below. Fully extended mature female larvae may reach 40.0 to 65.0 mm in length.

The mature larvae of *Zarhipis* are long, straight bodied, tapered at both ends, and somewhat flattened. The body is distinctly banded. The small head is prognathous and can be partially retracted within the thorax. The head bears three-segmented antennae, and simple eyes are absent. The first thoracic segment is longer than wide, sometimes longer than the last two segments combined. The short legs are four-segmented, including the claw. The 10-segmented abdomen lacks projections. Segments one through eight are nearly equal in size; segments nine and 10 are reduced, the last probably serving as an additional leg. Adult females are similar in external appearance to the larvae but can be readily distinguished by the presence of compound eyes and a genital opening on the underside of the ninth abdominal segment.

SIMILAR CALIFORNIA FAMILIES:
- net-winged beetles (Lycidae)—antennae not feathery; head and mandibles not visible from above; elytral ridges connected by distinct but less conspicuous cross veins (except in *Calochromus*)
- *Pterotus obscuripennis* (Lampyridae)—antennae 11-segmented, each segment with a single branch; mandibles not visible
- false soldier beetles (Omethidae)—antennae 11-segmented
- soldier beetles (Cantharidae)—antennae not feathery; mandibles not visible

CALIFORNIA FAUNA: Nine species in five genera.

The large size, contrasting colors, and strikingly featherlike antennae of *Zarhipis* males make these insects very conspicuous. They are also are weakly bioluminescent. The Western Banded Glowworm *(Z. integripennis)* (males 12.0 to 23.0 mm) (pl. 127) is found in western Oregon and Washington south through much of California into Baja California and southwestern Arizona. In California it is distributed in all regions below 6,000 ft except in the Great Basin and Colorado Desert. The color of the upper surface of the head varies from entirely orange to entirely black, except for a small orange area around the mouthparts. The pronotum is usually orange and is occasionally marked with black splotches. The elytra are black to reddish brown. The lower surface of the abdomen is usually orange, or orange with the last two segments black, or mostly reddish black overall. The larva preys on millipedes. A larva runs along side a potential victim and throws a coil of its body around the front of the millipede. Then, facing its victim, the larva immobilizes its prey by injecting enteric fluids through hollow, sickle-shaped mandibles into the millipede's neck. This fluid serves two functions: it quickly immobilizes the prey before it can release toxic chemical deterrents, and it initiates the digestive process. Previous speculation that the larva's bite severs the millipede's ventral nerve cord is apparently unfounded. The larva then buries the millipede, removes the head completely, and crawls into the headless millipede corpse and begins to feed on its internal organs. The larvae are active in January and February, and adult males are taken at lights April through June. *Zarhipis truncaticeps* (12.0 to 16.0 mm) has a black head with a light brown area adjacent to the mouthparts.

The pronotum is yellowish orange, while the elytra are black. The abdomen is always black from above, while below it varies from yellow with a blackish tip to entirely yellowish orange. The male is attracted to lights in the deserts of southern California and southwestern Arizona from March through May. *Zarhipis tiemanni* (17.0 mm) is known from the northern Mojave Desert, southern Nevada, and northeastern Arizona. It is found with *Z. integripennis* in eastern Kern County and is easily distinguished by the much shorter and more pointed elytra.

COLLECTING METHODS: Males are readily attracted to lights. Raking moist soil from underneath plants, where millipedes are active, sometimes reveals the larvae. The larvae are also attracted to boards and other flat objects deliberately placed on the ground, using them for shelter. Covered pitfall traps, used in conjunction with metal strips as drift fences, are also productive.

FIREFLIES and GLOWWORMS   Lampyridae

Fireflies, also called lightning bugs, are neither flies nor true bugs but are flat, soft-bodied beetles. Some adult females look more like grubs than beetles and are called glowworms. As in the glowworm family (Phengodidae), female larvae undergo a modified pupal stage before becoming sexually mature. Larviform females are distinguished from larvae externally by having compound eyes.

Most fireflies and glowworms live in the tropics and subtropics. About 150 species live in the United States, most in the south and east. Visitors from these areas often remark on the lack of lightning bugs or fireflies in California. Of the 16 species found in the state, only a few are bioluminescent, and none produce flashing lights for communication.

The light-producing organs of fireflies, if present at all, are whitish or yellowish and are located underneath the abdomen. These organs are supplied with air by numerous air tubes called tracheae. By regulating the oxygen supply to these organs, fireflies control the brightness and duration of their light. The color of the light varies among species from light green to orange and may be determined by temperature and humidity. Male fireflies have very large eyes adapted for locating light-producing females. The larvae of all species glow, but most adults do not.

Bioluminescence in fireflies is virtually 100 percent efficient, with almost all the energy that goes into the system given off as light. An incandescent light bulb is not nearly as efficient, with 90 percent of the electrical energy lost as heat. In fact, the light-producing organs of one firefly produces 1/80,000th of the heat produced by a candle flame of the same brightness.

Nocturnal, light-producing species spend their days hiding beneath bark or in leaf litter, or resting on leaves. Most fireflies with weak or no light-producing organs, such as *Ellychnia*, are active during the day and are generally found on flowers or on streamside vegetation, while others (e.g., *Pterotus*) are nocturnal. The feeding habits of adult fireflies are poorly known, and many species appear not to feed at all.

The predatory larvae attack snails and slugs, earthworms, and small insects such as cutworms. Chemicals are pumped through channeled mandibles to paralyze and liquefy prey, and the tissues are then swallowed.

IDENTIFICATION: Male fireflies and glowworms are elongate, flattened, and soft-bodied beetles ranging in length up to 16.0 mm. They are usually pale brown, black and reddish brown, or black with some red or pink markings. The head is concealed by the leading edge of the pronotum. The head is hypognathous, and the eyes are relatively large. The antennae are usually 10- (*Microphotus*) or 11-segmented and are threadlike, saw-toothed, or comblike. Some females (*Microphotus*) have only six or seven segments. The pronotum is flattened and distinctly margined at the sides. The scutellum is visible. The elytral margins are nearly parallel, widest at the middle, and almost or completely conceal the abdomen. The surface may have a few faint ribs running lengthwise. The tarsal formula is 5-5-5, with the next to last segment of each foot heart shaped. The claws are equal in size and simple. The abdomen has eight (males) or seven (females) segments visible from below.

Mature larvae are spindle shaped or straight sided or may resemble somewhat flattened pillbugs, with distinct armored segments. They vary in color from dark brown or black to pink. The head is partially retracted into the thorax, with a pair of three-segmented antennae and two pairs of simple eyes. The legs are five-segmented, including the claw. The 10-segmented abdomen appears nine-segmented from above and lacks projections. The last abdominal segment is located under the body and functions as an additional leg.

SIMILAR CALIFORNIA FAMILIES:

- net-winged beetles (Lycidae)—elytral ridges usually connected by distinct but less conspicuous cross veins (except in *Calochromus*)
- adult male glowworms (Phengodidae)—head exposed; abdomen slightly exposed; sicklelike mandibles clearly visible from above; antennae feathery, 12-segmented, each segment with two branches
- false soldier beetles (Omethidae)—labrum distinct; antennae separated by nearly twice the diameter of the antennal pits; abdomen without light-producing organs
- soldier beetles (Cantharidae)—head exposed

CALIFORNIA FAUNA: 18 species in seven genera.

The Pink Glowworm (*Microphotus angustus*) is found in the Sierra Nevada and the Coast, Transverse, and Peninsular Ranges and surrounding foothills from June through August. The bright pink larviform female (10.0 to 15.0 mm) (pl. 128) is found on rocks, walls, or posts at night. It hangs its head upward and curls its abdomen forward to reveal the continuous bright green glow of the light-producing organs to attract a mate. Within minutes, one or more males appear and attempt to mate with the signaling female. The pale brown adult male is small (6.0 to 10.0 mm), winged, and has large, bulging eyes. The male's light is weak and produced only when under duress.

Species of the genus *Ellychnia* do not bioluminesce. Their flat, black bodies are marked with bright rose-colored marks on the pronotum. The California Glowworm (*E. californica*) (9.5 to 16.0 mm) (pl. 129) is found throughout California. The larva believed to be associated with this species has bright pale green luminescent areas on the abdomen, with black back plates separated by pale membranes. It is active at night and preys upon snails. Adults of this and other species in the genus are found on flowers or low on grassy vegetation, particularly in moist habitats. Five species of *Ellychnia* are known from California.

Pyropyga nigricans (4.2 to 8.5 mm) is known from all parts of the United States, except in the southeast, and is the only representative of the genus in California. The body is entirely black, except the pronotum, which has a pale median strip and black borders. It is very similar to *Ellychnia*, but smaller.

Pterotus obscuripennis (9.5 to 12.0 mm) (pl. 130) is attracted to lights in spring and early summer in the chaparral and foothill regions throughout the central and southern parts of the state. The head, feathery antennae, and most of the tarsi are reddish brown, and the elytra are black. It is similar in appearance to species of *Zarhipis* (Phengodidae) but is readily distinguished by having 11-segmented antennae; the last seven segments each have a single, threadlike projection. This species also occurs in western Oregon and Washington.

COLLECTING METHODS: The bioluminescent female glowworms are found in oak woodlands and coniferous forests. Look for them just after dark on rocks, walls, and fence posts, or during the day under stones and other objects in or near wet habitats in June and July. Male *Microphotus* and *Pterotus* occasionally fly to lights. The following technique is useful for attracting males of *Microphotus* and *Pleotomus* (pl. 131). Attach red, yellow, and green light-emitting diodes (LEDs) to a piece of plywood. Align the LEDs so that they are in a line and 6 in. apart. Wire the LEDs in series and connect to an on/off switch, using two AA batteries as a power source. All of these electrical components may be purchased from an electronics store. Switch on the LEDs for random periods of 4 to 60 seconds. Activity periods for males may be relatively brief during the mating season (late spring through midsummer), so be sure to try this method at different times during spring and summer and throughout the night.

Both male and female *Ellychnia* and *Pyropyga* can be taken on tree trunks, rocks, and vegetation, or as they feed on flowers.

SOLDIER BEETLES Cantharidae

Soldier beetles are nocturnal or active during the day when they are frequently encountered on flowers and foliage in both wooded habitats and in open fields in spring and summer. In the drier regions of California they are usually found on streamside plants. They feed mostly on high-nutrient liquids drawn from nectar or insect prey. Carnivorous soldier beetles *(Cantharis, Pacificanthia,* and *Podabrus)*, especially those that feed on pestiferous aphids and their relatives, are considered to be of some use as biological control agents.

When disturbed, some soldier beetles will quickly withdraw their legs and drop to the ground and become lost in the tangle of plants and debris below. The contrasting red and black, bluish black, or gray colors of the adult beetles warn potential predators of their bad taste. The larvae, pupae, and adults all produce defensive secretions from their abdominal glands. Other beetle species may mimic distasteful soldier beetles to discourage attacks by predators.

The larvae are nocturnal and develop under bark or in damp areas beneath stones, logs, or other objects on the ground and are carnivorous or omnivorous. They attack caterpillars, maggots, and grasshopper eggs, whereas plant-feeding species graze on grasses. A few are minor pests on crops, particularly potatoes and celery. The larval stage may take one to three years.

The adults of *Cantharis, Chauliognathus,* and *Podabrus* occasionally suffer lethal infections of a zygomycetous fungal pathogen *(Eryniopsis)* known to attack insects. Dead beetles are often found in a death grip, their mandibles imbedded in the tissue of a leaf with their bodies twisted upward between extended wings.

IDENTIFICATION: Adult California soldier beetles are long, soft-bodied beetles that resemble fireflies (Lampyridae) and are usually dark brown to black, sometimes with yellow, orange, or red markings. They reach up to 20 mm in length. The head is usually prognathous, but sometimes hypognathous, and has long, threadlike, 11-segmented antennae. The flat, margined pronotum is usually broader than the head, usually wider than long, and partially conceals the head *(Cantharis, Chauliognathus,* and *Pacificanthia)* or not *(Podabrus)*. The sides of the pronotum in male *Silis* are deeply notched with angular processes near the hind angles. The scutellum is visible. The soft elytra are smooth, sometimes appearing velvety, parallel sided, and nearly or completely conceal the abdomen, except in the Malthininae that have short elytra. The tarsal formula is 5-5-5, with the fourth segment appearing heart shaped. The claws are equal in size and are simple, toothed, or lobed. The abdomen has seven (females and some males) or eight (most males) segments visible from below.

The mature larvae are yellowish or brownish, silverfish shaped, and straight bodied. They are clothed with a covering of fine hair that gives them a velvety appearance. The distinct head is prognathous and has a pair of large simple eyes. The antennae are three-segmented. The thoracic segments are nearly equal in size or gradually decreasing from front to back. The well-developed legs are five-segmented, including the claw. The 10-segmented abdomen does not have any projections, but there are paired glands on the first nine segments that produce defensive secretions. The last segment is small but visible from above.

SIMILAR CALIFORNIA FAMILIES:

- fireflies and glowworms (Lampyridae)—head covered by pronotum, labrum evident; abdomen sometimes with light-producing organs
- net-winged beetles (Lycidae)—labrum evident; elytral ridges connected by distinct but less-conspicuous cross veins (except in *Calochromus*)
- adult male glowworms (Phengodidae)—antennae feathery; mandibles distinct and sickle shaped
- false soldier beetles (Omethidae)—labrum distinct; tarsal segments three and four with two lobes underneath; upper abdominal plates without glandular openings
- blister beetles (Meloidae)—bodies more cylindrical; head with neck; tarsi 5-5-4
- false blister beetles (Oedemeridae)—prothorax without distinct side margins; tarsi 5-5-4
- some longhorn beetles (Cerambycidae)—tarsi appear 4-4-4 but are 5-5-5

CALIFORNIA FAUNA: Approximately 100 species in 10 genera.

The Brown Leatherwing Beetle *(Pacificanthia* [formerly *Cantharis*] *consors)* (12.0 to 20.0 mm) (pl. 132) is known only in California. The head is exposed and not covered by the pronotum, nor is it distinctly narrowed behind the large eyes. The pronotum is narrower than the elytra. The head, prothorax, and legs are reddish brown, and the knees and tarsi are blackish. The velvety elytra are covered with dense, erect hairs of uniform length. The adult is found under loose bark and is a common visitor to porch lights in late spring and early summer. It emits a musty odor when handled or crushed. The larva probably lives among plant litter and preys on small insects. The adult is a predator and has been observed feeding on the Citrus Mealybug *(Pseudococcis citri).*

In the genus *Cultellunguis* the head is exposed. One of the front claws is cleft; the upper blade of the claw is simple and slender, while the lower one is shaped like a pruning knife. *Cultellunguis americana* (8.0 mm) (pl. 133) has a yellowish to brownish yellow head and pronotum, the latter with black markings, blackish elytra, and uniformly brownish yellow legs. It occurs along the Pacific coast from San Diego to Oregon and Washington. This species is associated with Oregon white oak *(Quercus garryana)* in Oregon.

California has 33 known species of *Podabrus* (6.0 to 14.0 mm) (pl. 134). The head is not covered by the pronotum and is distinctly narrowed behind the eyes. The adults are largely predators of aphids but will also attack other soft-bodied plant-feeding insects. The velvety gray, brown, yellow, or pinkish larvae are also predaceous and may be abundant in leaf litter, especially in coniferous forests. The Downy Leather-winged Beetle *(P. pruinosus)* (9.0 to 14.0 mm) has black elytra with whitish hairs that give it a grayish blue appearance. The head, prothorax, margins of the abdominal segments, legs, and bases of the antennae are mostly yellow, but the legs are sometimes darker in specimens from northern California. The larva lives in the soil and is covered with pink, velvety hairs. This species is widely distributed, ranging from Oregon to southern California, and feeds on aphids.

COLLECTING METHODS: Soldier beetles are commonly encountered on foliage during the day in spring and early summer and are collected by hand or by sweeping low vegetation. A beating sheet is helpful in collecting specimens resting on the branches of shrubs and trees. *Podabrus* are often abundant on vegetation in meadows or near lakes, streams, and ponds. *Pacificanthia* are sometimes found under loose bark and are frequently attracted to lights at night. *Chauliognathus* are found on flowers and are known in California only from the eastern mountains of the Mojave Desert.

SKIN BEETLES   Dermestidae

Skin beetles are primarily scavengers that feed on materials high in protein, such as fur, feathers, and carcasses, as well as pollen and nectar. The adults and larvae of some species are important economic pests and cause millions of dollars worth of damage annually. Some species are used by natural history museums around the world to clean animal skeletons for use in research collections and exhibits, while others are considered serious museum pests. Others are associated with bird and mammal nests, where they feed on feathers, hair, and other organic debris. A few species feed on cork, seeds, grains, and other cereal products. Pollen feeders are frequently found on flowers or in the nests of bees and wasps. Household and museum pests, such as *Attagenus*

PLATES

Agyrtidae (pl. 40)
Amphizoidae (pl. 24)
Anobiidae (pls. 145–150)
Anthicidae (pls. 224–225)
Attelabidae (pl. 286)
Bostrichidae (pls. 139–144)
Bothrideridae (pls. 169–170)
Buprestidae (pls. 87–100)
Cantharidae (pls. 132–134)
Carabidae (pls. 4–20)
Cerambycidae (pls. 226–267)
Chrysomelidae (pls. 268–285)
Cleridae (pls. 155–159)
Coccinellidae (pls. 173–181)
Cucujidae (pl. 166)
Cupedidae (pls. 1–2)
Curculionidae (pls. 287–300)
Dascillidae (pls. 82–83)
Dermestidae (pls. 135–138)
Dryopidae (pl. 105)
Dytiscidae (pls. 25–30)
Elateridae (pls. 111–124)

continued ➤

Elmidae (pls. 101–104)
Endomychidae (pls. 171–172)
Erotylidae (pls. 167–168)
Eucnemidae (pls. 109–110)
Geotrupidae (pl. 53–55)
Glaphyridae (pl. 56)
Gyrinidae (pls. 21–22)
Haliplidae (pl. 23)
Heteroceridae (pls. 106–107)
Histeridae (pls. 36–39)
Hydrophilidae (pls. 31–35)
Lampyridae (pls. 128–131)
Lucanidae (pls. 47–49)
Lycidae (pls. 125–126)
Meloidae (pls. 210–223)
Melyridae (pls. 160–162)
Mordellidae (pl. 182)
Nitidulidae (pls. 163–164)
Oedemeridae (pls. 207–209)
Phengodidae (pl. 127)
Pleocomidae (pl. 52)
Psephenidae (pl. 108)
Rhipiceridae (pl. 84)
Ripiphoridae (pls. 183–184)
Rhysodidae (pl. 3)
Scarabaeidae (pls. 57–81)
Schizopodidae (pls. 85–86)
Silphidae (pls. 41–43)
Silvanidae (pl. 165)
Staphylinidae (pls. 44–46)
Tenebrionidae (pls. 189–206)
Trogidae (pls. 50–51)
Trogossitidae (pls. 151–154)
Zopheridae (pls. 185–188)

Cupedidae / Rhysodidae

Plate 1. *Prolixocupes lobiceps* (Cupedidae), 8.0–11.0 mm.

Plate 2. *Priacma serrata* (Cupedidae), 10.0–22.0 mm.

Plate 3. Wrinkled Bark Beetle (*Omoglymmius hamatus*, Rhysodidae), 6.2–6.8 mm.

Carabidae

Plate 4.
Omophron dentatus
(Carabidae)
5.1–7.1 mm.

Plate 5.
Black Calosoma
(*Calosoma semilaeve*,
Carabidae)
20.0–27.0 mm.

Plate 6.
Fiery Searcher
(*Calosoma scrutator*,
Carabidae)
30.0–35.0 mm.

Plate 7.
Scaphinotus punctatus
(Carabidae)
14.0–25.0 mm.

Carabidae

Plate 8. Mojave Giant Tiger Beetle *(Amblycheila schwarzi,* Carabidae), 22.0–28.0 mm.

Plate 9. California Night-stalking Tiger Beetle (*Omus californicus,* Carabidae), 12.0–21.0 mm.

Carabidae

Plate 10.
Pan-American Big-headed
Tiger Beetle
(*Tetracha carolina*,
Carabidae)
12.0–20.0 mm.

Plate 11.
Western Tiger Beetle
(*Cicindela oregona*,
Carabidae)
9.0–14.0 mm.

Plate 12.
California Tiger Beetle
(*Cicindela californica*,
Carabidae)
11.0–17.0 mm.

Plate 13.
White-striped Tiger Beetle
(*Cicindela lemniscata*,
Carabidae)
7.0–9.0 mm.

Carabidae

Plate 14.
Big-headed Ground Beetle
(*Scarites subterraneus*,
Carabidae)
16.0–25.0 mm.

Plate 15.
Pterostichus lama
(Carabidae)
18.0–30.0 mm.

Plate 16.
Anchomenus funebris
(Carabidae)
8.1–9.8 mm.

Plate 17.
*Laemostenus
complanatus* (Carabidae)
14.0 mm.

Carabidae

Plate 18. False Bombardier Beetle (*Chlaenius cumatilis*, Carabidae), 11.9–14.5 mm.

Plate 19. Tule Beetle (*Tanystoma maculicolle*, Carabidae), 9.0 mm.

Carabidae / Gyrinidae

Plate 20. *Brachinus favicollis* (Carabidae), 9.5–10.5 mm.

Plate 21. *Dineutus solitarius* (Gyrinidae), 9.0–10.0 mm.

Plate 22. *Gyrinus plicifer* (Gyrinidae), 5.0–6.5 mm.

Haliplidae / Amphizoidae / Dytiscidae

Plate 23.
Peltodytes simplex
(Haliplidae)
3.0–4.0 mm.

Plate 24.
Trout-stream Beetle
(*Amphizoa insolens*,
Amphizoidae)
10.9–15.0 mm.

Plate 25.
Stictotarsus striatellus
(Dytiscidae)
3.85–4.70 mm.

Plate 26.
Agabus regularis
(Dytiscidae)
9.2–11.3 mm.

Dytiscidae

Plate 27. *Rhantus gutticollis* (Dytiscidae), 9.8–13.0 mm.

Plate 28. Giant Green Water Beetle (*Dytiscus marginicollis*, Dytiscidae), 26.7–33.0 mm.

Dytiscidae / Hydrophilidae

Plate 29. Sunburst Diving Beetle (*Thermonectus marmoratus*, Dytiscidae), 10.0–15.0 mm.

Plate 30. *Eretes sticticus* (Dytiscidae) 12.7–17.0 mm.

Plate 31. *Berosus punctatissimus* (Hydrophilidae), 6.0–8.0 mm.

Hydrophilidae

Plate 32.
Hydrochara lineata
(Hydrophilidae)
13.5–17.0 mm.

Plate 33.
Tropisternus ellipticus
(Hydrophilidae)
8.0–12.0 mm.

Plate 34.
Giant Black Water Beetle
(*Hydrophilus triangularis*,
Hydrophilidae)
33.0–40.0 mm.

Plate 35.
Spotted Dung Beetle
(*Sphaeridium scarabaeoides*,
Hydrophilidae)
5.0–7.0 mm.

Histeridae

Plate 36.
Xerosaprinus sp.
(Histeridae)
1.5–4.5 mm.

Plate 37.
Saprinus lugens
(Histeridae)
4.5–8.0 mm.

Plate 38.
Iliotona cacti
(Histeridae)
4.5–7.5 mm.

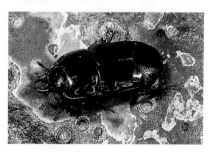

Plate 39.
Atholus bimaculatus
(Histeridae)
3.0–5.3 mm.

Agyrtidae / Silphidae

Plate 40.
Necrophilus hydrophiloides (Agyrtidae)
10.0–13.0 mm.

Plate 41.
Garden Carrion Beetle (*Heterosilpha ramosa*, Silphidae)
11.0–17.0 mm.

Plate 42.
Red and Black Burying Beetle (*Nicrophorus marginatus*, Silphidae)
13.9–22.0 mm.

Plate 43.
Nicrophorus guttula (Silphidae)
14.0–20.0 mm.

Staphylinidae

Plate 44. Hairy Rove Beetle (*Creophilus maxillosus*, Staphylinidae), 11.0–23.0 mm.

Plate 45. Devil's Coach Horse (*Ocypus olens*, Staphylinidae), 17.0–33.0 mm.

Staphylinidae / Lucanidae

Plate 46.
Pictured Rove Beetle
(*Thinopinus pictus*,
Staphylinidae)
12.0–22.0 mm.

Plate 47.
Oak Stag Beetle
(*Platyceroides agassizi*,
Lucanidae)
9.0–10.3 mm.

Plate 48.
Oregon Stag Beetle
(*Platycerus oregonensis*,
Lucanidae)
8.3–10.3 mm.

Plate 49.
Rugose Stag Beetle,
(*Sinodendron rugosum*,
Lucanidae)
11.0–18.0 mm.

Trogidae / Pleocomidae

Plate 50. *Omorgus suberosus* (Trogidae), 9.0–14.0 mm.

Plate 51. *Trox gemmulatus* (Trogidae), 9.0–12.0 mm.

Plate 52. Southern Rain Beetle (*Pleocoma australis*, Pleocomidae), 24.0–28.0 mm.

Geotrupidae / Glaphyridae

Plate 53.
Bolboceras obesus
(Geotrupidae)
6.5–12.0 mm.

Plate 54.
Bolbocerastes regalis
(Geotrupidae)
10.0–21.0 mm.

Plate 55.
Gopher Beetle
(*Ceratophyus gopherinus*,
Geotrupidae)
15.0–23.0 mm.

Plate 56. Bee Scarab
Lichnanthe apina,
(Glaphyridae)
9.7–13.5 mm.

Scarabaeidae

Plate 57.
European Dung
Beetle (*Aphodius
fimetarius*,
Scarabaeidae)
6.0–9.0 mm.

Plate 58.
Aphodius lividus
(Scarabaeidae)
3.0–6.0 mm.

Plate 59.
Ataenius platensis
(Scarabaeidae)
3.6–4.9 mm.

Plate 60.
Canthon simplex
(Scarabaeidae)
5.0–9.0 mm.

Scarabaeidae

Plate 61.
Brown Dung Beetle
(*Onthophagus gazella*,
Scarabaeidae)
8.0–13.0 mm.

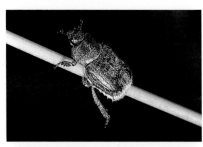

Plate 62.
Grape Vine Hoplia
(*Hoplia callipyge*,
Scarabaeidae)
5.0–11.0 mm.

Plate 63.
Serica perigonia
(Scarabaeidae)
7.0–9.0 mm.

Plate 64.
Dinacoma marginata
(Scarabaeidae)
15.1–21.1 mm.

Scarabaeidae

Plate 65.
Dusty June Beetle
(*Amblonoxia palpalis*,
Scarabaeidae)
16.0–26.0 mm.

Plate 66.
Ten-lined June Beetle
(*Polyphylla decemlineata*,
Scarabaeidae)
18.0–31.0 mm.

Plate 67.
Diplotaxis sierrae
(Scarabaeidae)
11.0–14.5 mm.

Plate 68.
Coenonycha testacea
(Scarabaeidae)
7.0–11.0 mm.

Scarabaeidae

Plate 69.
Dichelonyx valida vicina
(Scarabaeidae)
8.5–14.5 mm.

Plate 70.
Hairy June Beetle
(*Phobetus comatus*,
Scarabaeidae)
12.0–17.0 mm.

Plate 71.
Cotalpa flavida
(Scarabaeidae)
22.0–26.0 mm.

Plate 72.
Little Bear
(*Paracotalpa ursina*,
Scarabaeidae)
10.0–23.0 mm.

Scarabaeidae

Plate 73.
Paracotalpa puncticollis
(Scarabaeidae)
17.5–22.0 mm.

Plate 74.
Cyclocephala pasadenae
(Scarabaeidae)
12.0–14.0 mm.

Plate 75.
Western Carrot Beetle
(*Tomarus gibbosus obsoletus*, Scarabaeidae)
14.0–16.0 mm.

Plate 76.
Sleeper's Elephant Beetle
(*Megasoma sleeperi*,
Scarabaeidae)
24.8–30.5 mm.

Scarabaeidae

Plate 77. *Hemiphileurus illatus* (Scarabaeidae), 18.0–22.0 mm.

Plate 78. Green Fig Beetle (*Cotinis mutabilis*, Scarabaeidae), 20.0–30.0 mm.

Plate 79. *Euphoria fascifera trapezium* (Scarabaeidae), 12.8–13.1 mm.

Scarabaeidae / Dascillidae

Plate 80.
Cremastocheilus westwoodi (Scarabaeidae)
7.0–9.0 mm.

Plate 81.
Valgus californicus (Scarabaeidae)
8.3–9.0 mm.

Plate 82.
Dascillus davidsoni (Dascillidae)
8.0–20.0 mm.

Plate 83.
Anorus piceus (Dascillidae)
7.0–10.0 mm.

Rhipiceridae / Schizopodidae

Plate 84. *Sandalus cribricollis* (Rhipiceridae), 19.0 mm.

Plate 85. *Schizopus laetus* (Schizopodidae), 9.9–18.0 mm.

Plate 86. *Glyptoscelimorpha viridis* (Schizopodidae), 6.2–8.7 mm.

Buprestidae

Plate 87.
Polycesta californica
(Buprestidae)
9.0–19.5 mm.

Plate 88.
Spotted Flower Buprestid
(*Acmaeodera connexa*,
Buprestidae)
7.2–13.0 mm.

Plate 89.
Acmaeodera gibbula
(Buprestidae)
10.0–12.0 mm.

Plate 90.
Sculptured Pine Borer
(*Chalcophora angulicollis*,
Buprestidae)
22.0–31.0 mm.

Buprestidae

Plate 91. *Gyascutus dianae* (Buprestidae), 8.4–20.4 mm.

Plate 92. *Dicerca hornii* (Buprestidae), 14.0–22.0 mm.

Buprestidae

Plate 93.
Golden Buprestid
(*Cypriacis aurulenta*,
Buprestidae)
12.0–22.0 mm.

Plate 94.
Juniperella mirabilis
(Buprestidae)
21.0 mm.

Plate 95.
Trachykele nimbosa
(Buprestidae)
14.0–18.0 mm.

Plate 96.
California Flatheaded
Borer (*Phaenops
californica*, Buprestidae)
7.0–9.0 mm.

Buprestidae

Plate 97.
Anthaxia sp.
(Buprestidae)
4.0–8.0 mm.

Plate 98.
Chrysobothris monticola
(Buprestidae)
10.2–16.5 mm.

Plate 99.
Chrysobothris octocala
(Buprestidae)
10.2–17.0 mm.

Plate 100.
Agrilus walsinghami
(Buprestidae)
9.0–13.0 mm.

Elmidae

Plate 101.
Heterlimnius koebelei
(Elmidae)
2.0–2.5 mm.

Plate 102.
Optioservus quadrimaculatus
(Elmidae)
1.8–2.5 mm.

Plate 103.
Zaitzevia parvula
(Elmidae)
2.0–2.5 mm.

Plate 104.
Lara avara
(Elmidae)
6.0–7.0 mm.

Dryopidae / Heteroceridae

Plate 105. *Postelichus immsi* (Dryopidae), 5.9–8.0 mm.

Plate 106. *Lanternarius brunneus* (Heteroceridae), 3.0–5.0 mm.

Plate 107. *Neoheterocerus gnatho* (Heteroceridae), 4.0–7.0 mm.

Psephenidae / Eucnemidae

Plate 108. *Eubrianax edwardsii* (Psephenidae), 4.0–5.0 mm.

Plate 109. *Anelastes druryi* (Eucnemidae), 7.5–13.0 mm.

Plate 110. Dohrn's Elegant Euchemid Beetle (*Palaeoxenus dohrni*, Eucnemidae), 13.0–19.0 mm.

Elateridae

Plate 111.
Euthysanius lautus male
(Elateridae)
15.0–19.0 mm.

Plate 112.
Euthysanius lautus female
(Elateridae)
to 35.0 mm.

Plate 113.
Elater lecontei
(Elateridae)
21.0–31.5 mm.

Plate 114.
Megapenthes tartareus
(Elateridae)
9.0–13.0 mm.

Elateridae

Plate 115. *Ampedus cordifer* (Elateridae), 8.0–11.0 mm.

Plate 116. *Danosoma brevicornis* (Elateridae), 10.0–18.0 mm.

Elateridae

Plate 117.
Lacon sparsus
(Elateridae)
12.0–15.0 mm.

Plate 118.
Lacon rorulenta
(Elateridae)
11.0–16.0 mm.

Plate 119.
Sugar Cane Wireworm
(*Conoderus exsul*,
Elateridae)
9.5–11.0 mm.

Plate 120.
Western Eyed Click Beetle
(*Alaus melanops*,
Elateridae)
20.0–35.0 mm.

Elateridae

Plate 121.
Black and White Click
Beetle (*Chalcolepidius
webbi*, Elateridae)
25.0–38.0 mm.

Plate 122.
Athous axillaris
(Elateridae)
9.4–12.9 mm.

Plate 123.
Sugar-beet Wireworm
(*Limonius californicus*,
Elateridae)
8.0–12.0 mm.

Plate 124.
Pityobius murrayi
(Elateridae)
21.0–30.0 mm.

Lycidae / Phengodidae

Plate 125. *Dictyoptera simplicipes* (Lycidae), 6.5–12.5 mm.

Plate 126. *Plateros lictor* (Lycidae), 3.5–8.3 mm.

Plate 127. Western Banded Glowworm (*Zarhipis integripennis*, Phengodidae), males 12.0–23.0 mm.

Lampyridae

Plate 128.
Pink Glowworm female
(*Microphotus angustus*,
Lampyridae)
females 10.0–15.0 mm.

Plate 129.
California Glowworm
(*Ellychnia californica*,
Lampyridae)
9.5–16.0 mm.

Plate 130.
Pterotus obscuripennis
(Lampyridae)
9.5–12.0 mm.

Plate 131.
Pleotomus nigripennis
(Lampyridae)
15.0 mm.

Cantharidae

Plate 132. Brown Leatherwing Beetle (*Pacificanthia consors*, Cantharidae), 12.0–20.0 mm.

Plate 133. *Cultellunguis americana* (Cantharidae), 8.0 mm.

Plate 134. *Podabrus* sp. (Cantharidae), 6.0–14.0 mm.

Dermestidae

Plate 135.
Common Carrion Dermestid
(*Dermestes marmoratus*, Dermestidae)
10.0–13.0 mm.

Plate 136.
Buffalo Flower Beetle
(*Orphilus subnitidus*, Dermestidae)
3.0–4.0 mm.

Plate 137.
Varied Carpet Beetle
(*Anthrenus verbasci*, Dermestidae)
2.5–3.0 mm.

Plate 138.
Skin Beetle
(*Trogoderma simplex*, Dermestidae)
2.4–4.4 mm.

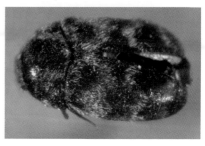

Bostrichidae

Plate 139.
Stout's Hardwood Borer
(*Polycaon stouti*,
Bostrichidae)
10.0–23.0 mm.

Plate 140.
Giant Palm Borer
(*Dinapate wrighti*,
Bostrichidae)
30.0–52.0 mm.

Plate 141.
Western Twig Borer
(*Amphicerus cornutus*,
Bostrichidae)
11.0–13.0 mm.

Plate 142.
Apatides fortis
(Bostrichidae)
9.0–20.0 mm.

Bostrichidae / Anobiidae

Plate 143.
Lead Cable Borer
(*Scobicia declivis*,
Bostrichidae)
3.5–7.0 mm.

Plate 144.
Southern Powder-post
Beetle (*Lyctus planicollis*,
Bostrichidae)
4.0–6.0 mm.

Plate 145.
White-marked Spider
Beetle (*Ptinus fur*,
Anobiidae)
2.0–4.3 mm.

Plate 146.
Ernobius montanus
(Anobiidae)
3.0–4.2 mm.

Anobiidae

Plate 147.
Xeranobium sp.
(Anobiidae)
6.0–9.0 mm.

Plate 148.
Drugstore Beetle
(*Stegobium paniceum*,
Anobiidae)
2.0–3.0 mm.

Plate 149.
Euvrilleta distans
(Anobiidae)
5.0–7.5 mm.

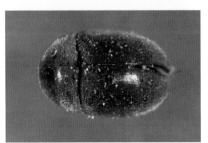

Plate 150.
Cigarette Beetle
(*Lasioderma serricorne*,
Anobiidae)
2.0–4.0 mm.

Trogossitidae

Plate 151.
Ostoma pippingskoeldi
(Trogossitidae)
5.3–10.5 mm.

Plate 152.
Calitys scabra
(Trogossitidae)
6.6–12.2 mm.

Plate 153.
Green Bark Beetle
(*Temnoscheila chlorodia*,
Trogossitidae)
9.0–20.0 mm.

Plate 154.
Tenebroides sp.
(Trogossitidae)
2.0–10.0 mm.

Cleridae

Plate 155.
Cymatodera oblita
(Cleridae)
9.0–13.0 mm.

Plate 156.
Phyllobaenus scaber
(Cleridae)
2.0–4.0 mm.

Plate 157.
Ornate Checkered Beetle
(*Trichodes ornatus douglasianus*, Cleridae)
5.0–15.0 mm.

Plate 158.
Enoclerus quadrisignatus
(Cleridae)
8.0–12.0 mm.

Cleridae / Melyridae

Plate 159.
Red-legged Ham Beetle
(*Necrobia rufipes*,
Cleridae)
3.5–7.0 mm.

Plate 160.
Malachius sp.
(Melyridae)
6.0–8.0 mm.

Plate 161.
Collops sp.
(Melyridae)
3.0–8.0 mm.

Plate 162.
Listrus sp.
(Melyridae)
2.6–3.3 mm.

Nitidulidae / Silvanidae

Plate 163. Pineapple Beetle (*Carpophilus humeralis*, Nitidulidae), 3.0–5.0 mm.

Plate 164. *Meligethes rufimanus* (Nitidulidae), 2.1–2.5 mm.

Plate 165. Saw-toothed Grain Beetle (*Oryzaephilus surinamensis*, Silvanidae), 1.7–3.3 mm.

Cucujidae / Erotylidae

Plate 166. Red Flat Bark Beetle (*Cucujus clavipes puniceus*, Cucujidae), 10.0–17.0 mm.

Plate 167. *Megalodacne fasciata* (Erotylidae), 9.0–15.5 mm.

Plate 168. Clover Stem Borer (*Languria mozardi*, Erotylidae), 4.0–9.0 mm.

Bothrideridae / Endomychidae

Plate 169.
Deretaphrus oregonensis
(Bothrideridae)
9.7–11.5 mm.

Plate 170.
Oxylaemus californicus
(Bothrideridae)
3.2–4.4 mm.

Plate 171.
Endomychus limbatus
(Endomychidae)
3.2–4.2 mm.

Plate 172.
Aphorista morosa
(Endomychidae)
6.2–7.2 mm.

Coccinellidae

Plate 173. Mealybug Destroyer (*Cryptolaemus montrouzieri*, Coccinellidae), 3.4–4.5 mm.

Plate 174. *Chilocorus orbus* (Coccinellidae), 4.0–5.1 mm.

Plate 175. Vedalia (*Rodolia cardinalis*, Coccinellidae), 2.6–4.2 mm.

Coccinellidae

Plate 176. Convergent Lady Beetle (*Hippodamia convergens*, Coccinellidae), 4.2–7.3 mm.

Plate 177. Two-spotted Lady Beetle (*Adalia bipunctata*, Coccinellidae), 3.5–5.2 mm.

Coccinellidae

Plate 178.
Seven-spotted Lady Beetle
(*Coccinella septempunctata*, Coccinellidae)
6.5–7.8 mm.

Plate 179.
Multicolored Asian Lady Beetle
(*Harmonia axyridis*, Coccinellidae)
4.8–8.0 mm.

Plate 180.
Myzia subvittata
(Coccinellidae)
5.7–8.0 mm.

Plate 181.
Ashy Gray Lady Beetle
(*Olla v-nigrum*, Coccinellidae)
3.7–6.1 mm.

Mordellidae / Ripiphoridae

Plate 182. *Mordella hubbsi* (Mordellidae), 3.5–4.5 mm.

Plate 183. *Macrosiagon cruenta* (Ripiphoridae), 5.0–8.0 mm.

Plate 184. *Ripiphorus rex* (Ripiphoridae), 9.0–11.0 mm.

Zopheridae

Plate 185.
Phellopsis obcordata
(Zopheridae)
10.0–15.5 mm.

Plate 186.
Zopherus granicollis granicollis (Zopheridae)
12.6–20.5 mm.

Plate 187.
Ironclad Beetle
(*Phloeodes pustulosus*,
Zopheridae)
15.0–25.0 mm.

Plate 188.
Plicate Ironclad Beetle
(*Phloeodes plicatus*,
Zopheridae)
12.0–16.0 mm.

Tenebrionidae

Plate 189. Desert Ironclad Beetle (*Asbolus verrucosus*, Tenebrionidae), 18.0–21.0 mm.

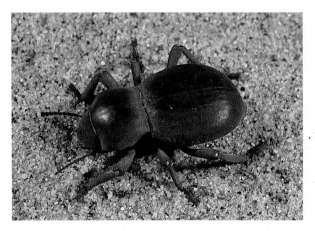

Plate 190. *Asbolus laevis* (Tenebrionidae), 17.0–20.0 mm.

Tenebrionidae

Plate 191.
Cryptoglossa muricata
(Tenebrionidae)
14.0–24.0 mm.

Plate 192.
Schizillus laticeps
(Tenebrionidae)
19.0–23.0 mm.

Plate 193.
Nyctoporis carinata
(Tenebrionidae)
11.0–18.0 mm.

Plate 194.
Coniontis sp.
(Tenebrionidae).

Tenebrionidae

Plate 195.
Eusattus dilatatus
(Tenebrionidae)
11.0–14.0 mm.

Plate 196.
Globose Dune Beetle
(*Coelus globosus*,
Tenebrionidae)
5.0–7.0 mm.

Plate 197.
Edrotes ventricosus
(Tenebrionidae)
6.4–10.0 mm.

Plate 198.
Megeleates sequoiarum
(Tenebrionidae)
7.0–9.5 mm.

Tenebrionidae

Plate 199.
Gigantic Eleodes
(*Eleodes gigantea*,
Tenebrionidae)
30.0–35.0 mm.

Plate 200.
Wooly Darkling Beetle
(*Eleodes osculans*,
Tenebrionidae)
12.0–16.0 mm.

Plate 201.
Eleodes nigropilosa
(Tenebrionidae)
8.0–12.0 mm.

Plate 202.
Argoporis bicolor
(Tenebrionidae)
8.3–14.0 mm.

Tenebrionidae

Plate 203.
Common Mealworm
(*Tenebrio molitor*,
Tenebrionidae)
12.0–18.0 mm.

Plate 204.
Scotobaenus parallelus
(Tenebrionidae)
17.0–21.0 mm.

Plate 205.
Coelocnemis californica
(Tenebrionidae)
18.0–24.0 mm.

Plate 206.
Iphthiminus serratus
(Tenebrionidae)
20.0–25.0 mm.

Oedemeridae

Plate 207. Wharf Borer (*Nacerdes melaneura,* Oedemeridae), 7.0–15.0 mm.

Plate 208. *Eumecomera cyanipennis* (Oedemeridae), 6.0–9.0 mm.

Plate 209. *Rhinoplatia ruficollis* (Oedemeridae), 5.0–12.0 mm.

Meloidae

Plate 210.
Pyrota palpalis
(Meloidae)
6.0–17.0 mm.

Plate 211.
Cordylospasta opaca
(Meloidae)
6.0–19.0 mm.

Plate 212.
Inflated Beetle
(*Cysteodemus armatus*,
Meloidae)
7.0–18.0 mm.

Plate 213.
Elegant Blister Beetle
(*Eupompha elegans*,
Meloidae)
7.0–13.0 mm.

Meloidae

Plate 214.
Phodaga alticeps
(Meloidae)
8.0–25.0 mm.

Plate 215.
Pleuropasta mirabilis
(Meloidae)
6.0–13.0 mm.

Plate 216.
Tegrodera erosa
(Meloidae)
13.0–29.0 mm.

Plate 217.
Punctate Blister Beetle
(*Epicauta puncticollis*,
Meloidae)
9.0–12.0 mm.

Meloidae

Plate 218.
Linsleya compressicornis neglecta (Meloidae)
9.0–11.0 mm.

Plate 219.
Lytta magister
(Meloidae)
16.5–33.0 mm.

Plate 220.
Lytta auriculata
(Meloidae)
6.0–19.0 mm.

Plate 221.
Lytta stygica
(Meloidae)
7.0–15.0 mm.

Meloidae / Anthicidae

Plate 222.
Meloe barbarus
(Meloidae)
5.0–14.0 mm.

Plate 223.
Nemognatha lurida apicalis (Meloidae)
7.0–15.0 mm.

Plate 224.
Tanarthrus alutaceus
(Anthicidae)
2.3–3.8 mm.

Plate 225.
Notoxus calcaratus
(Anthicidae)
2.65–4.15 mm.

Cerambycidae

Plate 226.
Parandra marginicollis
(Cerambycidae)
14.0–22.0 mm.

Plate 227.
Willow Root Borer
(*Archodontes melanopus aridus*, Cerambycidae)
21.0–47.0 mm.

Plate 228.
Pine Sawyer
(*Ergates spiculatus*,
Cerambycidae)
40.0–65.0 mm.

Plate 229.
Giant Mesquite Borer
(*Derobrachus geminatus*,
Cerambycidae)
32.0–70.0 mm.

Cerambycidae

Plate 230.
California Prionus
(*Prionus californicus*,
Cerambycidae)
24.0–55.0 mm.

Plate 231.
Hairy Pine Borer
(*Tragosoma depsarius*,
Cerambycidae)
18.0–36.0 mm.

Plate 232.
Arhopalus asperatus
(Cerambycidae)
17.0–31.0 mm.

Plate 233.
Asemum nitidum
(Cerambycidae)
15.0–20.0 mm.

Cerambycidae

Plate 234. *Centrodera spurca* (Cerambycidae), 20.0–30.0 mm.

Plate 235. *Leptura obliterata soror* (Cerambycidae), 9.0–17.0 mm.

Cerambycidae

Plate 236.
Yellow Velvet Beetle
(*Lepturobosca
chrysocoma*,
Cerambycidae)
9.0–20.0 mm.

Plate 237.
Judolia instabilis
(Cerambycidae)
6.0–15.0 mm.

Plate 238.
Ortholeptura valida
(Cerambycidae)
17.0–23.0 mm.

Plate 239.
Ribbed Pine Borer
(*Rhagium inquisitor*,
Cerambycidae)
9.0–21.0 mm.

Cerambycidae

Plate 240.
Stenostrophia tribalteata sierrae (Cerambycidae)
7.0–11.0 mm.

Plate 241.
Golden-winged Elderberry Borer
(*Desmocerus auripennis auripennis*, Cerambycidae)
23.0–30.0 mm.

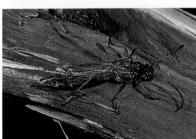

Plate 242.
Necydalis cavipennis (Cerambycidae)
13.0–24.0 mm.

Plate 243.
Lion Beetle
(*Ulochaetes leoninus*, Cerambycidae)
17.0–32.0 mm.

Cerambycidae

Plate 244.
Paranoplium gracile
(Cerambycidae)
12.0–24.0 mm.

Plate 245.
Brothylus gemmulatus
(Cerambycidae)
12.0–22.0 mm.

Plate 246.
Phoracantha recurva
(Cerambycidae)
14.0–30.0 mm.

Plate 247.
Lampropterus ruficollis
(Cerambycidae)
4.5–8.0 mm.

Cerambycidae

Plate 248.
Banded Alder Borer
(*Rosalia funebris*,
Cerambycidae)
23.0–40.0 mm.

Plate 249.
Callidium antennatum
(Cerambycidae)
9.0–14.0 mm.

Plate 250.
Mesquite Borer
(*Megacyllene antennata*,
Cerambycidae)
12.0–25.0 mm.

Plate 251.
Neoclytus balteatus
(Cerambycidae)
8.0–16.0 mm.

Cerambycidae

Plate 252.
Nautical Borer
(*Xylotrechus nauticus*,
Cerambycidae)
8.0–16.0 mm.

Plate 253.
Crossidius mojavensis
(Cerambycidae)
10.0–18.0 mm.

Plate 254.
*Crossidius suturalis
minutivestis*
(Cerambycidae)
14.0–19.0 mm.

Plate 255.
Plionoma suturalis
(Cerambycidae)
10.0–16.0 mm.

Cerambycidae

Plate 256.
Trachyderes mandibularis reductus (Cerambycidae)
17.0–32.0 mm.

Plate 257.
Tragidion armatum (Cerambycidae)
20.0–30.0 mm.

Plate 258.
Hairy Borer
(*Ipochus fasciatus*, Cerambycidae)
4.5–10.0 mm.

Plate 259.
Cactus Beetle
(*Moneilema semipunctatum*, Cerambycidae)
15.0–30.0 mm.

Cerambycidae

Plate 260.
Oregon Fir Sawyer
(*Monochamus scutellatus oregonensis*,
Cerambycidae)
13.0–27.0 mm.

Plate 261.
Spotted Pine Sawyer
(*Monochamus clamator latus*, Cerambycidae)
14.0–29.0 mm.

Plate 262.
Spotted Tree Borer
(*Synaphaeta guexi*,
Cerambycidae)
11.0–27.0 mm.

Plate 263.
Poliaenus obscurus ponderosae
(Cerambycidae)
6.0–9.5 mm.

Cerambycidae

Plate 264.
Acanthocinus princeps
(Cerambycidae)
13.0–24.0 mm.

Plate 265.
Coenopoeus palmeri
(Cerambycidae)
15.0–27.0 mm.

Plate 266.
Saperda calcarata
(Cerambycidae)
18.0–25.0 mm.

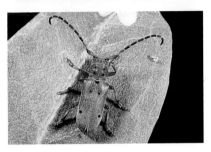

Plate 267.
Tetraopes femoratus
(Cerambycidae)
8.0–19.0 mm.

Chrysomelidae

Plate 268.
Asparagus Beetle
(*Crioceris asparagi*,
Chrysomelidae)
4.7–7.0 mm.

Plate 269.
Eggplant Tortoise Beetle
(*Gratiana pallidula*,
Chrysomelidae)
5.0 mm.

Plate 270.
Eucalyptus Beetle
(*Trachymela sloanei*,
Chrysomelidae)
6.0–7.0 mm.

Plate 271.
Green Dock Beetle
(*Gastrophysa cyanea*,
Chrysomelidae)
4.0–5.0 mm.

Chrysomelidae

Plate 272.
Trirhabda geminata
(Chrysomelidae)
5.5–7.0 mm.

Plate 273.
Trirhabda confusa
(Chrysomelidae)
7.0–8.5 mm.

Plate 274.
Elm Leaf Beetle
(*Xanthogaleruca luteola*,
Chrysomelidae)
6.0–6.5 mm.

Plate 275.
Western Spotted
Cucumber Beetle
(*Diabrotica
undecimpunctata*,
Chrysomelidae)
6.0–7.5 mm.

Chrysomelidae

Plate 276. Western Striped Cucumber Beetle (*Acalymma trivittatum*, Chrysomelidae), 4.0–6.0 mm.

Plate 277. Alder Flea Beetle (*Altica ambiens*, Chrysomelidae), 5.0–6.0 mm.

Plate 278. Dichondra Flea Beetle (*Chaetocnema repens*, Chrysomelidae), 1.5 mm.

Chrysomelidae

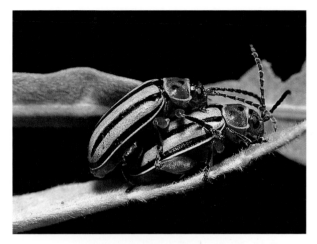

Plate 279. *Disonycha alternata* (Chrysomelidae), 6.8–7.7 mm.

Plate 280. Blue Milkweed Beetle (*Chrysochus cobaltinus*, Chrysomelidae), 6.5–11.5 mm.

Chrysomelidae

Plate 281. *Glyptoscelis albida* (Chrysomelidae), 6.0–10.5 mm.

Plate 282. Red and Yellow Leaf Beetle (*Cryptocephalus castaneus*, Chrysomelidae), 3.9–5.3 mm.

Plate 283. Red-shouldered Leaf Beetle (*Saxinis saucia*, Chrysomelidae), 4.0–5.0 mm.

Chrysomelidae / Attelabidae

Plate 284. Pea Weevil (*Bruchus pisorum*, Chrysomelidae), 4.0–5.0 mm.

Plate 285. Bean Weevil (*Acanthoscelides obtectus*, Chrysomelidae), 2.5–3.5 mm.

Plate 286. Western Rose Curculio (*Merhynchites wickhami*, Attelabidae), 4.5–5.5 mm.

Curculionidae

Plate 287. Rice Weevil (*Sitophilus oryzae*, Curculionidae), 2.0–3.5 mm.

Plate 288. Cactus Weevil (*Metamasius spinolae validus*, Curculionidae), 15.0–25.0 mm.

Plate 289. Yucca Weevil (*Scyphophorus yuccae*, Curculionidae), 8.0–24.0 mm.

Curculionidae

Plate 290. Tule Billbug (*Sphenophorus aequalis picta*, Curculionidae), 9.0–21.0 mm.

Plate 291. California Acorn Weevil (*Curculio occidentis*, Curculionidae), 5.0–10.0 mm.

Curculionidae

Plate 292. Vegetable Weevil (*Listroderes costirostris*, Curculionidae), 6.4–10.0 mm.

Plate 293. *Trigonoscuta morroensis* (Curculionidae), 6.0–7.6 mm.

Plate 294. Fuller's Rose Weevil (*Naupactus godmanni*, Curculionidae), 5.0–9.0 mm.

Curculionidae

Plate 295. *Ophryastes desertus* (Curculionidae), 11.0–23.0 mm.

Plate 296. *Otiorhynchus meridionalis* (Curculionidae), 7.0–10.0 mm.

Curculionidae

Plate 297.
Apleurus albovestitus
(Curculionidae)
11.8–21.4 mm.

Plate 298.
Red Turpentine Beetle
(*Dendroctonus valens*,
Curculionidae)
5.7–9.5 mm.

Plate 299.
Fir Engraver
(*Scolytus ventralis*,
Curculionidae)
4.0 mm.

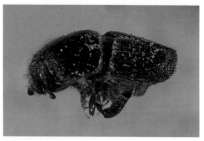

Plate 300.
Pine Engraver
(*Ips pini*,
Curculionidae)
3.5–4.2 mm.

and *Anthrenus*, resemble small lady beetles (Coccinellidae) mottled with black, brown, tan, and white hairs or scales.

Adults feed largely on pollen and nectar but will enter homes and other buildings in spring and summer to lay their eggs on animal products. The larvae scavenge all kinds of protein materials. They develop in dark, undisturbed places and are responsible for damaging woolen materials, carpets, silk products, dried meats, and museum specimens, including collections of insects. The pupa is usually formed within the skin of the last larval instar.

The clusters of bristly hairs found on the bodies of the larvae serve as an irritating deterrent to predatory mammals, reptiles, and birds. Located on the membranes between some of the upper abdominal plates, these hairs can be raised and spread into a protective fan as a defense against predators. This same defense system is also thought to entangle the mouthparts of ants and other small arthropod predators. The hairs are found in house dust, are known to cause human allergies, and have been linked to asthma attacks.

With the increased use of synthetics in the manufacture of furniture and carpets, skin beetles are not as common in homes as they used to be. However, they will occasionally attack synthetic fibers stained with sweat, urine, food, or drink. Household reinfestations of skin beetles may be the result of undetected natural reservoirs such as bird, mammal, or paper wasp nests. Spider webs, windowsills, and light fixtures containing dead insects are breeding grounds for some species of skin beetles. Even some insulation materials wrapped around electrical lines hidden in walls are potential breeding sites for these pests.

IDENTIFICATION: Skin beetles are unique among beetles in that most of them have a simple eye located between the compound eyes. Adults are compact, nearly oval or round beetles, ranging in length up to 12.0 mm. The head is hypognathous and capable of being retracted into the prothorax. The five- to 11-segmented antennae are distinctly clubbed and attached in front of the eyes. The pronotum is broader than long and narrowed toward the head. The side margins of the prothorax are distinct, or at least finely ridged toward the back. The scutellum is usually visible. The elytra are smooth or distinctly ribbed, clothed with hair or scales, and completely conceal the abdomen. The tarsal formula is 5-5-5, with claws equal in size and simple. The abdomen usually has five segments visible from below.

Mature skin beetle larvae are long or almost oval and nearly cylindrical, or oval and flattened. The dark body is covered with clumps of short or long bristly hairs. The distinct head is hypognathous and has three-segmented antennae. Zero, three, or six pairs of simple eyes are present. The thorax is short. The legs are five-segmented, including the claw. The abdomen is nine- or 10-segmented, with a pair of projections present dorsally on the ninth segment in *Dermestes* and *Orphilus,* but absent in all other genera.

SIMILAR CALIFORNIA FAMILIES:

- deathwatch beetles (Anobiidae)—antennae longer and club, if present, lopsided
- pill beetles (Byrrhidae)—antennal club, if present, formed gradually
- plate-thigh beetles (Eucinetidae)—antennae threadlike
- wounded-tree beetles (Nosodendridae)—legs with tibiae expanded (see *Orphilus subnitidus*)

CALIFORNIA FAUNA: 49 species and one subspecies in 11 genera.

Like other species of *Dermestes,* the Common Carrion Dermestid *(D. marmoratus)* (10.0 to 13.0 mm) (pl. 135) is associated with the latter stages of decomposition of a carcass. This large skin beetle moves with quick, jerky movements. The shoulders of the elytra are distinctly marked with large patches of whitish gray scales. Both the adult and larva feed on dried flesh. Pupation takes place underneath or near the food supply. The cosmopolitan Hide Beetle *(D. maculatus)* (5.5 to 10.0 mm) is smaller and black, with slender, whitish hairs scattered over the elytra. These and other species of *Dermestes* are sometimes mass reared in museums to clean the skeletons of vertebrate specimens. Three other species of *Dermestes* are recorded in California.

The black, somewhat oval Buffalo Flower Beetle *(Orphilus subnitidus)* (3.0 to 4.0 mm) (pl. 136) is found throughout North America. Unlike other species of skin beetles, *Orphilus* species lack hair or scales on their bodies. The adult is commonly found during summer on various flowers. The larva has been found in rotten wood and has been reared from dead insects found inside a building. The systematic relationships of *Orphilus* are not clearly understood, and this genus has also been placed with the wounded-tree beetles (Nosodendridae).

Beetles of the genus *Attagenus* are brown or black and sparsely clothed with hair. Three species occur in the state. The adult Black

Carpet Beetle *(A. megatoma)* (2.8 to 5.0 mm) is reported from bird, Alfalfa Leafcutting Bee *(Megachile rotundata)*, and mud dauber wasp *(Sceliphron)* nests. The larva is known to infest barley, cottonseed meal cake, and mixed feeds. It occasionally attacks woolens in homes and infests cereal products in mills.

Anthrenus beetles are covered in multicolored scales. Some species are serious pests in insect collections that are not properly sealed or protected with fumigants. Females lay their eggs on dried proteinaceous materials on which their larvae will later feed. The time before hatching is more dependent upon temperature than relative humidity. The pupa remains almost totally within the confines of the last larval skin. Adults are found on flowers, where they feed on pollen and nectar. The Bird Nest Carpet Beetle *(A. lepidus)* (1.9 to 3.0 mm) is found throughout North America. It is mottled brown and white. The mature and bristled, black larva is about 3.0 mm in length and is found in the nests of various birds and occasionally attacks fabrics in homes. The cosmopolitan Furniture Carpet Beetle *(A. flavipes)* (2.0 to 3.5 mm) attacks upholstered furniture, hair padding, feathers, and woolens, as well as bookbindings and brushes made of natural fibers. The Museum Beetle *(A. museorum)* (2.2 to 3.6 mm) is nearly cosmopolitan. Its golden brown larva feeds on furs, woolen materials, animal mounts, and dried insects. The adult is mottled to varying degrees with tan and gray scales. The Varied Carpet Beetle *(A. verbasci)* (2.5 to 3.0 mm) (pl. 137) is distributed nearly worldwide and is often found in granaries and flour mills. It is readily distinguished from all other *Anthrenus* in North America by its long, narrow scales. The scales are usually white or yellowish and arranged in a zigzag pattern on a dark brown to black background. The larva develops in a wide variety of foods and is a pest in insect collections. It feeds on the insect remains in spider webs and animal nests and apparently also preys on the eggs of the Gypsy Moth *(Lymantria dispar)*. Eight species of *Anthrenus* live in California.

Of the eight species of California *Trogoderma,* only the Skin Beetle *(T. simplex)* (2.4 to 4.4 mm) (pl. 138) is considered a pest. It occasionally infests stored products and is found in every state west of the Rocky Mountains and is widespread in California. This beetle is black with wavy, dull reddish brown bands. The adult feeds on the pollen of various flowers. It apparently breeds in the nests of solitary bees *(Osmia)* and mud dauber wasps.

COLLECTING METHODS: Smaller flower-visiting species are swept from flowers or collected by hand during the spring and summer months. Species of *Anthrenus* are found indoors on windowsills of homes and other buildings with infested animal products. All stages of *Dermestes* are found on or underneath carcasses, whereas the adults are occasionally taken at lights.

BOSTRICHID BEETLES Bostrichidae

Bostrichid beetles are also known as powder-post beetles and twig borers or branch borers. They develop either in living trees or dead wood. The tunneling activity of both the adults and larvae often reduces the wood to powder. In fact, the name "powder-post beetle" is derived from the way one group of bostrichids (formerly Lyctidae) leaves behind a tube, or "post," of fine, powdery frass inside tunneled wood. Other bostrichid larvae leave galleries tightly packed with coarser dust and frass mixed with wood fragments. The closely placed exit holes of some bostrichid species have suggested another common name, the "shot-hole borers."

Bostrichid beetles are especially fond of dead branches and fire-killed hardwoods. Some species prefer old wood, whereas others attack cut and seasoned wood. The tunneling activities of some larvae are particularly damaging to old dwellings and furniture. Others mine through living limbs of weakened cultivated trees or tunnel through green shoots of living plants, and some tropical species attack felled timber and bamboo. Because of their tendency to bore into wood products, bostrichid beetles are widely distributed around the world through commerce. Some species cause considerable damage to stored products, especially dried roots and grains, whereas others prefer to feed on woody fungi.

Many bostrichids have a special symbiotic relationship with bacteria. The bacteria are kept in special organs inside the midgut and help with the digestion of wood. The larvae are especially tolerant of extremely low moisture environments.

IDENTIFICATION: Bostrichid beetles are elongate and cylindrical, or somewhat flattened in overall body shape and are usually black, dark brown, or reddish brown. They range in length from 2.0 to 23.0 mm (one genus to 52.0 mm). The head is usually hypognathous, not generally visible from above (it is visible in lyctine bostrichids and *Polycaon*), and may or may not be covered by the pronotum. The eight- to 11-segmented antennae are straight and tipped with a two- to four-segmented club. The pronotum is somewhat square, the rounded margins with or without fine teeth. The front of the pronotum may be rough and rasplike and is sometimes armed with small horns. The scutellum is visible. The elytra are parallel sided and variously modified with lines of punctures, ridges, or apical spines and completely conceal the abdomen. The elytral surface is coarsely or finely punctate. The apices of the elytra may or may not slope sharply downward to give the beetle an abruptly "cut-off" look. The tarsal formula is 5-5-5, with each claw equal in size and toothed. The abdomen has five segments visible from below.

The mature larvae are cream or dull white, nearly cylindrical, and C shaped. The distinct, dark head can be partially retracted within the thorax. The antennae are three-segmented, and one pair of simple eyes is either present or absent. The thoracic segments are usually enlarged, while the smaller abdominal segments are approximately equal in length. The small legs are five-segmented, including the claw. The abdomen is 10-segmented and lacks projections.

SIMILAR CALIFORNIA FAMILIES:

- Rugose Stag Beetle (*Sinodendron rugosum*, Lucanidae)—antennae lamellate; head prognathous; male horned
- some deathwatch beetles (Anobiidae)—antennae sawtoothed or fan shaped
- cylindrical bark beetles (Colydiinae, Zopheridae)—mandibles concealed
- bark beetles and ambrosia beetles (Curculionidae)—antennal club compact

CALIFORNIA FAUNA: 27 species in 18 genera.

Of the two species of *Polycaon* in California, Stout's Hardwood Borer *(P. stouti)* (10.0 to 23.0 mm) (pl. 139) is the most common. It occurs in British Columbia, Washington, Oregon, California, and Arizona. Its large, conspicuous head is not obscured by the pronotum, and the body is entirely black. This species attacks fruit and nut trees, eucalyptus *(Eucalyptus)*, maple *(Acer)*, oak *(Quercus)*, California laurel *(Umbellularia californica)*, Pacific madrone *(Arbutus menziesii)*, manzanita *(Arctostaphylos)*, western sycamore *(Platanus racemosa)*, and other native trees. It is also reported to damage cured and treated hardwoods in lumberyards and in buildings in mountainous regions of the state. Adults may appear inside new buildings, where they have emerged from infested timbers or crates. Although rarely a pest, this beetle is recorded emerging from wood cabinets more than 20 years after they were installed. Since it does not attack finished wood, this type of infestation demonstrates that the wood was infested before manufacture. It is nocturnal and readily attracted to lights.

The shiny black Giant Palm Borer *(Dinapate wrighti)* (30.0 to 52.0 mm) (pl. 140) sometimes has a reddish tinge. The mouthparts, legs, and underside are dark reddish brown. It is the largest bostrichid in the world and was once considered extremely rare, being prized by collectors in the late 1800s and early 1900s. Later, it was discovered that this spectacular beetle could be found in numbers in the desert oases of Baja California and the Colorado Desert, where California fan palms *(Washingtonia filifera)* occur. The female tunnels into the crown of the palm to lay 400 to 500 eggs. The pale yellow larva bores through the trunks of dead and dying palms to feed and pupate, a process that may take several years. It is not unusual to hear both adults and larvae chewing inside palm trunks from several feet away. Adults emerge at night from June through August, leaving behind dime-sized emergence holes in the palm's trunk. This beetle has become a pest of planted palms in Arizona and California, including date palms and street trees. A second species, *D. hughleechi,* was described from mainland Mexico in 1986.

The Western Twig Borer *(Amphicerus cornutus)* (11.0 to 13.0 mm) (pl. 141) is found in southern California. It is cylindrical in shape and dark brown or black. The head is hypognathous, and the pronotum has a row of fine teeth located at the fore portion of the lateral margin. Each segment of the antennal club has longitudinal depressions. The elytra end with blunt teeth at their tips. The larva feeds on oak, mesquite *(Prosopis),* and other hardwoods. The adult attacks small living branches of fruit trees, causing the branches to break and die. Populations are seldom large enough to cause serious harm to fruit trees. This species is also known from Arizona, New Mexico, Texas, and Mexico.

Apatides fortis (9.0 to 20.0 mm) (pl. 142) is similar in appearance to *Amphicerus,* except the segments of the antennal clubs have round depressions. They riddle dead mesquite branches in the Sonoran Desert of Arizona and the adjacent Colorado Desert in California. Their chewing inside the branches can be heard at some distance from the host tree. The adult is frequently taken at lights during summer and fall.

The adult Lead Cable Borer *(Scobicia declivis)* (3.5 to 7.0 mm) (pl. 143) is dark reddish brown to black with reddish mouthparts, legs, and eight-segmented antennae. It is similar in appearance to *Dinapate* but much, much smaller. The Lead Cable Borer is unusual in that it often bores into the lower side of lead-lined telephone cables at the hanger supports, but it does not feed on lead. High temperatures apparently stimulate this beetle, and fires in areas with large populations are often followed by outbreaks of cable-boring activity. It is also known as the Short-circuit Beetle because its burrowing activity allows moisture into the cable, resulting in the short-circuiting of wires and interruption of electrical service. Both the adult and larva normally attack oak, maple, eucalyptus, fig *(Ficus)*, and other trees. It occasionally causes damage in wineries by boring through wooden wine casks and storage tanks.

Spotted Limb Borers *(Psoa maculata)* (7.0 to 15.0 mm) are bronze and clothed in gray hairs. The elytra are blackish or bluish green with four or more reddish, yellow, or white spots. It breeds only in dead twigs of trees and shrubs. Eggs are laid on dying branches of plants, including oak, apple *(Malus)*, grapes *(Vitis)*, white sage *(Salvia apiana)*, and chaparral broom or coyote brush *(Baccharis pilularis)*. Branches hollowed by the feeding activities of the immature beetles are sometimes filled with both larvae and pupae. The adult is alert and flies readily when alarmed. This beetle is active from spring through early summer

Three species of *Lyctus* are recorded in the state. The Southern Powder-post Beetle *(L. planicollis)* (4.0 to 6.0 mm) (pl. 144) is black with sparse, yellowish white hairs scattered over the body. This species is widespread throughout the southern and western United States. Its presence is indicated by small, round emergence holes in wood ringed with fine white powder. It damages a wide range of seasoned hardwoods, including flooring, timbers, plywood, tool handles, gunstocks, furniture, and antiques.

COLLECTING METHODS: Adults of some species are readily attracted to lights. Others are netted on the wing during spring and summer flights or collected by beating infested dead branches. Some species feed on fungi and are hand picked from the host. Wood- and bark-feeding species are gathered from infested materials stored in rearing chambers. Check wooden furniture for exit holes and powdery residue. Emerging beetles are attracted to sunny windows and are found lying dead on windowsills.

DEATHWATCH BEETLES and SPIDER BEETLES — Anobiidae

When viewed from above, the hood- or bell-shaped pronotum of most deathwatch beetles usually conceals the head. The name "deathwatch" beetle is attributed to the behavior of the European Furniture Beetle *(Anobium punctatum)*. During the mating season the male strikes the top of its head to the adjacent pronotal margin to produce a series of audible clicks. These clicks, emanating from the infested furniture surrounding hushed, deathbed vigils were thought by some to warn of impending death. Other species of deathwatch beetles are also known to make clicking noises as part of their courtship behavior.

Deathwatch beetles and spider beetles include several important pests of stored products, including the Cigarette Beetle *(Lasioderma serricorne)* and the Drugstore Beetle *(Stegobium paniceum)*. They cause considerable economic damage by infesting drugs, tobacco, seeds, spices, cereals, and leather.

The larvae of most deathwatch beetles attack hardwoods and softwoods, boring into bark, dry wood, twigs, seeds, woody fruits, and galls. Species of *Ernobius* attack pine cones. A few species feed on woody fungi and puffballs or attack the young stems and shoots of growing trees. Like the bostrichids (Bostrichidae), some deathwatch and spider beetles store symbiotic yeastlike organisms in special pouches in their midgut that aid with the digestion of wood.

Species of deathwatch beetles, spider beetles, and bostrichid beetles are sometimes referred to as "powder-post beetles." The tiny tunnels or "shot holes" of the larvae are about 2 mm in diameter and contain pellets smaller than those of termites.

Spider beetles, once considered to be a separate family (Ptinidae), often have inflated elytra conspicuously wider than the pronotum, giving them a spiderlike appearance. Their small bodies are usually covered to various degrees with setae or scales. Although some are woodborers, most spider beetles feed on accumulations of plant and animal materials. Some species are common in bird or mammal nests, whereas others feed on pollen in the nests of solitary bees or animal dung. In homes, spider beetles and their larvae are found in flour and other cereal products, wool, and other similar dried plant or animal materials.

IDENTIFICATION: California spider beetles and deathwatch beetles are humpbacked and globular to long and cylindrical in shape. They are tan, brown, or blackish. Some species have patches of scales or fine setae. They range in length up to 9.0 mm. The head is hypognathous, partially withdrawn into the prothorax, and covered from above partially or completely by the leading edge of the pronotum. The 10- or 11-segmented antennae are sometimes tipped with a one- to three-segmented club. In males the club segments are elongate, giving the club an asymmetrical appearance. The pronotum is as wide or wider than the head, broadly oval, or almost square or is bell or hood shaped; the side margins are variable. The scutellum is usually visible. The elytral surface is grooved or pitted, smooth, or rough, and the elytra completely conceal the abdomen. The tarsal formula is 5-5-5, with each claw equal in size, small, and simple. The abdomen has five segments (four in *Gibbium psylloides*) visible from below.

The mature larvae are white or whitish, straight sided, nearly cylindrical, and weakly to strongly C shaped. The distinct head is hypognathous and has one- or two-segmented antennae. One pair of simple eyes is either present or absent. The thoracic segments are nearly equal in size. The legs are five-segmented, including the claw. The 10-segmented abdomen lacks any projections. The last abdominal segment is greatly reduced.

SIMILAR CALIFORNIA FAMILIES:

- bostrichid beetles (Bostrichidae)—antennae short with compact club
- skin beetles (Dermestidae)—antennae with symmetrical club
- minute tree fungus beetles (Ciidae)—antennae with symmetrical club
- bark beetles (Scolytinae, Curculionidae)—antennae with symmetrical club

CALIFORNIA FAUNA: 150 species in 47 genera.

Spider beetles are so named because of their small, round bodies and six spidery legs. A pair of long antennae completes their eight-legged appearance. They range in color from pale to blackish brown. Two species of *Mezium* occur in California, both with the head and prothorax clothed in scales. The American

Spider Beetle *(M. americanum)* (1.5 to 3.5 mm) is a distinctive and cosmopolitan species that is encountered only rarely in California. It occurs in homes, warehouses, and mills, where it breeds primarily in dried animal materials. It will also attack plant products such as tobacco seed, cayenne pepper, opium, and grain. The head, pronotum, legs, and bases of the elytra are covered with dense, brownish scales. The pronotum is bulging on either side of the furrow running its entire length. Most of the elytral surface is black and shining, with a few scattered hairs. The abdomen is narrow, filling only one-third of the elytral space. A second species, *M. affine* (2.3 to 3.5 mm), is similar in appearance but lacks the bulging pronotum on either side of the pronotal furrow. Another similar species, the Shiny Spider Beetle *(Gibbium psylloides)* (1.7 to 3.2 mm) is completely shiny above, lacking any scales.

At least 15 species of *Ptinus* spider beetles have been recorded from California. The Brown Spider Beetle *(P. clavipes)* (2.3 to 3.2 mm) is widely distributed in the state. It is a scavenger throughout the home, feeding on books, feathers, rodent droppings, sugar, legumes, seeds, animal feeds, and all kinds of grains. The White-marked Spider Beetle *(P. fur)* (2.0 to 4.3 mm) (pl. 145) is a reddish brown beetle covered in yellowish hairs. The male is smaller and more slender than the female. The female has two white patches located near the base and tip of each elytron. These patches are sometimes joined to form bands. The females of both species are wingless. *Ptinus californicus* (4.0 to 5.0 mm) is distributed throughout much of the state and is found in the nests of ground-nesting bees.

The genus *Trigonogenius* has 11-segmented antennae, a hidden scutellum, and rows of pits on its elytra. *Trigonogenius globulus* (2.5 mm) is the only species in the genus found in California. This brown beetle is densely clothed above and below with yellowish scales. It is found throughout the state in homes and has been recorded in large numbers breeding on manure, beet seeds, and oats.

Sphaericus gibboides (1.8 to 2.2 mm) is the sole representative of the genus in the state. It also has 11-segmented antennae, but the pits on the elytra are scattered and not arranged in rows. It is dull brown and covered above with yellowish or brown scales that lie almost flat on the body surface. It is found along the coast of California, where it is known to infest stores of seeds and spices.

It also attacks herbarium and insect specimens in collections.

The species presented below are not spiderlike at all. Sixteen species of *Ernobius* inhabit California. They are small, reddish brown beetles with bulging eyes and are elongate, straight sided, and clothed with short, fine hair. The sides of the prothorax are sharply margined or keeled. The larvae live in dry conifer twigs, dead cones, under bark, or in fungal galls on conifers. *Ernobius montanus* (3.0 to 4.2 mm) (pl. 146) is pale brown with 11-segmented antennae; the ninth segment is never shorter than segments six through eight combined. It is found in the Transverse and Peninsular Ranges of southern California, where it attacks the cones and twigs of Coulter, Jeffrey, single-leaf pinyon, and ponderosa pine *(Pinus coulteri, P. jeffreyi, P. monophylla,* and *P. ponderosa)*. *Ernobius melanoventris* (2.5 to 4.2 mm) is found throughout the coniferous forests of California, Idaho, Oregon, and Washington, where it attacks the cones of Jeffrey and ponderosa pine. As its name suggests, the abdominal segments are usually black but are sometimes mottled black and brown.

The Deathwatch or Furniture Beetle *(Anobium punctatum)* (2.7 to 4.5 mm) apparently arrived from Europe as a stowaway in imported furniture. It is now established along the Pacific and Atlantic coasts of North America. In Europe it is considered a serious and widespread pest, as it attacks well-seasoned softwoods and hardwoods and damages furniture, timbers, paneling, flooring, and other interior woodwork. It also attacks the bindings of books. However, it causes only occasional damage in the United States. The adult beetle is cylindrical and brown and is sparsely clothed with short hairs.

The Pacific Powder-post Beetle *(Hemicoelus gibbicollis)* (2.5 to 5.5 mm) damages buildings all along the Pacific coast, from Alaska to California. Structural timbers and subflooring made from both hardwoods and softwoods, especially beneath older buildings, are attacked and sometimes destroyed by large numbers of these beetles. The prothorax is distinctly narrower than the elytra, which are distinctly grooved and clothed with short, yellowish hair.

Thirteen species of *Xeranobium* (6.0 to 9.0 mm) (pl. 147) are found in Arizona, California, Nevada, Oregon, and southwestern Texas. Antennal segments four through eight are serrate to pectinate, and the side of the pronotum has a distinct margin only near the base. The hind tarsus is nearly as long or longer than the hind tibia. Little is known of their biology other than that the larvae of two species develop in iodine bush *(Allenrolfea occidentalis),* and the adults are sometimes attracted to lights. Eight species occur in California.

The Drugstore Beetle *(Stegobium paniceum)* (2.0 to 3.0 mm) (pl. 148) is cosmopolitan and attacks spices, herbs, legumes, biscuits, and candy. The beetle is straight sided, light reddish brown, and clothed with short, bent, and scattered erect setae. The last three antennal segments are gradually expanded to form a loose club.

Of the 13 species of *Euvrilleta,* five occur in California. *Euvrilleta distans* (5.0 to 7.5 mm) (pl. 149) is elongate, uniformly brown, and occurs in the foothills of southern California. It is distinctly pubescent with faintly grooved elytra. Both the procoxae and mesocoxae are touching. It is occasionally attracted to lights.

The bane of cigar aficionados, the Cigarette Beetle *(Lasioderma serricorne* (2.0 to 4.0 mm) (pl. 150) is also a serious pest of spices, legumes, grains, and cereal products. This oval beetle is light reddish brown and clothed in short, dense hair. The antennae are saw-toothed in appearance. This species lacks the longitudinal grooves of the elytra that are present in the Drugstore Beetle.

COLLECTING METHODS: Deathwatch beetles are usually collected in small numbers, either at lights, or by beating, sweeping vegetation, or rearing from infested wood and fungus. Spider beetles living indoors are found in stored organic products or in related debris. Outdoors a few might be taken with similar techniques used for deathwatch beetles, as well as in pitfall traps or animal nests. Disturbed deathwatch and spider beetles withdraw their legs, remain very still, and are easily overlooked. Check exposed or infested stored products for adults and larvae.

BARK-GNAWING BEETLES   Trogossitidae

Bark-gnawing beetles are a relatively small family of beetles, with less than 60 species known in the United States. Previously known as the Ostomidae or Ostomatidae and frequently misspelled as "Trogositidae," the Trogossitidae consists of long and cylindrical to somewhat flattened, or broadly oval and distinctly flat beetles. Most North American species are found under the bark of dead trees, logs, and stumps. The longer, more cylindrical species are predators of wood-boring beetles and their larvae. At least one California species, *Eronyxa expansus* (4.3 to 5.7 mm), may be a predator of the Incense-cedar Scale *(Xylococculus macrocarpae)*. On the other hand, the widespread Cadelle *(Tenebroides mauritanicus)* feeds on a variety of stored organic products as well as the larvae of other grain-infesting insects. Most of the broad, flat genera, such as *Calitys* and *Ostoma,* are associated with fungi or dried vegetable matter.

IDENTIFICATION: Adult California bark-gnawing beetles range in length from 2.3 to 20.0 mm and come in two forms. The first group is broadly oval, flattened, and compact. Their body is dull and distinctly sculptured. The second group is made up of species that are long, cylindrical to flat beetles whose head and prothorax are loosely attached to the rest of the body, so the beetle appears slightly "narrow waisted." Their body surface is smooth, more or less shiny and black, brown, blue, or bluish green. The head of bark-gnawing beetles is prognathous or slightly hypognathous. The clubbed antennae consist of 10 to 11 segments. The antennal club is asymmetrical, often bulging to one side. The rectangular pronotum is broader than the head and wider than long or nearly square. The triangular scutellum is visible, sometimes only barely so. The elytra completely conceal the abdomen. The tarsal formula is 5-5-5, with the first segment usually very small. The claws are equal in size and are simple. The abdomen has five visible sterna with the sutures distinct.

The mature larvae are creamy white or pinkish, elongate, and almost cylindrical or slightly flattened. The distinct head has three-segmented antennae and is hypognathous. Two or five pairs of simple eyes are present or absent. The prothorax is covered with a single (e.g., *Temnoscheila* and *Tenebroides*) or a pair *(Calitys)* of armored plates. The prothorax is enlarged in *Calitys* and *Ostoma.* The usually well-developed, widely separated legs are five-segmented, including the claw. The 10-segmented abdomen may have pairs of fleshy bumps and appears nine-segmented from above. The ninth abdominal segment bears a pair of upturned, hooklike projections. The last segment is reduced and located beneath the body.

SIMILAR CALIFORNIA FAMILIES:
- some ground beetles (Carabidae)—antennae threadlike
- cylidrical bark beetles (Colydiinae, Zopheridae)—mandibles not visible from above

CALIFORNIA FAUNA: 31 species in 10 genera.

The genus *Ostoma* is known from North America and Europe. Three species are found in California. Their broad, oval body is somewhat flattened and their elytra have six distinct ridges. The margins of the pronotum and elytra are distinctly spread out and flattened. The elytral ridges of *O. pippingskoeldi* (5.3 to 10.5 mm) (pl. 151) are broken and studded with small bumps, while the surface has reddish patches. This species is known west of the Great Plains, extending from British Columbia to California, Arizona, and New Mexico. The larva and adult are found in association with fungus (*Fomes, Polyporus;* Basidomycetes, Polyporaceae) under the bark of fir *(Abies)*, pine *(Pinus),* Douglas-fir *(Pseudotsuga menziesii),* and hemlock *(Tsuga).*

The four species of *Calitys* are distributed in Africa, Europe, and North America. The parallel elytral margins are bumpy and expanded laterally, with one continuous ridge and two rows of small, buttonlike bumps, or tubercles, on each elytron. Two species are found in California. *Calitys scabra* (6.6 to 12.2 mm) (pl. 152) is distributed in Europe and North America, where it is known throughout Canada and the western and northern United States. *Calitys* is associated with fungus *(Polyporus)* under pine bark.

Of the six species of *Temnoscheila* (previously known as *Temnochila*) found in California, the Green Bark Beetle *(T. chlorodia)* (9.0 to 20.0 mm) (pl. 153) is the best known. It is found west of the Great Plains and is widely distributed in the mountains, deserts, and valleys of California during spring and summer. It is long, slender, and shiny metallic green, blue green, or occasionally purple. The larva is slender and somewhat flattened in appearance. Both adult and larva are predators, hunting under the bark and in galleries of fallen trees for wood-boring insects. The Green Bark Beetle is an important predator of several bark- and wood-boring beetles and their larvae. The Green Bark Beetle larva enters the galleries of bark beetles (Scolytinae, Curculionidae) and consumes the eggs, larvae, pupae, and newly emerged adults. Efforts to mass rear *Temnoscheila* to control forest pests have so far proven unsuccessful. The adult and larva are found under the bark of ponderosa pine *(Pinus ponderosa),* white fir (*A. concolor*), incense-cedar *(Calocedrus decurrens),* and coast live oak *(Quercus agrifolia).* The adult can deliver a painful nip when handled carelessly.

Tenebroides (pl. 154) is also predatory and found throughout western North America, often under the bark of conifers recently killed by bark beetles. They are similar to *Temnoscheila*, but these shiny black beetles are flatter and somewhat shorter. *Tenebroides corticalis* (4.3 to 9.7 mm) is found coast to coast across the northern United States. It occurs in coniferous forests throughout California. The cosmopolitan Cadelle *(Tenebroides mauritanicus)* (6.5 to 10.0 mm) is one of the largest beetles infesting granaries and pantries. It feeds on stored grain insects and occasionally damages grain. It will sometimes burrow into wooden grain bins or other nearby wooden fixtures, causing them to collapse. Three other species are found in the state.

COLLECTING METHODS: Look for bark-gnawing beetles during the day beneath bark of dead conifers and broadleaf trees. At night some species are found crawling on dead branches, logs, and stumps. *Temnoscheila* and *Tenebroides* are occasionally attracted to lights, whereas *Calitys* and *Ostoma* are found beneath bark of conifers in association with fungus.

CHECKERED BEETLES Cleridae

Checkered beetles are somewhat robust or long, hairy, and "bug-eyed" insects that are sometimes brightly marked with distinct patterns. Some strikingly colored species that are active during the day are thought to mimic wasps, flies, and other beetles. These species are boldly marked with black and red or yellow patterns or are uniformly shiny green or blue.

Checkered beetles are found in a wide variety of habitats, from flowers and beetle-infested trees to dried carcasses and stored foods. Many species are associated with trees and shrubs, where they are found under bark, in the tunnels of various borers, in galls, or on leaves and branches.

Both larvae and adults prey on insects, especially beetles. The larvae of checkered beetles associated with dead wood prey on the immature stages of bark beetles (Scolytinae, Curculionidae), longhorn beetles (Cerambycidae), and metallic wood-boring beetles (Buprestidae). Adults (e.g., *Enoclerus*) run up and down branches hunting adult beetle prey and searching for mates and egg-laying sites. The drab and nocturnal *Cymatodera* lay their eggs and feed at night, hiding during the day beneath bark or other debris. The larvae of *Trichodes* develop in the nests of bees and wasps, whereas those of *Aulicus* attack the underground egg pods of grasshoppers. The predatory *Necrobia* frequents stored food products infested with insects and mites.

IDENTIFICATION: California checkered beetles are long and slender, or robust and flattened beetles typically clothed in moderately long, erect hairs and range in length up to 20.0 mm. The head is hypognathous and is usually as wide or wider than the prothorax. The antennae have eight to 11 segments and are usually abruptly clubbed but are sometimes threadlike, saw-toothed, or comblike. The pronotum is longer than wide and narrower than the base of the elytra. The scutellum is visible. The elytra almost or completely conceal the abdomen. The tarsal formula is 5-5-5, with all claws equal in size and simple or toothed. The abdomen has five or six segments visible from below.

The mature larvae are variously colored and mottled with black or shades of brown and are long, slender, and straight bodied. Most are uniform in width, whereas others are markedly thicker at the middle of the abdomen. The distinct head has a pair of well-developed three-segmented antennae and is prognathous. There are zero to five pairs of simple eyes. The prothorax usually has a large armored plate across its back divided by a faint line. The widely separated legs are five-segmented, including the claw. The 10-segmented abdomen may have fleshy folds or wrinkles and is long and straight, or slightly expanded toward the end. The ninth segment has a plate across the back, usually with long and hooked or short, blunt projections.

SIMILAR CALIFORNIA FAMILIES:
- soldier beetles (Cantharidae)—antennae long, threadlike
- some bostrichid beetles (*Psoa*, Bostrichidae)—maxillary palps not expanded
- soft-winged flower beetles (Melyridae)—pronotum margined; antennae usually saw-toothed
- narrow-waisted bark beetles (*Elacatis*, Salpingidae)—head prognathous
- antlike flower beetles (Anthicidae)—head with neck
- antlike leaf beetles (Aderidae)—head with neck

CALIFORNIA FAUNA: Approximately 71 species in 19 genera.

Nineteen species of *Cymatodera* are found in California, and examination of the male genitalia is often necessary to identify the species. The nocturnal adults are usually brown, but some California species may have faint yellowish markings on the elytra. The body is widest toward the rear. *Cymatodera californica* (20 mm) is the largest bark-gnawing beetle in the state and is commonly found at lights. The adult has been found on the dead branches of coast live oak *(Quercus agrifolia)*. The immature of the flightless *C. ovipennis* (7.0 to 11.0 mm) is found in the larval tunnels of the small, flightless *Ipochus fasciatus* (Cerambycidae) on lemonadeberry *(Rhus integrifolia)*. It has also been reared from pine cones infested with various species of moth larvae and from twigs of elderberry *(Sambucus)* used as nests by solitary bees. *Cymatodera oblita* (9.0 to 13.0 mm) (pl. 155) is widespread in the western United States and occurs throughout California, where it is readily attracted to lights at night. *Cymatodera* larvae prey on the immature stages of wood-boring beetles by digging through the sawdust-packed tunnels of infested wood to search for larvae and pupae.

Adult *Phyllobaenus* (2.0 to 4.0 mm) are small beetles with distinctly bulging eyes and resemble ants in their form and behavior. Adults are found on flowers, bushes, trees, and logs, where they attack certain bark beetles *(Ips)* and other small borers. Thirteen species are known from California. *Phyllobaenus merkeli* is a predator of *Ips* in Jeffrey Pine *(Pinus jeffreyi)* and *Phloeosinus* (Scolytinae, Curculionidae) infesting cypress *(Cupressus)*. The larva and adult of *Phyllobaenus scaber* (pl. 156) feed on numerous small insects. The larva hunts in abandoned moth larva galls on chaparral broom or coyote brush *(Baccharis pilularis)*, eating the pupae of parasitic wasps, larvae of book lice, and eggs of spiders. Before pupating, the larva spins a loose, silken shelter and is one of the few beetles known to spin silk.

Most species of *Trichodes* develop in the nesting cells of bees. Three species and two subspecies occur in California. The wasp-colored Ornate Checkered Beetle *(T. ornatus)* (5.0 to 15.0 mm) and its subspecies is widely distributed in western North America and is common throughout California in spring and summer. It is found on a wide variety of flowers, including wild onion *(Allium)*, brodiaea *(Brodiaea)*, yucca *(Yucca)*, and various members of the sunflower family. The elytra are variably patterned with dark blue, blue, or greenish bands with yellow or faintly orange yellow to red patches. The bands sometimes cover the basal two-thirds of the elytra or are reduced to an oval spot just behind the middle. Although it feeds primarily on pollen, the adult will readily capture and devour insect prey, particularly other small, flower-visiting beetles. Mating takes place on flowers, after which the female sometimes consumes the male. The eggs are laid in the flower heads. The newly hatched larvae attach themselves to visiting leaf-cutter bees or vespid wasps and hitch a ride back to the unsuspecting host's nest, where they may take from one to four years to develop. Two recognized subspecies occur in the state. *Trichodes o. douglasianus* (pl. 157) is widespread except in the southeastern part of the state, where *T. o. tenuosus* lives. *Trichodes o. tenuosus* is smaller, parallel sided, and often marked with red, a color never found on *T. o. douglasianus*.

Species of the genus *Aulicus* are patterned in red and black or bluish black markings that may serve as some kind of warning coloration. Six species live in California. *Aulicus terrestris* (10.0 to 12.5 mm) lives on the ground and occurs along the interior of the southern Coast Ranges in central California. The elytra are black with a pair of large reddish orange humeral spots, each broadly joined at the sides with a large reddish orange subapical spot. The ventral segments of the abdomen are edged in red. The elongate, pink larva feeds upon the egg masses of the Lubber Grasshopper *(Esselenia vanduzeei)*, whereas the adult attacks moth caterpillars. This beetle and grasshopper are found in only a few localities in central California and always together. *Aulicus bicinctus* (6.5 to 7.3 mm) is similar to *A. terrestris* but more robust, with four smaller, reddish orange spots that are not connected along the margin. The ventral segments of the abdomen are bluish black throughout. It probably has a similar biology as *A. terrestris* and is known only from the San Gabriel and San Bernardino Mountains, where it has been collected on interior live oak *(Q. wislizenii)* and chamise *(Adenostoma fasciculatum)*. Both of these beetles are active in spring.

The genus *Enoclerus* is found in both deciduous and coniferous forests and also preys on bark beetles (e.g., *Dendroctonus, Ips,* and *Scolytus,* Curculionidae). The larvae feed upon the immature stages, whereas the adults feed on adult bark beetles and other small insects. They can capture and kill prey many times their own size. Adults are found on summer nights running about on infested trees looking for food and mates. Occasionally they are found at lights. Thirteen species live in the state. The Red-bellied Clerid *(E. sphegeus)* (8.5 to 12.0 mm) is an important predator of *Dendroctonus* beetles that attack conifers, especially *D. monticolae, D. pseudotsugae,* and *D. ponderosae*. This species has a yellow face, blackish body with a slight metallic luster, a broad, gray stripe across the middle of the elytra, and a bright red abdomen. Eggs are laid at the entrance of bark beetle galleries, which the larva enters in search of its prey. As the *Dendroctonus* adults begin to emerge, the mature red-bellied clerid larvae leave the bark beetle galleries to burrow into the soil at the base of the tree to pupate. The adult is typically active during summer and fall. The Black-bellied Clerid *(E. lecontei)* (9.0 to 14.0 mm) is black with gray markings on the elytra, especially at the rear. It is widespread throughout the west, and its life cycle is closely tied to that of its bark beetle prey. *Enoclerus quadrisignatus* (8.0 to 12.0 mm) (pl. 158) is found throughout much of the eastern and southwestern United States during summer. In California this species is found in the southern deserts, as well as dry scrub and pine forest in the adjacent foothills. The head, thorax, and basal third of the elytra, as well as the underside, are dull red. The remaining two-thirds of the coarsely and densely punctate elytra are black with two yellowish white bands narrowly interrupted at the elytral suture. This nocturnal species is sometimes attracted to lights and is a voracious predator of bark beetles and their broods.

Members of the genus *Necrobia* are primarily scavengers of dry carrion and other dried organic matter. The three California species are cosmopolitan. The Red-legged Ham Beetle *(N. rufipes)* (3.5 to 7.0 mm) (pl. 159) infests dried and smoked meats. It is dark metallic blue or green with reddish brown legs. Most of the damage is done by the larva, which prefers boring into the fattiest portions. This beetle also attacks cheese, hides, and dried carrion such as old road kills and Egyptian mummies. A similar species, *N. violacea* (3.0 to 5.0 mm), is entirely metallic blue or green, with dark brown or black legs and antennae. The Red-shouldered Ham Beetle *(N. ruficollis)* (3.5 to 6.2 mm) is also metallic blue, but the legs, prothorax, and bases of the elytra are brownish red. It occurs mainly on animal products but is also found on stored plant products, where it feeds on other pests in addition to the plant materials themselves.

COLLECTING METHODS: Flower-visiting *Trichodes* and *Phyllobaenus* are easily collected by hand or gathered in a sweep net. Search trunks, limbs, and twigs of trees infested with wood-boring and bark beetles both day and night for *Enoclerus*. Some *Cymatodera* and *Enoclerus* are readily attracted to lights. Beating small trees and woody shrubs will yield specimens of *Phyllobaenus*. In addition to checking infested meats and stored products, species of *Necrobia* are sometimes encountered on carcasses.

SOFT-WINGED FLOWER BEETLES — Melyridae

Soft-winged flower beetles feed on both plant and animal materials. Many species gather in large aggregations on conifers or flowering plants as they feed on pollen and look for mates. Some species *(Collops* and *Malachius)* feed on pollen and nectar or prey on small arthropods. Recent studies of the genus *Collops* suggest that they might be important predators of crop pests in alfalfa, cotton, and sorghum fields.

Two of the more conspicuous California genera *(Collops* and *Malachius)* have eversible vesicles, normally hidden balloonlike structures on the undersides of the prothorax and abdomen that expand like air bags. They are thought to be scent organs that produce odors to deter would-be predators. Species in both genera are often brightly colored to warn potential predators.

Species in the genus *Listrus* feed on the pollen and nectar of at least 26 plant families and are considered important pollinators. Their hairy bodies are often covered with pollen grains, which are then carried to other flowers.

The biology of larval soft-winged flower beetles is poorly known. The scant information available suggests that they live in leaf litter or under bark, probably feeding on detritus, fungi, and the eggs and larvae of various small arthropods. The larvae of *Malachius* have been found under bark in association with the galleries of wood-boring beetles. *Malachius bipustulatus* larvae, found in Europe and Asia, have been collected in sagebrush stems *(Artemisia)* inhabited by the larvae of tumbling flower beetles *(Mordellistena,* Mordellidae). Some, such as the larvae of *Listrus* and *Collops,* prefer sandy soils. Larval *Collops* are known to attack the larvae of several important weevil crop pests (Curculionidae). The larvae of *Listrus* have been found at the bases of plants growing in coastal dunes.

IDENTIFICATION: California soft-winged flower beetles are slender or robust, somewhat flattened to strongly convex beetles. They are often brightly colored in red and blue or mottled gray. The head is nearly as wide as the prothorax, with eyes bulging. The antennae are 11-segmented but appear 10-segmented in *Collops* and related genera where the second segment is small and partly hidden within the first segment. The antennae are threadlike, saw-toothed, or comblike. The pronotum is often squarish and is usually broader than long. The scutellum is visible. The elytra are soft or not, do not conform to the contours of the abdomen, and completely cover the abdomen or are somewhat cut off, exposing some of the abdominal segments. The tarsal formula is 5-5-5, but males of *Collops* are 4-5-5, with equal claws simple or toothed. The abdomen has five or six segments visible from below, sometimes with eversible vesicles along the sides.

The mature larvae are pale, long, and almost cylindrical or slightly flattened. The thoracic segments are only slightly wider than the head and the abdomen. The distinct head has short, three-segmented antennae and usually four to five pairs of simple eyes. The prothorax often has one to three pairs of armored plates or dark spots on the back, while the remaining segments usually have only a pair. The 10-segmented abdomen appears nine-segmented from above. The ninth segment usually has a pair of curved or hooked projections. The last segment is reduced and often located underneath the body.

SIMILAR CALIFORNIA FAMILIES:

- some soldier beetles (Cantharidae)—antennae long, threadlike; tarsi 5-5-5
- some leaf beetles (Chrysomelidae)—antennae long, threadlike; tarsi appear 4-4-4 but are 5-5-5
- checkered beetles (Cleridae)—clypeus not distinct; antennae usually clubbed, occasionally saw-toothed or threadlike
- some sap beetles (*Meligethes,* Nitidulidae)—antennae distinctly clubbed
- some fruitworms (*Xerasia,* Byturidae)—antennae distinctly clubbed

CALIFORNIA FAUNA: 435 species in 24 genera.

The genus *Malachius* (6.0 to 8.0 mm) (pl. 160) includes 17 species in California, plus several undescribed species. The antennae are distinctly 11-segmented, with the second segment clearly visible. The tarsi of both the male and the female are five-segmented. Species are shiny black or metallic blue with red or reddish orange markings. They are often found on flowers in spring and summer.

The genus *Collops* (3.0 to 8.0 mm) (pl. 161) contains some of the largest beetles in the entire family. Of the 30 species recorded from the United States, most occur in the west. Thirteen species are known from California. The antennae appear 10-segmented but are actually 11-segmented. The second segment is small and hidden from view. In males, the third antennal segment is expanded and scooped out. During courtship preceding mating, the enlarged segments are held together and receive the females' mouthparts. In males, the front tarsi are four-segmented, and the remaining tarsi are five-segmented. Many of the species in this genus are predators and are considered beneficial to some crops, such as alfalfa *(Medicago sativa)*. *Collops bipunctatus* (6.5 to 8.0 mm) is known from moderate elevations (5,000 to 6,000 ft) along the eastern slope of the Sierra Nevada.

More than 100 species of the genus *Listrus* (pl. 162) (formerly known as *Amecocerus*) are found in North America, most in the western United States and Canada. These small (2.6 to 3.3 mm), gray beetles are often variously marked with light or dark scaly spots. The side margins of the prothorax are finely serrated. These beetles are frequently found packed by the dozens in a wide variety of flowers during spring and summer, feeding on nectar and pollen. Because their hairy bodies are readily covered with pollen, species of *Listrus* are important pollinators. They are among the most poorly known beetles in North America and are very difficult to identify to species.

COLLECTING METHODS: Although a few species fly to lights at night, most species of soft-winged flower beetles are collected by hand from flowers or by beating and sweeping vegetation.

SAP BEETLES   Nitidulidae

The sap beetle family contains approximately 200 species in 29 genera in North America. They are small and found in a wide variety of habitats. Sap beetles have one of the most catholic feeding habits among all beetles. Adults are sometimes found in flowers, but most are encountered around sap flows or in decaying wood, rotting fruit, carrion, and fungi. Some adults feed on seeds, whereas others live under bark and prey on bark beetles (Curculionidae). The larvae of *Cybocephalus* feed on scale insects. A few species of *Epuraea*, *Carpophilus*, and *Aethina* are associated with bee nests, and members of *Amphotis* are associated with ants.

Some species of *Carpophilus* are pests of stored products,

especially dried fruit. The amount of fruit consumed by the adults and larvae is negligible, but the produce is soiled with their waste and no longer suitable for market. These and other genera also inoculate food stores with molds and other organisms that hasten their decomposition. Other sap beetles, such as some *Glischrochilus, Epuraea,* and *Carpophilus,* are known to carry and spread fungal diseases of plants.

The larvae of some European species of *Pityophagus* feed on bark beetles, a habit that may be shared by some of their American relatives. Larvae of some species have modified glands that secrete a mucouslike substance that is believed to function in aspects of life associated with fungi and potentially as a deterrent to predators. Most species pupate in the ground.

IDENTIFICATION: California sap beetles are small, ranging in length from 1.0 to 7.0 mm. The body is usually oval in outline and slightly humpbacked or flattened. The head is prognathous (except in *Cybocephalus*) and bears 11-segmented antennae that usually end in a three-segmented club. The scutellum is visible. The elytra are usually short, appearing cut off or separately rounded at the tips, and usually do not cover the entire abdomen. The tarsal formula is 5-5-5, with a small fourth segment, except in one genus *(Cybocephalus)* where it is 4-4-4. The abdomen typically has five segments visible from below.

The mature larvae are long, slender, and nearly cylindrical in cross section and have straight or slightly curved bodies. The distinct head is somewhat smaller than the prothorax and slightly flattened. It has well-developed, three-segmented antennae and two to four pairs of simple eyes. The prothorax usually has an armored plate across the back, while the remaining segments may or may not have plates or spots. The short, yet well-developed, legs are five-segmented, including the claw. The abdomen appears nine-segmented from above, but the tenth segment is small and hidden from above and is sometimes modified into a leglike structure. The ninth segment always has a pair of simple to elaborate projections.

SIMILAR CALIFORNIA FAMILIES:

- false clown beetles (Sphaeritidae)—elytra lined with rows of punctures and cover all but last abdominal segment
- clown beetles (Histeridae)—antennae elbowed; elytra short, lined with rows of punctures, and cover all but last two abdominal segments

- round fungus beetles (Leiodidae) — elytra grooved; antennae gradually clubbed
- some rove beetles (Staphylinidae) — antennae usually not distinctly clubbed; elytra with rows of punctures or distinctly ridged
- water scavenger beetles (Hydrophilidae) — long maxillary palps; weakly developed antennal club
- marsh beetles (Scirtidae) — antennae threadlike to sawtoothed; elytra completely covering abdomen
- pill beetles (Byrrhidae) — antennae threadlike or gradually clubbed, club with three to seven segments
- some bark-gnawing beetles (Trogossitidae) — antennal club loose; side margin of pronotum sharp; tarsal claws long
- wounded-tree beetles (Nosodendridae) — tibiae broadly expanded; elytra completely covering abdomen
- skin beetles (Dermestidae) — body usually covered with scales or hairs
- short-winged flower beetles (Kateretidae) — antennal club less compact; elytra always short and exposing at least two abdominal segments
- pleasing fungus beetles (Erotylidae) — elytra completely covering abdomen
- shining flower beetles (Phalacridae) — dorsal surface shiny; antennal club less compact
- minute fungus beetles (Corylophidae) — head often not visible from above
- hairy fungus beetles (Mycetophagidae) — body covered with short hairs; antennal club less compact
- tortoise beetles (Cassidinae, Chrysomelidae) — head not visible from above; antennae threadlike or gradually clubbed

CALIFORNIA FAUNA: 53 species in 21 genera.

Species of *Carpophilus* are some of the most commonly encountered sap beetles in California. They are oval in outline and have short, truncate elytra exposing the last two or three abdominal segments. Some species are pests, infesting dried fruits and grains. They become active as early as February and actively search for rotting fruit in orchards, where they find mates and lay hundreds of translucent, whitish eggs. They can be quite abundant in backyards with fruit trees. One of the most common and important pests of dried fruits is the Dried Fruit Beetle *(C. hemipterus)* (2.1 to 4.0 mm), which is distributed throughout the state. It infests figs, dates, raisins, and numerous other kinds of decomposing fruits and vegetables. This dull or shiny black beetle has shortened elytra, with each elytron marked with one large and one small, yellowish orange spot; the smaller spot is on the shoulders and the larger, wavy spot more or less at the apex of the elytra. The complete life cycle may take as little as 15 days during summer. The Pineapple Beetle *(C. humeralis)* (3.0 to 5.0 mm) (pl. 163), also called the Yellow-shouldered Sap Beetle, apparently originated in southern Africa and is now distributed worldwide. It is shiny brown to black, with a pale spot at the base of each elytron near the humerus. The antennae are reddish with a blackish club, while the legs are blackish or reddish black. It is an occasional pest of stored products, especially figs and other dried fruits. The Cactus Flower Beetle *(C. pallipennis)* (2.5 to 4.0 mm) is brown to reddish brown with dull yellow elytra and coarsely pitted head, pronotum, and elytra. It is found inside cholla cactus flowers *(Opuntia)* in large numbers in the Colorado and Mojave Deserts. Other species of *Carpophilus* are also known from *Opuntia*, but they are not as strikingly colored. The larva develops in old, decaying cholla fruits and other organic matter and occasionally infests stored foods such as corn.

Three species of *Meligethes,* commonly called pollen beetles, are known in California. They are most often encountered in large numbers in a variety of flowers, where they feed on both pollen and petals. *Meligethes* larvae also feed on pollen. Most species are shining black, but a few have a metallic green or blue luster. One species, *M. nigrescens* (1.8 to 2.3 mm), is shining black without a metallic luster. It has grayish yellow legs and antennae and is very common throughout the Pacific coast states, across southern Canada, and into the northern and central United States. It also occurs in much of Europe, North Africa, and the Middle East. Another species, *M. rufimanus* (2.1 to 2.5 mm) (pl. 164), is found throughout California and much of North America. The otherwise black body has elytra with a slight metallic reflection, and the legs are dark brown to black.

COLLECTING METHODS: Look inside flowers, especially those of cacti and daisies in the deserts and chaparral areas. Sweeping through blooming vegetation will also yield specimens. Search decaying fruits in the wild or those associated with orchards for those species that infest fruits. Inspect sap flows, fruiting bodies of mushrooms, and rotting fruit. Pitfall traps baited with fermenting malt or molasses are also effective for species known to occur on fungi, including subterranean species. Place about a quarter of an inch of malt or molasses in a metal or plastic container and add enough water so that the mixture is about an inch deep. Add a couple pinches of yeast to the solution. A few species are attracted to carcasses in advanced stages of decay. Several species are attracted to lights. Many species may be sifted from leaf litter and loose bark. The most effective way to collect a wide variety of taxa is to use a flight-intercept trap in forested areas.

SILVAND FLAT BARK BEETLES — Silvanidae

The biologies of the vast majority of American silvanid flat bark beetles, especially those species not considered of any economic importance, are unknown. Many are found under loose tree bark, where they are thought to feed on fungi. Others are found on plants or in plant debris and also on fungi. Several species in the genera *Ahasverus, Cathartus, Oryzaephilus,* and *Nausibius* are pests of stored grains, grain products, nuts, and spices.

IDENTIFICATION: The small size, somewhat flattened bodies, and distinctly margined, wavy, or toothed sides of the pronotum will distinguish silvanid flat bark beetles from most other California beetles. Adults are elongate and parallel sided to slightly egg shaped in outline. Most are brownish or blackish, sometimes with conspicuous hairs. They range in length up to 6.0 mm. The head is broad and usually distinctly narrowed behind the eyes. The head is prognathous. The 11-segmented antennae are either long and threadlike, with cylindrical segments and an inconspicuous club (*Uleiota*), or shorter, with beadlike segments and a distinct three-segmented club (*Carthartosylvanus, Nausibius, Oryzaephilus,* and *Silvanus*). The prothorax is longer than wide, with distinctly wavy or toothed side margins. The scutellum is short but visible. The elytra are long, distinctly pitted or rough, and cover the abdomen completely. The tarsal formula is 5-5-5 (4-4-4 in *Uleiota*), with claws equal and simple. The abdomen has five distinctly visible segments.

The mature yellowish or whitish larvae are elongate, somewhat flat, and straight bodied. The distinct head is broad and prognathous. The antennae are usually long and two- (*Carthar-*

tosylvanus, Nausibius, Oryzaephilus, and *Silvanus)* or three-segmented *(Uleiota).* There are five to six pairs of simple eyes. The thoracic segments are roughly equal in size. The well-developed legs are three-segmented, including the claw, and are close together *(Carthartosylvanus, Nausibius, Oryzaephilus,* and *Silvanus)* or widely separated *(Uleiota).* Each of the 10 abdominal segments is approximately equal in length. The ninth segment has a pair of long, whiplike projections or small bumps.

SIMILAR CALIFORNIA FAMILIES:
- flat bark beetles (Cucujidae)—more flattened; antennae not clubbed; pronotum usually wider than long to almost square; tarsi 5-5-4 in males
- lined flat bark beetles (Laemophloeidae)—head and pronotum with a distinct line or fine ridge near sides
- root-eating beetles (Monotomidae)—antennae 10-segmented with a one- or two-segmented club
- jugular-horned beetles (Prostomidae)—large, forward projecting mandibles; antennae weakly clubbed; sides of pronotum not margined; tarsi 4-4-4

CALIFORNIA FAUNA: Seven species in five genera.

Oryzaephilus is distinguished from other small, straight-sided beetles by the six broad, sawlike spines on each side of its prothorax. Two of the several species that are pests of stored products are established in California. The omnivorous Saw-toothed Grain Beetle *(O. surinamensis)* (1.7 to 3.3 mm) (pl. 165) infests cereals, bread, pasta, nuts, cured meats, sugar, and many other products. Dried fruits, especially raisins, are usually infested only after they have been stored for long periods of time. Their small, slender, flattened bodies permit their passage through narrow cracks and crevices, including the openings in poorly sealed food packages. The adult is typically found in the field feeding on fallen, moldy, drying fruits such as figs. The Merchant Grain Beetle *(O. mercator)* (2.2 to 3.1 mm) seems to prefer nuts and nut products. It is similar in appearance but has slightly longer elytra, larger eyes, and a narrow "temple" behind the eye, less than one-third of the eye width. In the Saw-toothed Grain Beetle, the eyes are relatively smaller, with the temples at least one-half the eye width.

COLLECTING METHODS: Look for silvanid flat bark beetles under loose bark, in decaying plant materials, especially rotting fruit, or in infested stored products.

FLAT BARK BEETLES   Cucujidae

Flat bark beetles live up to their name. These long, incredibly flat beetles and their larvae are adapted for crawling in the narrow spaces between the bark and wood of logs and stumps. Little is known of their biology, but the adults and larvae of *Cucujus* are predators of small insects. Only two genera of this small family occur in North America, both of which live in California.

IDENTIFICATION: Adult California flat bark beetles are flat and rectangular in overall body shape and are red *(Cucujus)* or brown *(Pediacus)*. The head is broad and bulging behind the eyes *(Cucujus)* or not *(Pediacus)*. The head is prognathous. The antennae are beadlike and consist of 11 segments. The pronotum is shorter than wide or almost square. The scutellum is visible. The elytra are parallel sided and completely conceal the abdomen. The elytral surface is flat and finely pitted. The tarsal formula of the male is 5-5-4, the female 5-5-5. The tarsal claws are equal and simple. The abdomen has five segments visible from below.

The mature larvae are elongate, flat, and amber colored. The distinct head is broader than the rest of the body. The antennae are long and three-segmented. The simple eyes are absent. The thoracic segments are wider than long and nearly the same size. The long and widely separated legs are five-segmented, including the claw. Each of the 10 abdominal segments is approximately equal in length. The ninth segment has a pair of distinct, forklike projections supported by a short *(Cucujus)* or long *(Pediacus)* stalk.

SIMILAR CALIFORNIA FAMILIES:
- lined flat bark beetles (Laemophloeidae)—head and pronotum with a distinct line or fine ridge near sides
- silvanid flat bark beetles (Silvanidae)—less flattened; antennae clubbed or threadlike; pronotum usually longer than wide; tarsi 5-5-5 in both sexes

CALIFORNIA FAUNA: Two species in two genera.

The Red Flat Bark Beetle *(Cucujus clavipes puniceus)* (10.0 to 17.0 mm) (pl. 166) is readily distinguished from all other California beetles. It is a distinctively flat, fire-engine red beetle with black eyes, tibia, feet, and antennae. The head of the equally flattened, amber-colored larva is broader than the rest of the body. The Red Flat Bark Beetle is distributed in coniferous forests throughout the western United States and Canada, from Alaska southward. It is widespread in pine forests throughout California. All stages of this beetle are found under the bark of conifers and deciduous trees. Little is known of its biology, other than the larva and adult are predators of wood-boring beetles. Gut content analysis has also revealed that they consume bits of plant material and fungi. This species usually overwinters in larval form, but some adults are capable of surviving extended freezing temperatures. The subspecies *C. c. clavipes* is found from the Great Plains eastward. The remaining 11 species of *Cucujus* occur in Europe and Asia.

Pediacus depressus (2.8 to 4.5 mm), the only other species of California flat bark beetle, is readily distinguished from *Cucujus* by its brown color and much smaller size. It is also found under the bark of coniferous trees in the Cascades and Sierra Nevada. Recent studies indicate that the North American population may be distinct from European populations, requiring description as a new species. Additional new species from the Coast and Transverse Ranges also await description.

COLLECTING METHODS: Look for adult and larval flat bark beetles under the bark of downed logs and stumps, especially conifers such as pine *(Pinus)*. They sometimes occur under the bark of deciduous trees.

PLEASING FUNGUS BEETLES Erotylidae

The habits of the pleasing fungus beetles are poorly known. The larvae and adults are usually found with basidiomycetous fungus under the bark of rotten wood or in the soil on fungi associated with roots. They usually feed on the fruiting fungal bodies. Adults usually lay their eggs on their fungal host. The larvae feed either by burrowing through fungal tissues or by grazing the surface. Some California species are active at night (e.g., *Megalodacne*), whereas others *(Triplax)* are active during the day.

Recently, the lizard beetles (Languriidae) were included with the pleasing fungus beetles as a separate subfamily (Languriinae). The lizard beetles are found worldwide but are best represented in the tropics. Adults of the genus *Languria* are frequently found on flowers. Their larvae bore within the stems of plants, especially asters and legumes. Their feeding activities seldom kill the plant but do reduce its vitality, weakening stems and causing them to break off just above where the egg was laid. Adults feed on the pollen and leaves of the larval host plant.

One widespread species in North America, the Clover Stem Borer *(L. mozardi)* is a minor pest of alfalfa *(Medicago sativa)* in the southwest and red clover *(Trifolium pratense)* in the east. Another California species, *Cryptophilus integer*, is occasionally a pest of stored grains.

IDENTIFICATION: California pleasing fungus beetles are long and somewhat oval in shape, while lizard beetles are usually straight sided. They range up to 15.5 mm in length. Pleasing fungus beetles are usually reddish brown or black, sometimes with bright reddish orange markings *(Megalodacne)*, whereas Lizard beetles are often contrastingly colored, with the head and prothorax red and the elytra black. The hypognathous head is inserted into the pronotum. The antennae are 11-segmented and are tipped by a distinctly abrupt and flat three-segmented club. The pronotum is variable in shape and distinctly margined. The scutellum is visible. The elytra are smooth, sometimes with lines of pits, and completely conceal the abdomen. Legs have a tarsal formula of 5-5-5; the fourth segment is sometimes reduced, thus appearing 4-4-4. The first three tarsal segments are more or less broad, with brushy pads below. Claws are equal in size and simple. The abdomen has five segments visible from below.

The mature larvae are light colored, pale brown, elongate, and almost straight sided, cylindrical, or spindle shaped. The distinct head is prognathous or somewhat hypognathous and bears three-segmented antennae. Five or six pairs of simple eyes are present. The short to moderately long legs are five-segmented, including the claw. The 10-segmented abdomen appears nine-segmented from above. The ninth segment has a short pair of projections. The last segment is small and located mostly underneath the body.

SIMILAR CALIFORNIA FAMILIES:

- metallic wood-boring beetles or jewel beetles (Buprestidae)—antennae are not clubbed
- click beetles (Elateridae)—antennae are not clubbed
- handsome fungus beetles (Endomychidae)—front angles of pronotum distinct and extended forward; tarsi appear 3-3-3 but are 4-4-4
- sap beetles (Nitidulidae)—elytra usually short, leaving abdomen partially exposed
- short-winged flower beetles (Kateretidae)—elytra usually short, leaving abdomen partially exposed

- lady beetles (Coccinellidae)—antennae not as distinctly clubbed; tarsi appear 3-3-3 but are 4-4-4
- some darkling beetles (Tenebrionidae)—antennae not distinctly clubbed; tarsi 5-5-4
- some zopherid beetles (*Hyporhagus*, Zopheridae)—antennae fit into grooves under prothorax; tarsi 5-5-4
- some leaf beetles (Chrysomelidae)—antennae threadlike or gradually clubbed

CALIFORNIA FAUNA: 12 species in six genera.

Four species of *Dacne* occur in California. They are distinguished from other small (less than 6.0 mm) pleasing fungus beetles in the state by a tarsal formula that is clearly 5-5-5. *Dacne californica* (1.9 to 3.5 mm) is a somewhat cylindrical, reddish brown to black beetle. In most individuals the head, pronotum, and a spot near the base of each elytron are a lighter reddish orange. It is distributed throughout western North America, from British Columbia to Baja California. In California this species occurs in all regions except the deserts. It is found on fungus associated with a variety of soft- and hardwoods and is collected in leaf litter at the base of manzanita *(Arctostaphylos)*, oak *(Quercus)*, and various conifers. The larva burrows into the tissues of various bracket fungi, including *Lentinellus, Pholiota, Pleurota,* and *Polyporus.*

Megalodacne fasciata (9.0 to 15.5 mm) (pl. 167) is a relatively large and conspicuously colored species normally found east of the Rocky Mountains and south into Mexico along the Gulf Coast. It was accidentally introduced into northern California, where it has become established in Butte, Glenn, Placer, and Tehama Counties. This shiny black beetle is oblong-ovate and has two distinct and irregular reddish patches on each elytron. Each shoulder patch contains a round, black spot. In California both the adult and the larva have been found feeding on bracket fungi growing on the trunks of almond trees or beneath the bark of rotten oak logs. It is sometimes attracted to lights. Large numbers of adults have been found overwintering beneath bark in eastern United States.

The two California species of *Triplax* may be distinguished from other pleasing fungus beetles in the state by having strongly reduced fourth tarsal segments, appearing 4-4-4. *Triplax californica californica* (2.7 to 5.0 mm) is found from British Columbia to Nevada and California, where it is widespread except in the deserts. The body of this species is black, but the head, prothorax, and legs (except the middle and hind coxae) are reddish yellow. It is active during the day and is found on a variety of fungi on both deciduous and evergreen logs. The subspecies, *T. c. antica,* has a dark pronotum and is found east of the Sierra Nevada.

Three species of *Languria* are known from California. The Clover Stem Borer *(L. mozardi)* (4.0 to 9.0 mm) (pl. 168) is the most widespread of all lizard beetles in North America, ranging from parts of Canada to northern Mexico. It has a red head, five-segmented antennal club, red scutellum (in most specimens), and black elytra, and the abdomen is red with the last three segments usually black. This species attacks alfalfa and is sometimes considered a minor pest. The larva molts four times and works up and down inside the stem, feeding on the pith. The adult is typically active in the morning, feeding mainly on pollen, but it will also attack the leaves. *Languria californica* (6.5 to 9.0 mm) is active May through June and is widespread along the coast of California, where it has been found on milkvetch *(Astragalus)*. It also has a red head and five-segmented antennal club, but the scutellum, elytra, and underside are completely black. The abdomen is more heavily pitted than in the other two species of *Languria* found in the state. The largest species, *L. convexicollis* (9.0 to 11.5 mm), lives in the deserts of California and also occurs in Arizona and British Columbia. The head is black or black and red, and the antennal club is six-segmented. The scutellum, elytra, and underside are all black. It is active May through August and has been found on the flowers of the prickly poppy *(Argemone munita).*

COLLECTING METHODS: Look for pleasing fungus beetles on or near woody and soft fungi on standing and downed logs. They are also found beneath the bark of trees in relatively moist woods, especially in wet, wooded canyons. Sift beetles from moist duff beneath trees or extract them by placing leaf litter samples in a Berlese funnel. A few species may come to lights. Lizard beetles are found on prickly poppies, alfalfa, and other leguminous flowers and are collected by hand or by sweeping.

BOTHRIDERID BEETLES Bothrideridae

Most bothriderid beetles are external parasites on the larvae and pupae of wood-boring beetles such as deathwatch and spider (Anobiidae), bostrichid (Bostrichidae), metallic wood-boring (Buprestidae), longhorn (Cerambycidae), and weevils or snout

(Curculionidae) beetles. A few species feed on fungi, including those cultivated by ambrosia beetles (Platypodinae, Curculionidae). Somewhat flattened species are usually found under bark, whereas more cylindrical species tend to live in the tunnels of their beetle hosts. All seem to prefer older, drier trees.

Bothriderids are currently of little or no economic importance, but studies on their parasitic habits may reveal the importance of their roles in the natural control of wood-boring beetles in managed forests.

IDENTIFICATION: California bothriderids are long and narrow, at least three times longer than wide. The broad, flat head is prognathous and has clubbed antennae with 10 *(Oxylaemus)* or 11 *(Deretaphrus* and *Sosylus)* segments. The attachment point of the antennae to the head is visible from above, and the antennal club is one- *(Oxylaemus),* two- *(Sosylus),* or three-segmented *(Deretaphrus).* The pronotum is longer than wide. The scutellum is visible. The elytra are pitted and with or without ridges running their length. The tarsal formula is 4-4-4. The abdomen has five segments visible from below.

Of the genera present in California, only the larvae of *Sosylus* have been described. First-instar larvae, or triungulins, are long, spindle shaped, very flat, and more heavily armored than later stages. The mandibles are curved and sickle shaped. The legs are five-segmented with well-developed tibiae. The 10-segmented abdomen lacks hooks or projections. Mature larvae are elongate, almost cylindrical to strongly flattened, and almost parallel sided or with the abdomen slightly enlarged. The head is distinct and prognathous. The antennae are two-segmented. One pair of simple eyes is present. The small, widely separated legs are five-segmented. The 10-segmented abdomen lacks hooks or projections on the ninth segment.

SIMILAR CALIFORNIA FAMILIES:

- wrinkled bark beetles (Rhysodidae)—antennal segments beadlike, not clubbed; tarsi 5-5-5
- powder-post beetles (Lyctinae, Bostrichidae)—head hypognathous; tarsi 5-5-5
- small bark-gnawing beetles (Trogossitidae)—mandibles prominent in front; tarsi 5-5-5
- cylindrical bark beetles (Colydiinae, Zopheridae)—antennal insertions hidden from above; tarsi 4-4-4
- small darkling beetles (Tenebrionidae)—antennal insertions hidden from above; tarsi 5-5-4

CALIFORNIA FAUNA: Three species in three genera.

Deretaphrus oregonensis (9.7 to 11.5 mm) (pl. 169) is a long, slender, black beetle, about four times longer than wide, with short, thick 11-segmented antennae. The pronotum has a long median depression running lengthwise that is interrupted at about the first third of its length, just below the head. Each elytron has four distinct ridges running nearly their entire length, with two rows of distinct pits between the ridges. The adult is found throughout the state under loose, dry bark of dead Jeffrey pine *(Pinus jeffreyi)*. It has also been found at night wandering about on dead trunks of pines and deciduous hardwoods. The larva attacks the larvae of a longhorn borer, *Asemum striatum* (Cerambycidae), as well as other longhorn beetles, metallic wood-boring beetles (Buprestidae), and some of the larger bark beetles *(Dendroctonus,* Curculionidae). The larva constructs a shell in which to pupate.

Sosylus dentiger (4.5 to 5.0 mm) is brownish to black and about four-and-one-half times longer than wide. The antennae are 11-segmented. The elytra have ridges running their length that are connected at their tips and lack columns of deep distinct pits in between. Other species of *Sosylus* attack platypodine ambrosia beetles (Platypodinae, Curculionidae) in their tunnels at the bases of coniferous trees.

Oxylaemus californicus (3.2 to 4.4 mm) (pl. 170) is a reddish brown, cylindrical beetle about three times longer than wide. The antennae are 10-segmented. The elytra lack ridges but have distinct columns of somewhat elongate pits. It is fairly common in the Pacific Northwest and occurs as far south as the San Bernardino Mountains. It has been collected flying at dusk in spring, under the bark of various conifers, and at lights.

COLLECTING METHODS: Peeling back bark is probably the most productive method for collecting bothriderids. Since dead trees dry out from the top down, driving many wood-boring species down the trunk, look for predatory bothriderids in the bases and roots of standing dead or dying trees. All California species are occasionally collected at lights.

HANDSOME FUNGUS BEETLES Endomychidae

About 45 species of handsome fungus beetles live in the United States. They are typically found in decaying wood, beneath logs, or under bark, where they feed on the spores and microhyphae of fungi. Some species reflex bleed when alarmed, secreting a noxious fluid from the leg joints as a defensive measure, just like lady beetles (Coccinellidae). Handsome fungus beetles are seldom of any economic importance, but at least one species, the Hairy Cellar Beetle *(Mycetaea subterranea),* is considered a minor pest in granaries and warehouses, where it infests moldy stored grain products.

IDENTIFICATION: Handsome fungus beetles are distinguished from other California beetle families by the two lines or grooves on the pronotum. Adults are broadly rounded, oval, or elongate-oval, and sometimes moderately flattened. They range in size up to 8.0 mm. The head is withdrawn into the prothorax and slightly bent downward but is otherwise prognathous. The antennae are 11-segmented with a distinct, yet loose, three-segmented club. The pronotum is distinctly broader than the head and has a pair of distinct lengthwise impressions or grooves; the front angles are distinct and project forward. The scutellum is visible. The elytra are irregularly pitted, sometimes hairy, and completely conceal the abdomen. The tarsal formula is 4-4-4 but may appear 3-3-3. The first two tarsal segments are usually strongly lobed beneath, and the tarsal claws are equal in size and simple. The abdomen has five or six segments visible from below. Color is usually black with reddish or pale markings.

The mature larvae are brownish or grayish and are elongate, somewhat spindle shaped, or nearly cylindrical, whereas others are flattened or may resemble a "roly-poly" or pillbug (a crustacean). The distinct head has three-segmented antennae, the second segment long and conspicuous, and the third relatively short. The head is usually prognathous. Most have two to four pairs of simple eyes. The thoracic and abdominal segments of some may appear rough or somewhat spiny. The legs are four-segmented. The abdomen is 10-segmented but appears nine-segmented from above, sometimes with protruding bumps on the sides. The ninth segment usually has projections.

SIMILAR CALIFORNIA FAMILIES:

- lady beetles (Coccinellidae)—antennae not as distinctly clubbed; pronotal grooves lacking
- pleasing fungus beetles (Erotylidae)—front angles of pronotum not extended forward; elytra with rows of punctures; tarsi appear 5-5-5

- minute bark beetles (Cerylonidae)—never black; last segment of palps small or pointed; elytra almost always with rows of punctures
- some leaf beetles (Chrysomelidae)—tarsi appear 4-4-4 but are 5-5-5

CALIFORNIA FAUNA: 11 species in nine genera.

Two species of *Endomychus* are found in the United States, one of which is found in the west. The third tarsal segment is very small and the tarsi appear 3-3-3 but are actually 4-4-4. When threatened they will bleed reflexively from their leg joints. *Endomychus limbatus* (3.2 to 4.2 mm) (pl. 171) has shiny reddish yellow elytra with black spots just before the base and apex. The base of the pronotum does not have a membrane. It is known from British Columbia and Alberta to Washington, Idaho, and Oregon, and south to California and Nevada. It lives in mountainous regions of California and has been found beneath the bark of a standing dead pine *(Pinus)*.

Aphorista morosa (6.2 to 7.2 mm) (pl. 172) lives in the coniferous forests of California and ranges northward into southeastern Oregon. The head is reddish brown, while the elytra and center of the pronotum are almost black. The base of the pronotum has a membrane. The margins of the pronotum and appendages and underside are reddish yellow. The adult is found under the bark of conifers. Two species of *Aphorista* occur in the state.

COLLECTING METHODS: The most effective way to collect handsome fungus beetles is by sifting rotten wood and leaf litter. They can also be located by searching host fungi on rotten wood or under bark. A few species may be attracted to lights.

LADY BEETLES Coccinellidae

Lady beetles, also known as ladybird beetles or ladybugs, are among the most recognized and beloved insects in the world. The bright, contrasting red and black coloration of some species has attracted people's attention for centuries, although only a fraction of the world's nearly 6,000 species are so colored. A considerable body of folklore is associated with them. In the Middle Ages, certain species were dedicated to the Virgin Mary and named Beetles of Our Lady. They are widely equated with good luck and are often associated with happy events such as weddings. Consequently, in many parts of the world, harming a lady beetle is thought to bring bad luck.

Lady beetles are surprisingly diverse in their habits. Some species become dormant during hot or cold weather. A few migrate to the mountains, where they assemble in small clusters or huge aggregations. Some species, particularly in the tropics, are plant feeders and crop pests. A small number feed on molds and fungi. However, the vast majority of lady beetles are predatory and feed on a variety of plant pests such as aphids, scale insects, mealy bugs, and even mites. As such, they are considered among the most beneficial of insects and are commonly released as biological control agents.

Since the late 1800s, more than 100 exotic species of lady beetle have been introduced into California to control various pests. Some of these became permanently established. Unfortunately, a number of established species appear to have contributed to the displacement of native lady beetles, posed a threat to other nontarget insects through direct predation, or created a public nuisance by aggregating inside homes and on buildings. The Multicolored Asian Lady Beetle *(Harmonia axyridis)* is one such example.

Lady beetles themselves are attacked by a number of insect predators, but they are well defended against many birds, mammals, ants, and other generalist predators. The bright colors of many advertise the presence of repellent chemicals in their bodies. The yellowish fluid released from their "knees" (leg joints) during reflex bleeding contains toxic alkaloids that render these beetles distasteful to their attackers. Although most lady beetles will exhibit this defensive behavior when disturbed, the toxicity of the fluid varies greatly. Most of the less pungent species are drab, but some falsely advertise themselves as unpalatable by adopting the colors and patterns of the more toxic species. Lady beetles also serve as models for other beetles and beetlelike insects to mimic.

The Ladybird Beetle Parasite *(Dinocampus coccinellae)* is a braconid wasp that attacks many species of ladybirds in California but is found most commonly on the Convergent Lady Beetle *(Hippodamia convergens)*. The female wasp inserts an egg into the larva, pupa, or adult lady beetle. The egg soon hatches, and the tiny wasp larva feeds on the lady beetle's internal tissues.

When the wasp larva reaches its full size, it bores a hole through its host. The lady beetle is often weakened or killed by this ordeal. The wasp spins a cocoon beneath the lady beetle's body, using silk to bind itself between a twig and the ladybird. The adult wasp emerges from its cocoon about a week later. In spring and summer, the bodies of parasitized Convergent Lady Beetles are found on various plants in natural areas and gardens. A related wasp parasite, *D. terminatus,* feeds only on the Ashy Gray Lady Beetle *(Olla v-nigrum).*

IDENTIFICATION: California's lady beetles are typically oval or round in outline and weakly hemispherical to strongly humpbacked in profile. The antennae are seven- to 11-segmented and form a gradual club. The head is often deeply inserted into the prothorax and usually not clearly visible from above. The scutellum is visible. The elytra are smooth, often shiny, and cover the abdomen completely in most species. The legs generally have a tarsal formula that appears 3-3-3 but is actually 4-4-4. The true third segment is tiny and hidden at the base of the fourth segment. Tarsal claws are simple or toothed. The abdomen has five or six segments easily seen from underneath; a seventh segment is rarely visible. The first abdominal segment usually has at least one distinct line immediately behind the attachment of the hind legs.

The mature larvae are frequently marked with white, yellow, orange, or red against a gray, blue, violet, or black ground color. Many are covered with a fleecy white wax. They are elongate and frequently bear spines, bumps, or wrinkles. The distinct head is prognathous and has antennae with one to three segments. There are zero to three pairs of simple eyes. The prothorax has two or four armored plates across the back, while the remaining two segments have only two each. The legs are long, particularly in active predatory species, and are five-segmented, including the claw. The 10-segmented abdomen is broadest behind the thorax and gradually tapers toward the rear. The last segment is short, narrow, and in some species, used as a false leg.

SIMILAR CALIFORNIA FAMILIES:

- round fungus beetles (Leiodidae)—antennal club abrupt; elytra often grooved
- marsh beetles (Scirtidae)—antennae threadlike or saw toothed
- shining flower beetles (Phalacridae)—antennal club abrupt; tarsi appear 4-4-4 but are 5-5-5; no distinct markings on the elytra

- pleasing fungus beetles (Erotylidae)—antennal club abrupt
- handsome fungus beetles (Endomychidae)—front angles of pronotum distinctly pointed forward
- minute fungus beetles (Corylophidae)—antennal club abrupt
- leaf beetles (Chrysomelidae)—tarsi appear 4-4-4 but are 5-5-5.

CALIFORNIA FAUNA: Approximately 180 species.

The Mealybug Destroyer *(Cryptolaemus montrouzieri)* (3.4 to 4.5 mm) (pl. 173) is a dull black beetle with silver pubescence and reddish head, prothorax, abdomen, and tips of the elytra. The larva is yellow and clothed in long waxy threads, somewhat resembling its prey, the mealybugs. This species was introduced into California in 1892 and has since become widely established in the state. It feeds on many kinds of mealybugs.

Axion plagiatum (5.0 to 7.0 mm) is shiny black and oblong and has a large spot on each elytron. In the San Francisco Bay area, the spot is yellowish, but elsewhere it is red. The underside of its thorax is black, and the underside of its abdomen is yellow or red. It is found on western sycamore trees *(Platanus racemosa)* infested with the tiny Sycamore Scale insect *(Stomacoccus platani)*. It is also found on coast live oak *(Quercus agrifolia)* and is known to congregate in tree hollows and acorn caps.

Chilocorus orbus (4.0 to 5.1 mm) (pl. 174) is a hemispherical, shiny black beetle that is rounded in outline. It bears a single red spot on each elytron. Its underside is black except for the abdomen, which is yellow or red. The spiny, black larva has a yellow band that crosses the body just in front of the middle. This widespread species feeds on a variety of scale insects, particularly those that infest orchards and ornamental trees. A related species, *C. fraternus* (3.4 to 5.1 mm) of the Great Central Valley, is very similar in appearance and can be distinguished from it only by examination of the male genitalia.

The history of modern biological control began with the introduction of the Vedalia *(Rodolia cardinalis)* (2.6 to 4.2 mm) (pl. 175) into southern California in 1888 to control the Cottony Cushion Scale *(Icerya purchasi)*. Both insects are native to Australia. The elongate, nearly parallel-sided adults are dark red with swirling black markings on their elytra that are somewhat obscured by a covering of fine hairs, giving them a grayish appearance. The female tends to be redder than the more blackish male.

The genus *Hippodamia* is well represented in the state. They are distinguished by their somewhat flattened, elongate appearance and their tarsal claws that are toothed or forked. Some species of *Hippodamia* are so similar in appearance to each other that identification requires an examination of the male genitalia by a specialist. The Convergent Lady Beetle *(H. convergens)* (4.2 to 7.3 mm) (pl. 176) is one of the most common lady beetles in North America and is found year-round. It ranges throughout the entire United States to Ontario and British Columbia in Canada. It has also been introduced to the Antilles and Central and South America. The black pronotum sports two long, whitish spots that converge toward each other at the back. Each elytron usually has seven discrete black spots, but this pattern varies; some individuals are spotless or nearly so. The somewhat long and flattened larva is a velvety black or gray with small orange spots on the first and fourth abdominal segments. The pupa is orange with black spots and found attached to plants, walls, and fence posts. In early spring, adult females begin to lay eggs on plants with ample supplies of aphids, producing several generations by summer. As summer temperatures increase and food supplies dwindle, the adults take wing and fly up into cooler mountain canyons to feed. In fall, as temperatures begin to drop and days become shorter, they fly to higher elevations and join communal groups on bare soil, trees, rocks, and logs, apparently drawn together by aggregation pheromones. Thousands to millions of beetles congregate in these groups. Those that matured in the Great Central Valley fly up to the Sierra Nevada, while those in coastal southern California travel to the Transverse and Peninsular Ranges. These aggregations are found throughout winter and early spring, sometimes covered with snow. These beetles are often collected and sold in garden centers for pest control, but when released into the garden, they usually fly away and do little to control local plant pests. As spring approaches, the beetles become more active, mate, and travel from their overwintering sites back to lower elevations to reproduce. In southern California, many overshoot the land and, aided by southward-traveling winds, are blown out to sea or are deposited on the beaches. These beached lady beetles become news items nearly every year. The Five-spotted Lady Beetle *(H. quinquesignata)* (4.0 to 7.0 mm) may or may not have convergent pale spots on the pronotum. The elytra are orange red, and the elytral markings are quite variable. They range from individuals that lack black markings to those that have three distinct black bands across the elytra. This species is also widespread in the state.

The Two-spotted Lady Beetle *(Adalia bipunctata)* (3.5 to 5.2 mm) (pl. 177) has a black head and red orange elytra. Its black prothorax has an irregular white margin; in some individuals, there appears to be a black M shape on a white background. The body is black beneath. Each wing cover typically has one round, black spot in the middle, but this pattern can vary. This common species feeds on aphids and is widely distributed in temperate North America and Europe, as well as Argentina and Chile.

The California Lady Beetle *(Coccinella californica)* (5.1 to 6.8 mm) lacks any markings on its red orange elytra. Its black head has two widely separated pale spots and a black pronotum with white markings on the front toward the sides. The elytra usually lack markings, except on some individuals inhabiting the Channel Islands. This common species lives only along the California coast, where it feeds on aphids. Adults often congregate in higher elevations. The European Seven-spotted Lady Beetle *(C. septempunctata)* (6.5 to 7.8 mm) (pl. 178) was introduced into California in 1979 following its introduction into New Jersey. It also has a black head with two separated white spots, and the elytra have a total of seven black spots. This species is well established and may be displacing the California Lady Beetle in some areas.

The Multicolored Asian Lady Beetle *(Harmonia axyridis)* (4.8 to 8.0 mm) (pl. 179) is a large and spectacular species now well established throughout southern and central California. Its common name reflects the extreme diversity of colors and patterns it exhibits. In some forms the pronotum is dull white marked with a black M shape. The elytra range from pale yellow orange to a bright red orange or black, with up to 20 spots. The mature dark gray to black larvae have cone-shaped spiny projections, with an orange line running down each side of the body. Originally from Japan, Korea, and the former U.S.S.R., it was introduced into Louisiana during the late 1970s and early 1980s and has since spread throughout the southeast. In California it was released in Visalia in 1991 and spread statewide in about two years. It typically overwinters in sheltered places in lower elevations, including in and around homes and other buildings. It sometimes occurs in such numbers to be a nuisance. This beetle is a voracious predator of aphids and psyllids. In southern California it is known to feed on the Red Gum Eucalyptus Psyllid *(Glycaspis brimblecombii)*, a common serious pest of eucalyptus trees *(Eucalyptus)*.

Myzia are relatively large, cream-colored lady beetles that live in forested areas. *Myzia interrupta* (6.5 to 8.0 mm) is oval in outline, and the lateral border of the elytra is slightly expanded. The head lacks any markings, while the pronotum has three light brown markings near the base, and the elytra usually have light brown stripes. *Myzia subvittata* (5.7 to 8.0 mm) (pl. 180) is less oval and is somewhat pointed toward the rear. The head has a dark spot, and the dark brown pronotum is fringed with a broad, yellowish white border. The margins of the elytra are broadly expanded and flattened, particularly toward the front. Both of these species are frequently encountered at lights.

The nearly hemispherical Ashy Gray Lady Beetle *(Olla v-nigrum)* (3.7 to 6.1 mm) (pl. 181) is distributed throughout North America south to Argentina. It has two distinct color forms. The most common is ashy gray above with black spots on the pronotum and elytra. The underside is tan, beige, or light brown. The less common form is all black above with white markings on the head and pronotum and a prominent red spot on each elytron; the underside is dark but the abdomen is red. In some the tarsi are red as well. It resembles *Chilocorus* and *Axion,* but unlike those genera, it has white at the leading edge of its head and prothorax. It is common on willow *(Salix)* trees in parks and riparian areas and feeds on psyllids and aphids.

COLLECTING METHODS: Search for lady beetles by visually inspecting flowering plants (especially those covered with aphids), sweep-netting plants such as oak and willow, and beating woody vegetation. A few species are attracted to lights at night.

TUMBLING FLOWER BEETLES — Mordellidae

Aptly named, tumbling flower beetles jump by kicking out with their legs, causing them to bounce and tumble unpredictably when disturbed as they attempt to avoid predators and escape collectors. They are also quick to fly when approached.

North America has more than 200 species, nearly three-quarters of which are in the genus *Mordellistena*. They are small, mostly black, wedge-shaped beetles commonly found on flowers, often in great numbers in spring and summer. Adults apparently feed on pollen and nectar of a variety of plants. There appears to be little relationship between adult feeding preferences and larval food plants. In spite of reports to the contrary, there is no direct evidence that tumbling flower beetle larvae prey on insects. Instead, they feed on galls, fungi, rotten wood, and inside the stems of numerous herbaceous plants and shrubs.

IDENTIFICATION: The humpbacked and wedge-shaped body form, long, narrow, pointed abdomen, and jumping behavior are distinctive. Adults are mostly black and 7.0 mm or less in length. The head is hypognathous and the short, 11-segmented antennae are saw-toothed and gradually clubbed, or threadlike. The pronotum is small, narrowed toward the head, and distinctly margined on the sides. The scutellum is visible. The elytra are smooth and narrowed behind. The elytral surface is clothed with fine hair and is frequently ornamented with lighter colored hair arranged in the form of lines, bands, or spots. The elytra completely conceal all but part of the last, slender abdominal segment. The tarsal formula is 5-5-4, with the claws equal in size and toothed. The abdomen has five segments visible underneath, the last long, narrow, and pointed.

The mature larvae are white, long, more or less straight sided or slightly widest at the middle, somewhat cylindrical, with the body straight or slightly curved. The distinct head is hypognathous and bears very short two- or three-segmented antennae. The prothorax is somewhat enlarged, while the remaining two segments are smaller and somewhat equal in size. The legs are very short, with two or three indistinct segments, and no distinct claw. The 10-segmented abdomen appears nine-segmented from above; segments one through six have pairs of wartlike bumps on the back. The ninth abdominal segment has a single, simple spine or a pair of small projections. The last segment is small and more visible underneath the body.

SIMILAR CALIFORNIA FAMILIES:

- wedge-shaped beetles (*Macrosiagon,* Ripiphoridae)—last abdominal segment not long and pointed

CALIFORNIA FAUNA: 21 species in two genera.

The hind legs of *Mordella* lack ridges but have distinct bumps on the tibiae. Of the three species of *Mordella* known in California, *M. grandis* (5.2 to 7.0 mm) is the largest and is widely distributed throughout the state. It is black with scattered brownish hairs. This species also occurs in British Columbia, Idaho, and

Oregon. *Mordella albosuturalis* (3.25 to 5.0 mm) is black with scattered reddish brown hairs with a narrow strip of golden hairs running along the elytral suture. It is found in northern California northward through Oregon to British Columbia, and eastward to Nevada and Utah. *Mordella hubbsi* (3.5 to 4.5 mm) (pl. 182) is also black, but the front and middle femora are reddish brown. It is found in the mountainous regions of the state.

Adult tumbling flower beetles of the genus *Mordellistena* (2.0 to 4.0 mm) are frequently found on flowers, especially California buckwheat *(Eriogonum fasciculatum)*. The 18 California species are distinguished from those of *Mordella,* the only other genus in the state, by having one to six oblique to distinct ridges on their hind legs. Larval host records do not exist for California species, but larvae are recorded elsewhere as feeding on grasses, legumes, asters, and other plants. There is at least one report of the larvae infesting hemp *(Cannabis)*. Most species are black, but some are uniformly light colored or with bands, stripes, or spots.

COLLECTING METHODS: Tumbling flower beetles are wary and escape easily, but they can be picked up on flowers by hand, scooped into a container, or collected with a sweep net. Malaise and flight-intercept traps produce the greatest diversity of species, especially those that do not visit flowers. A few are sometimes attracted to lights.

RIPIPHORID BEETLES — Ripiphoridae

The ripiphorid beetle family contains about 425 species worldwide, with more than 50 species occurring in North America. Ripiphorid beetles have hypermetamorphic life cycles, where the larvae start out as active, silverfishlike triungulins and later transform into sedentary grubs. Early instars feed internally on the larvae of other insects, whereas the later stages feed externally on their hosts. Females of the genus *Ripiphorus* lay their eggs on flowers. Upon hatching, the triungulin attaches itself to a solitary bee as it visits flowers and is carried back to the bee's nest. Species of the genus *Macrosiagon* parasitize wasps, whereas other genera attack cockroaches and wood-boring beetles.

Adult ripiphorid beetles are ephemeral, living for only a day or two. Consequently, the information on their lives is fragmentary. Males and females rest on low grasses or flowers or meet in mating swarms. The comblike antennae of the males presumably increase their ability to locate females emitting sexual pheromones. Females are often seen on flowers of buckwheat *(Eriogonum)*, milkweed *(Asclepias)*, and various asters.

IDENTIFICATION: California ripiphorid beetles are wedge shaped with black and orange, red, or yellow coloration. They range from 4.0 to 12.0 mm in length. The head is hypognathous. The usually 11-segmented antennae are fan or comblike. The pronotum is narrowed behind the head and is without lateral margins. The scutellum is usually covered, at least in part, by the extended margin of the pronotum. The elytra are smooth, without grooves, and almost cover the abdomen *(Macrosiagon)* or are reduced to scalelike plates *(Ripiphorus)*. Legs have a tarsal formula of 5-5-4, with the claws equal in size and comblike or toothed. The abdomen has five visible segments.

The minute (0.45 to 0.95 mm) first larval instar is a triungulin. The heavily armored, spindle-shaped body expands greatly after feeding on the host larva. The distinct head is large and prognathous. The antennae are three-segmented with a terminal stylus, and there are five pairs of simple eyes. The thoracic segments are distinct. The five-segmented legs, including the claws, are long and slender. The 10-segmented abdomen does not have any projections. Later instars (two through six) are C shaped, lightly armored, and blind. The head is hypognathous. The antennae and legs are greatly reduced, and the thoracic and abdominal segments have cone-shaped horns.

SIMILAR CALIFORNIA FAMILIES:

- twisted-winged parasites (order Strepsiptera)—eyes stalked; elytra short and knoblike; abdomen pointed
- tumbling flower beetles (Mordellidae)—elytra cover most of the abdomen; antennae threadlike or saw-toothed; last abdominal segment acutely pointed

CALIFORNIA FAUNA: Approximately 18 species in two genera.

Macrosiagon is the most commonly encountered genus of ripiphorid beetles in the state. Their elytra nearly conceal the long but telescoping abdomen. Little is known of the life history of the California species of *Macrosiagon*. Elsewhere they have been reported to parasitize wasps of the families Tiphiidae, Scoliidae, Vespidae, and Sphecidae. *Macrosiagon cruenta* (5.0 to 8.0 mm) (pl. 183) is known from Canada and is widespread in the United States. It is found throughout California during late spring and summer on the flowers of buckwheat, milkweed, salt marsh fleabane *(Pluchea odorata),* and baccharis *(Baccharis emoryi).*

The 12 California species of *Ripiphorus* are characterized by short, scalelike elytra that expose the short, broad abdomen. *Ripiphorus smithi* (4.0 to 7.0 mm) lives in the Great Central Valley. It parasitizes ground-nesting bees that burrow into hard-packed soil. The adult is short-lived, with the male living less than a day. Males emerge first, followed by the females. After mating, the male flies away, leaving the female to lay its eggs on the flowers of alkali mallow *(Malvella leprosa).* Less than 10 eggs are laid at a time, with more than 800 eggs produced during the female's short life. Upon hatching, the triungulin attaches itself to the bees that visit the flower in search of pollen and nectar. The triungulin is carried back to the nest. It penetrates the body of the bee's larva to feed internally. It molts again to assume a more grublike form. Eventually it leaves the larva's body to feed externally. The host larva remains alive until the beetle larva has nearly completed its life cycle. The pupal stage lasts about two weeks. *Ripiphorus rex* (9.0 to 11.0 mm) (pl. 184) ranges from northern Mexico, Texas, New Mexico, and Utah to central and southern California. It is black with yellow antennae (in the male), legs, and elytra, and a variegated yellow and black abdomen. It is unique among the species of *Ripiphorus* in having a serrated outer edge of the middle tibiae.

COLLECTING METHODS: Sweep or examine flowers and low vegetation closely during the warmer parts of the year, particularly those attractive to bees and wasps. Ripiphorid beetles are rare in collections partly because of their short flight period, although they may be quite abundant locally.

ZOPHERID BEETLES   Zopheridae

The Colydiidae and Monommatidae families are now placed as subfamilies in the Zopheridae. The zopherid beetles *Phloeodes* and *Zopherus,* sometimes called "ironclads" because of the incredibly hard exoskeletons, are the most conspicuous California species. These flightless beetles lack hind wings, and their thick elytra are more or less fused along the elytral suture. When disturbed, they tuck in their legs, drop to the ground, and play dead, resembling a piece of bark.

One of the best known ironclads is *Z. chilensis.* It ranges from southern Mexico to Venezuela, where it is found on the bark of dead trees in search of fungi to eat. In Mexico, the beetle, popularly known as the *ma'kech,* is decorated with brightly colored glass beads and fixed to a small chain or tether and pinned to clothing as a reminder of an ancient Yucatecan legend. The beetle's thick, hard exoskeleton affords protection from desiccation and, if properly cared for, these living jewels may live many months or more on a diet of rotten wood, cereal, and apples. Tourists sometimes return to California with these gaudily decorated beetles illegally. From time to time they are sold legally in the state as a novelty.

California *Phloeodes* and *Zopherus* are found in a variety of woodland, forest, or desert habitats. They are found under bark or in dead wood, tunneling in rotten or sound logs and stumps of various conifers and hardwoods, where they feed on the fruiting bodies of tough, fleshy, or woody fungi. Their larvae have very small legs and are lightly armored, both apparent adaptations for boring through reasonably hard wood. They feed in dead wood that has been attacked by white-rot fungi.

Although the adults are outwardly similar to some darkling beetles (Tenebrionidae), they lack the defensive glands that produce foul-smelling odors. Also, larval zopherid beetles have subtle yet distinct differences in the structure of their mouthparts and abdominal projections that distinguish them from darkling beetles.

IDENTIFICATION: Adult California zopherid beetles are elongate and parallel-sided black or blackish gray beetles. They range in length from 2.0 to 25.0 mm. The head is prognathous. The antennae are moderately to abruptly clubbed and consist of nine to 11 segments. The pronotum is fitted with special grooves on the upper or lower surface to receive the antennae. The scutellum is visible or hidden. The elytra are usually parallel sided and completely conceal the abdomen. The elytral surface is often roughened by raised bumps and deep pits or has sharp, well-defined ribs. The tarsal formula is 4-4-4 or 5-5-4. The claws are equal in size and simple. The abdomen has five visible sterna that are mostly fused together and have indistinct sutures. The first two, three, or four abdominal segments are sometimes fused together. The last abdominal segment may have a crescent-shaped groove or a pair of grooves.

The mature larvae are long and parallel sided and are nearly cylindrical or flattened. They are lightly sclerotized and white or pale creamy white. The head, thoracic segments, and legs are pale tan, while the claws, urogomphi, and mouthparts are black. The distinct head may have simple eyes or not, and it has a pair of three-segmented antennae. The well-developed legs are short and five-segmented. The abdomen is 10-segmented, with the ninth segment bearing a pair of projections.

SIMILAR CALIFORNIA FAMILIES:

- darkling beetles (Tenebrionidae) — species generally more rounded in outline; many with defensive secretions; many with abdomen with three fused abdominal segments instead of four; species often move about the ground freely rather than associated with trees or fungus; antennae less often inserted in grooves

CALIFORNIA FAUNA: 37 species in 17 genera.

Of the five species of *Usechus,* three live in Japan. The other two species are both found in Washington, Oregon, and California. The adults are brown, roughly sculptured, and covered with short, reddish or yellowish scalelike hair. Both adults and larvae are found with fungus and under moldy bark of dead logs and stumps in moist, wooded areas in the northern and central portions of the state, especially along the coast. The 11-segmented antennae fit into grooves along the upper front margins of the pronotum. The sides of the pronotum are almost straight and parallel to each other. The elytra are distinctly ribbed with double rows of pits between the ribs. *Usechus lacerta* (3.1 to 5.5 mm) is blackish and does not have small elytral processes at the bases of each elytron that extend forward over the pronotum. The larva is found in fungus under the bark of oaks *(Quercus)* and big-leaf maple *(Acer macrophyllum)*. *Usechus nucleatus* (2.9 to 4.8 mm) is brownish black to blackish yellow brown. The bases of the elytra have short processes that extend forward over the base of the pronotum. A similar California zopherid, *Usechimorpha barberi* (3.2 mm), is more rectangular and has distinctly clubbed antennae, and the sides of the pronotum are more angular and bulging at the middle.

The genus *Phellopsis* occurs in China, Japan, and the Russian Far East. One species, *P. obcordata* (formerly *P. porcata* in California) (10.0 to 15.5 mm) (pl. 185), occurs in North America. It has a transcontinental distribution in old-growth forests from Alaska to Newfoundland. In the eastern United States it ranges southward into Tennessee and North Carolina through the Appalachian highlands. In the west its southern range includes the Rocky Mountains of Idaho and Montana, as well as the Sierra Nevada in California. It is often found under bark or on fungi on downed logs. This species is distinguished from other large (10 mm or greater) California zopherids by having 11-segmented antennae.

The genus *Zopherus* is distributed primarily in the southern United States and Central America. Their antennae appear nine-segmented (club segments nine through 11 are fused), and the femora and tibiae are lined with rows of golden hairs. Deep grooves on the underside of the prothorax provide receptacles for the antennae. Five species and one subspecies occur in California. *Zopherus tristis* (10.5 to 22.0 mm) is found under the bark of desert tamarisk *(Tamarix)* and other trees in the Colorado Desert in spring and summer. *Zopherus granicollis granicollis* (12.6 to 20.5 mm) (pl. 186) is known from Death Valley and the Peninsular and Transverse Ranges. The adult is active throughout the year. The larva bores into the root crowns of dead single-leaf pines *(Pinus monophylla)* and Jeffrey pines *(P. jeffreyi)*. Adults have been collected beneath the bark of these trees and also at the bases of large California junipers *(Juniperus californica)* at night. *Zopherus g. ventriosus* (14.5 to 21.0 mm) is similar in appearance but has finer elytral tubercles and a dull luster. This species is found in the western foothills of the southern Sierra Nevada and Saline Valley in Inyo County. *Zopherus opacus* (13.0 to 20.3 mm) has the dullest surface luster of all the California species and is known from higher elevations of the White Mountains and the eastern Sierra in Inyo and Mono Counties. *Zopherus sanctaehelenae* (13.0 to 21.0 mm) is known only from populations in Napa County in northern California and is active in spring and summer. *Zopherus uteanus* (14.0 to 22.0 mm) is found in the Granite Mountains of eastern San Bernardino County. It also occurs in northern Arizona, southern Nevada, and southern Utah.

The genus *Phloeodes* has 10-segmented antennae with a one-segmented club and lacks rows of golden setae on the inner margins of the legs. When disturbed, adult *Phloeodes* feign death, drawing their legs and antennae into specially fitted grooves on their body. The larvae have been found on rotten logs of various trees. Fourteen species have been described from California, but only three of these names represent distinct species. The Ironclad Beetle *(P. pustulosus)* (15.0 to 25.0 mm) (pl. 187) is found throughout California. It is black or grayish black, with numerous shiny and irregularly spaced knobs resembling pustules on the elytra. The very tips of the elytra barely have traces of pale "crust," or vestiture. They are often found walking on the ground or hiding beneath the bark of oak logs and stumps where they are thought to feed on fungus in the rotting wood. The Diabolical Ironclad Beetle *(P. diabolicus)* is similar in size and color, but the last fifth of each elytron has a large and distinct patch of pale crusty vestiture. It is found in central and southern California, where it feeds on tough woody fungi growing on decaying stumps of coast live oak *(Q. agrifolia)*. The elytra of both species may have distinctive black velvety spots and crescents. The Plicate Ironclad Beetle *(P. plicatus)* (12.0 to 16.0 mm) (pl. 188) lacks tubercles on its back. The tips of the elytra resemble the folds of a curtain, or are plicate. They live beneath the bark of oak trees throughout the state. Adult and larval *Phloeodes* are recorded from cottonwood *(Populus)*, oak, and the introduced mulberry *(Morus)* but are likely to be associated with other trees as well.

The sole California monommatine, *Hyporhagus gilensis* (4.5 to 7.5 mm), is compact, oval in outline, and dull black in color. It is attracted to lights in dry foothill areas surrounding the southern deserts and has been found in the rotting stems of yuccas *(Yucca)*.

COLLECTING METHODS: Zopherid beetles are found beneath the bark of dead pines, oaks, cottonwoods, and other trees with fungal infestations throughout the year. Both *Phloeodes* and *Zopherus* are sometimes found walking on trails in oak woodlands or pine forests late in the afternoon or evening during late spring and summer. *Phloeodes* and *Zopherus* are long-lived in captivity and thrive on a diet of rotten wood, sliced apples, and oatmeal. *Hyporhagus gilensis* is attracted to lights in summer and is associated with rotting yucca stalks. Adult colydiines are usually found under or on bark, in the galleries of other wood-boring beetles, and at light.

DARKLING BEETLES   Tenebrionidae

Usually black and hard bodied, darkling beetles are among the most conspicuous beetles in California, especially in the deserts.

On cool or overcast days, or at night, they are often active and abundant, foraging for food. Most of these beetles scavenge both plant and animal matter. The head-standing stink beetles *(Eleodes)* are among some of the most familiar members of the family. Some long-lived beetles, either in captivity or subjects of mark and recapture surveys, are known to live 17 years or more.

Darkling beetles are found in nearly all habitats, except aquatic, and are particularly abundant in drier regions, especially in coastal and desert dune habitats. Many California species are sand dune obligates, requiring the fine, loose sand of beaches and deserts to burrow, feed, and breed. Steady plodders, these dune dwellers often leave long and familiar tracks winding over the fine sand.

To conserve water and regulate their body temperatures, about half of the species have sealed cavities beneath their wing covers to reduce the amount of water lost as they breathe. They also take shelter inside rodent burrows, among leaf litter, or beneath stones, bark, or logs during the hottest parts of the day.

At dusk or in the evening, darkling beetles are often found crawling on downed logs, on standing tree trunks, or over the ground. Species associated with trees are found either under bark or in galleries and tunnels carved out by wood-boring insects. Some prefer living in rotten wood, where they feed on fungi. Still other species graze upon the surfaces of lichens, algae, and mosses growing on the surfaces of bark and rocks.

Soil-dwelling larvae and adults in drier regions feed on the seeds, roots, and detritus of many kinds of plants. Chaparral and desert dwellers sometimes use the seed-strewn mounds of black *(Messor)* and red *(Pogonomyrmex)* harvester ants like a buffet, scavenging bits of plant detritus.

The long, cylindrical larvae are sometimes called false wireworms because of their resemblance to larval click beetles (Elateridae). The front pair of legs of soil and sand-dwelling larvae is often much thicker and more heavily armored than the remaining legs. The immature stages of the mealworms *Tenebrio* and *Zophobas*, sold in pet shops and bait stores, are familiar examples of darkling beetle larvae. Both larvae and adults are seldom of any negative economic importance, although some people consider their sheer numbers around homes and agricultural endeavors a nuisance.

About half of the species of darkling beetles have specialized abdominal glands and chambers that produce and store noxious chemicals, especially quinones. These and other chemicals, including organic acids, render the beetles unpalatable to most predators. The black color shared by many of these distasteful species serves as a warning to potential predators and is thought to be an example of Müllerian mimicry, a phenomenon where unrelated species advertise their possession of some defensive capability with a common physical feature or behavior.

IDENTIFICATION: California darkling beetles are usually black or dark brown, ranging in length up to 30.0 mm or more. They are extremely variable in shape, ranging from elongate and somewhat cylindrical, to oval or round and strongly humpbacked or hemispherical. The head is usually prognathous but is occasionally hypognathous. The antennae are normally 11-segmented, with the segments beadlike or gradually clubbed. The antennae are attached below the expanded rim that extends around the front of the head. The eyes are usually notched. The pronotum is either broadly rounded or hatchet shaped and is usually distinctly margined on the sides. The scutellum is visible. The elytra completely conceal the abdomen and are smooth, pitted, bumpy, grooved, or ridged. Species with elytra partially or completely fused have their flight wings reduced or absent. The legs are variable, from stout and specialized for digging, to somewhat slender and adapted for rapid movement. The tarsal formula is 5-5-4, rarely 4-4-4 *(Anchomma)*, with claws equal in size, usually toothed but sometimes comblike. The abdomen has five segments visible from below; the first three are fused together.

The mature larvae are whitish or reddish to yellowish brown and are usually long and cylindrical, or slightly flattened, and are straight bodied. Their body segments are tough and thickly armored. The distinct head is prognathous and bears two- or three-segmented antennae. Up to five pairs of simple eyes are present or absent, sometimes represented only by dark spots. The prothorax is larger than the remaining two thoracic segments. The five-segmented legs, including the claw, are well developed. The 10-segmented abdomen appears nine-segmented from above. The ninth segment ends with either projections, spines, or dense hairs. The last segment is reduced in size and is more easily visible beneath the body.

SIMILAR CALIFORNIA FAMILIES:
- ground beetles (Carabidae)—tarsi 5-5-5; first abdominal segment divided by hind coxae

- trout-stream beetles (Amphizoidae)—tarsi 5-5-5; found in or near water
- false darkling beetles (Melandryidae)—winged; pronotum often with two impressions at base
- pleasing fungus beetles (Erotylidae)—tarsi 5-5-5; antennal club abrupt; elytra sometimes with bright-colored markings
- minute bark beetles (Cerylonidae)—very small; tarsi 4-4-4; antennal club abrupt
- zopherid beetles (Zopheridae)—rough or ribbed elytra generally parallel sided in outline; pronotum often with distinct grooves underneath to receive the nine-, 10-, or 11-segmented antennae; abdomen usually with two to four fused abdominal segments

CALIFORNIA FAUNA: Approximately 445 species in 110 genera.

In California, the rough-looking Desert Ironclad Beetle *(Asbolus verrucosus)* (18.0 to 21.0 mm) (pl. 189) is widespread throughout the Colorado and Mojave Deserts, where it prefers sandy soils. The smooth-bodied *A. laevis* (17.0 to 20.0 mm) (pl. 190) is restricted to sand dunes primarily in the Colorado Desert. Both species emerge at dusk to scavenge plant and animal materials until the following morning. With their coating of bluish gray wax secreted by their extremely tough exoskeletons and penchant for playing dead when disturbed, they have earned the names "blue death feigners" and "ghost beetles." They are active year-round, but seem to have two peaks of activity, one in spring and the other in late summer. These darkling beetles are very long-lived. An individual of *A. verrucosus* is recorded to have lived 17 years in captivity. Two other species of *Asbolus* also occur in the state. *Asbolus* does not produce an odor when disturbed.

In cross section, the antennae of *Cryptoglossa* are broadly oval, and like *Asbolus,* their eyes are not divided into two regions. *Cryptoglossa* (formerly *Centrioptera*) *muricata* (14.0 to 24.0 mm) (pl. 191) is the most common of the three species found in California. It is similar in appearance to *A. verrucosus* but is more elongate and almost parallel sided. It is known primarily from the coarse-grained sandy soils of the Colorado and Mojave Deserts but also occurs in some of the drier regions along the western edge of the San Joaquin Valley. It also occurs in Arizona and Utah, as well as in Baja California and Sonora.

Schizillus is very similar to *Asbolus* and *Cryptoglossa,* but the eyes of *Schizillus* are completely divided by a strip of exoskeleton called the canthus. The most common species, *S. laticeps* (19.0 to 23.0 mm) (pl. 192), occurs in the Colorado and Mojave Deserts, including the adjacent foothills and desert mountain ranges, as well as the Big Panoche area in the San Joaquin Valley.

Four species of *Nyctoporis* are recorded from California. *Nyctoporis carinata* (11.0 to 18.0 mm) (pl. 193) is found throughout the foothills, mountains, and valleys of the state, but not in the deserts. Its coarsely sculptured body resembles those of zopherid beetles (Zopheridae). The pronotum is coarsely pitted, and the elytra have lines of distinctly raised bumps, with distinct ridges on each side. The tarsi are clothed underneath with a thick, yellow pile.

Coniontis (pl. 194) are stout, oval, and smooth darkling beetles commonly encountered throughout California in spring and summer, except in the deserts. All the species are similar in appearance and are very difficult to identify. Adults have front tarsi cylindrical in cross section and plain, nearly cylindrical front tibiae that are sometimes hooked at the tip. *Coniontis abdominalis* (13.5 to 19.0 mm) occurs primarily in the southern Coastal Ranges and the interior foothills of the Transverse Ranges. More than 40 species are known from the state.

Adult *Eusattus* typically forage on the surface throughout the year and are regularly parasitized by tachinid flies and braconid wasps. The front tarsi are cylindrical, with the first segment more than twice the length of the second, and the outer margin of the foretibia is greatly expanded. The underside of the prothorax is almost bald, except for the fringe of hairs next to the edge. Nine species and two subspecies are known from California. Three widespread species are commonly encountered. *Eusattus difficilis* (9.0 to 13.0 mm) is found throughout arid and semiarid habitats in southern California, from Inyo and Kern Counties in the north to central Baja California in the south. *Eusattus muricatus muricatus* (11.0 to 13.0 mm) ranges discontinuously in sandy habitats throughout most of the western United States, including eastern and southern California. This black, humpbacked beetle, whose rear margin of the prothorax is trimmed in short, fine, golden hairs, is widely distributed among isolated sand dunes in California's Great Basin, Mojave Desert, and Colorado Desert. The last antennal segment of this species is rounded and nearly symmetrical, in contrast to the angulate, noticeably asymmetrical last segment of *E. difficilis*. *Eusattus dilatatus* (11.0 to 14.0 mm) (pl. 195) is restricted to wind-blown sand formations from Riverside County south to Puerto Penasco, Sonora, Mexico. The middle and hind tibiae of this species are strongly bowed rather than nearly straight as in the other two species.

The genus *Coelus* includes a small number of species restricted primarily to coastal sand dunes along the Pacific coast of North America. The first segment of the front tarsi is flat, like a spatula, and extends beyond the second segment. These small, round, flightless beetles spend much of their time burrowing beneath the cover of various coastal plants, including sand verbena *(Abronia),* beach-bur *(Ambrosia chamissonis),* and saltbush *(Atriplex leucophylla).* At night or during cool foggy days, adults may burrow out into open sand, as evidenced by the shallow furrows. Therevid fly larvae, ant lions, as well as some mammals and reptiles, probably prey on *Coelus* larvae. Tachinid flies also occasionally parasitize them. The Globose Dune Beetle *(C. globosus)* (5.0 to 7.0 mm) (pl. 196) inhabits foredunes and sand hummocks immediately bordering the coast from Bodega Bay Head to Ensenada, Baja California, and all of the Channel Islands except San Clemente Island. The decline of its habitat has resulted in its consideration for federal listing as threatened or endangered by the U.S. Fish and Wildlife Service. The more widespread Ciliated Dune Beetle *(C. ciliatus)* (5.0 to 7.0 mm) inhabits sand dunes and other sandy soils from Vancouver, British Columbia, south to Ensenada, Baja California, mostly along the immediate coast.

Asidina confluens (18.5 to 26.0 mm) is a large, yet somewhat uncommon, beetle found in Arizona, Baja California, and Sonora. Each elytron has two distinct ridges running lengthwise along the body. In California it is found throughout the Colorado and Mojave Deserts, where it wanders about at dusk from August through November, but intact dead specimens are often found through April.

Stenomorpha (15.0 to 33.0 mm) is found throughout much of western North America, with 12 species found in California. The convex pronotum is wider than long, with the base strongly curved. The side margins of the somewhat inflated elytra are evenly rounded. Each elytron has three, weak ridges running its length. In some years *Stenomorpha* is especially common and found by the hundreds in the southern San Joaquin *(S. mckittricki)* and western Antelope *(S. costipennis)* Valleys in spring.

The genus *Edrotes* is restricted to the desert regions of North

America. All six species stridulate by rubbing their finely ridged hind femora against the minute saw-toothed edges of the elytra. The purpose of sound production of these energetic little beetles remains unknown. *Edrotes ventricosus* (6.4 to 10.0 mm) (pl. 197) has a total of eight distinct rows of erect hairs on its shiny black elytra. It is distributed in Nevada, southern Idaho, southwestern Utah, western Arizona, Sonora, and Baja California. In California it is widely distributed in both the Mojave and Colorado Deserts. Adults are collected under stones or as they wander about in sandy soil or dune habitats year-round. They are primarily active in late winter and spring and feed on salt grass *(Distichlis spicata)*, cheat grass *(Bromus tectorum)*, Russian thistle *(Salsola tragus)*, and wild onion *(Allium)*.

Megeleates sequoiarum (7.0 to 9.5 mm) (pl. 198) is the only species in the genus and is known only from California. Its brownish gray body is rough, resembling a small hide beetle (Trogidae) without lamellate antennae. The pronotum of both sexes has broadly expanded side margins, but only the male has a pair of raised bumps or distinct horns near the leading edge. Fully developed horns may reach half the length of the pronotum. The eyes are completely divided, and the last antennal segment is set partially within the preceding segment. It is found under the bark of firs *(Abies)* and other coniferous trees, where it presumably feeds on fungus (e.g., *Fomes*). *Megeleates* is found primarily in the northern Coastal Ranges and the higher elevations of the Sierra Nevada and Transverse Ranges.

Species of *Phaleria* are oval and quite variable in their coloration, ranging from yellowish to black. The Kelp Beetle *(P. rotundata)* (5.3 to 6.8 mm) is the only species in California and is dull, not shining, lacking any markings on its elytra. It is active throughout the year and is sometimes found in large numbers on the beach in or under seaweed. It ranges from San Francisco to Millers Landing, Baja California.

When disturbed, many species of *Eleodes* first stand on their heads and then discharge noxious chemical compounds. These compounds act as repulsive agents and are laced with hydrocarbons and acids that help the quinones penetrate the outer waxy layer of the exoskeleton of predatory arthropods. This defensive behavior and accompanying arsenal has earned them the names acrobat, clown, or stink beetles. Remains of the normally black beetles left in the desert sun for years will eventually bleach out, turning brown or reddish brown. The Armored Darkling Beetle *(E. armatus)* (24.0 to 35.0 mm) is generally nocturnal during summer but is active during the day and at dusk during cooler months. It feeds on flowers or scavenges plant detritus, dung, and even dead insects. It is found throughout the warmer, drier regions of California, including the San Joaquin Valley, western Great Basin, and the Mojave and Colorado Deserts. Active throughout the year, this species reaches its peak in fall. Each femur is armed with a stout spine. Raising its abdomen high in the air, an Armored Darkling Beetle is capable of sending a spray of repugnant fluid several inches. In spite of its chemical defenses, hairy scorpions and some rodents prey upon these beetles. The Gigantic Eleodes *(E. gigantea)* (30.0 to 35.0 mm) (pl. 199) is a large, smooth, elongate, and moderately shining species common in the state. The Wooly Darkling Beetle *(E.* [previously in the genus *Cratidus*] *osculans)* (12.0 to 16.0 mm) (pl. 200) also produces smelly defensive secretions. This distinctive beetle lives primarily in the coastal foothills and valleys and is easily recognized by the erect, reddish brown hair on the elytra. The larva lives in dry soil and plant litter, where it feeds on detritus. At times it is commonly encountered ambling along paths and roads in the evening and early morning hours. Another hairy species, *E.* (previously in the genus *Amphidora*) *nigropilosa* (8.0 to 12.0 mm) (pl. 201), is covered in black hairs and can be quite numerous under stones and boards along the central and southern coast. *Eleodes clavicornis* (10.2 to 13.0 mm) is a smaller species commonly found in coastal dunes under vegetation, from Humboldt County to Ventura County.

Argoporis bicolor (8.3 to 14.0 mm) (pl. 202) is black, moderately to strongly shining, and somewhat parallel sided. The mouthparts, antennae, and undersides are dark reddish, as are the tibiae and tarsi. The femora are bright red and, in the male, have two strong spines just before the tibiae. It lives primarily in the Colorado and Mojave Deserts.

The Common Mealworm *(Tenebrio molitor)* (adult length 12.0 to 18.0 mm) (pl. 203), also known as the Yellow Mealworm, is originally from Europe. There it sometimes infests stored grain and grain products in mills and warehouses but seldom becomes a household pest. Both the adult and larval Common Mealworm are nocturnal, inhabiting undisturbed accumulations of grain in dark corners or beneath bags of feed. It also attacks damp grain stored in bins. The elongate adult is dark brown to black with distinctly grooved elytra. The adult resembles a small *Zophobas*, but the undersides of its feet have only sparse, mostly dark-colored hairs.

The Super Mealworm is the larva of the large, black beetle *Zophobas morio* (22.0 to 27.0 mm), a native of Central and South America. Like the Common Mealworm, the Super Mealworm is frequently sold as live food or fishing bait. Although the adult resembles that of a large Common Mealworm, it is easily distinguished by having a dense pad of yellowish hairs beneath its feet.

All four species of *Scotobaenus* are found in California. They are shiny black and strongly oval beetles with heads set partially within the prothorax. Their eyes are kidney shaped, with the upper and lower lobes equal or approximately equal in size. The most widespread species, *S. parallelus* (17.0 to 21.0 mm) (pl. 204), is found from Oregon south to Baja California.

Other large and flightless "stink beetles" belong to the genus *Coelocnemis*. These beetles are found in conifer and oak-conifer woodlands throughout the state. In the southwestern deserts their distribution is restricted to a few of the higher mountain ranges. Like *Eleodes*, *Coelocnemis* species stand on their heads and emit defensive secretions from the tip of their abdomen. Members of this genus are readily distinguished from the similar-appearing *Eleodes* by possessing two rows of fine, golden brown hairs on the lower half of the inside margin of the hind tibiae. They are most often found beneath stones, debris, and loose bark or wandering around on the ground at night. Five species of *Coelocnemis* are found in California. The most common species are *C. magna* (19.0 to 31.0 mm) and *C. californica* (18.0 to 24.0 mm) (pl. 205). Both species are widespread in California except in the Great Central Valley and the desert lowlands. *Coelocnemis magna* ranges from the southwestern part of the state northward along the coastal mountains and foothills and along the lower elevations of the Sierra Nevada. It also occurs in Arizona, New Mexico, and Baja California. *Coelocnemis californica* ranges throughout the Pacific Northwest, south to the mountains throughout California, and east along the Rocky Mountains to southern Utah. The sides of the prothorax are smooth in *C. magna*, while *C. californica* has distinct pits or wrinkles.

The genus *Iphthiminus* contains five species in western North America, three of which occur in California. Of these, *I. serratus*

(20.0 to 25.0 mm) (pl. 206) is the most common. It is abundant in the coniferous forests from British Columbia south to the Sierra Nevada of California. It feeds on rotten wood, especially Douglas-fir *(Pseudotsuga menziesii)*. The adults are dull black, somewhat flattened, leggy beetles. They closely resemble *Coelocnemis* but lack the rows of golden brown hairs on the hind tibiae.

COLLECTING METHODS: Most darkling beetles are easy to collect and keep alive. Many species are found during the day, hiding beneath stones, debris, bark, or fungi. In the mountains some species are found on trees, logs, or stumps after dark. Dune dwellers along the coast and in the deserts are collected at dusk or after dark as they wander over the surface in search of food. Also along the coast, *Phaleria* can be found in or under kelp washed up on the beach. Desert species are attracted to trails of oatmeal and are then collected by hand throughout the night. Pitfall traps with runners and/or baited with oatmeal will also produce specimens.

FALSE BLISTER BEETLES Oedemeridae

The biology and ecology of the small family of false blister beetles have long been neglected because these beetles are neither prized by collectors nor, except for one species, considered to be of economic importance. They have earned the name "false blister beetles" not only because of their appearance, but also because some species are capable of producing blisters when pinched or squashed on the skin. Like blister beetles (Meloidae), the bodily fluids of false blister beetles contain the blistering agent cantharidin. Reactions vary from nothing at all, to mild and painless blistering, to extremely painful wheals that are slow to heal. Many blistering species have contrasting warning, or aposematic, colors to signal potential predators.

False blister beetles are most abundant along the coast and in moist, wooded canyons in California. Adults are frequently found visiting flowers, where they feed on nectar and pollen. They are also found on foliage, as well as in moist, rotten logs and are sometimes attracted to lights at night. The larvae develop in moist, decaying logs, stumps, and roots of conifers and other trees. They will also attack driftwood if it is of the right consistency.

IDENTIFICATION: False blister beetles are elongate, slender, and soft-bodied beetles ranging in length from 5.0 to 18.0 mm. The color ranges from metallic blue green, black, brown, or gray to yellowish brown, sometimes with yellow, red, or orange markings. The head is hypognathous and is usually longer than wide. The threadlike antennae are 11-segmented, appearing 12-segmented in some males. The pronotum covers the head slightly, is broader in the front than in the back, and lacks margins on the sides. The scutellum is visible. The elytra, which are wider than the base of the pronotum, are long, parallel sided, finely ridged, and almost or completely conceal the abdomen. The tarsal formula is 5-5-4, and the claws are equal in size and are simple or toothed. The abdomen has five segments visible from underneath.

The mature larvae are white, elongate, and nearly cylindrical, with a straight or slightly curved body. The distinct head is prognathous and has well-developed, three-segmented antennae. The thorax is short and slightly wider than the abdomen. The short legs are moderately or widely separated and are five-segmented, including the claw. The 10-segmented abdomen appears nine-segmented from above and usually lacks projections on the ninth segment. The last segment is short, wide, and more easily visible from below.

SIMILAR CALIFORNIA FAMILIES:

- soldier beetles (Cantharidae)—pronotum margined; tarsi 5-5-5
- blister beetles (Meloidae)—head usually wider than pronotum and with a distinct neck; pronotum never broader in the front than in back
- longhorn beetles (Cerambycidae)—tarsi appear 4-4-4 but are 5-5-5

CALIFORNIA FAUNA: 20 species in nine genera.

The Wharf Borer *(Nacerdes melaneura)* (7.0 to 15.0 mm) (pl. 207) is cosmopolitan and the only species in the family considered a pest. The larva damages ship's timbers, old pilings, and posts. It is found on or near the coast or along rivers or estuaries. This species resembles a longhorn beetle (Cerambycidae), with antennae nearly half the length of the body. It is usually brownish yellow to dull yellow with the elytral tips black. The legs vary from brownish yellow to black. In males the last antennal segment is constricted, making the antennae appear 12-segmented.

Xanthochroa californica (8.0 to 12.0 mm) lives in the northern Coast Ranges, where it is active May through July. It is occasionally attracted to light. This distinctive beetle has metallic blue green elytra with orange head, thorax, and femora. Three additional species of *Xanthochroa* occur in the state. Both *Xanthochroa* and *Nacerdes* have front tibiae tipped with only a single spur.

Ditylus quadricollis (13.0 to 23.0 mm) occurs in the Sierra Nevada and has been reared on wet and rotten Englemann spruce *(Picea engelmannii)* and western red cedar *(Thuja plicata)*. The adult is black, somewhat like a ground beetle (Carabidae) in appearance, and clothed in short, dark setae. It overwinters as an adult and has been found under the bark of conifers. This species also occurs throughout the Pacific Northwest and Idaho.

Eumecomera are found on flowers. They have front tibiae tipped with two spurs, simple claws, and mandibles that are divided at their tips. All three species occur in California. *Eumecomera cyanipennis* (6.0 to 9.0 mm) (pl. 208) is entirely metallic blue or blue green and occurs only in California. It is found in the southern Coast and Transverse Ranges. Another species, *E. bicolor* (7.0 to 11.0 mm), has a metallic blue head above and an orange thorax with two spots at the bases of the legs. The rest of the body is metallic blue. It occurs in the northern Coast Ranges and southeastern Oregon. *Eumecomera obscura* (8.0 to 13.0 mm) is simlar to *E. bicolor* but is black instead of metallic blue, and the last abdominal segment is yellow. It is found in the Coast Ranges and also occurs in Arizona, Colorado, New Mexico, Texas, and Utah.

Oxacis pallida (5.0 to 12.0 mm) is found on tumbleweed, also known as Russian thistle *(Salsola tragus)*, in the Colorado Desert and southern Great Central Valley in July. It also occurs in Arizona, New Mexico, Texas, and Utah. It varies in color from pale to black and appears powdery because of a coating of coarse hairs. *Oxacis bicolor* (6.0 to 9.0 mm) is found infesting flooring and woodwork afflicted with dry rot in buildings and has been reported from deep in mines in California and Colorado, presumably where the larvae were developing in timbers used to support the tunnels. Miners have complained that the bite of the beetle causes "severe inflammation and swelling." Three additional species in this genus are recorded in California, all with stout, blunt mandibles and simple tarsal claws.

Rhinoplatia ruficollis (5.0 to 12.0 mm) (pl. 209) is known from the western Great Basin in Nevada and California. It is active March through June in the Colorado and Mojave Deserts and in the western Transverse Ranges. It has been found in the flowers of plants in the mustard family. This bluish black to black beetle has reddish orange legs and antennae and is moderately clothed in dense, gray, velvety setae. The only other California species, *R. mortivallicoa* (9.0 to 14.0 mm), is known from Death Valley. This beetle is bluish black with reddish orange head, antennae, thorax, and tarsi. The middle and hind femora and tibiae are blackish orange. Both species have a long head with a distinct snout.

COLLECTING METHODS: Sweeping vegetation, especially flower heads and leaves, can be productive for collecting adults. Also look for them at lights along beaches and the shores of estuaries, where there are plenty of moist driftwood and fallen logs.

BLISTER BEETLES　　　　　　　　　　　Meloidae

Blister beetles are of particular interest because of their medical, veterinary, and agricultural importance. They contain cantharidin, a chemical that causes irritation and blistering to sensitive tissues. In Europe, cantharidin was collected from the blister beetle *Lytta vesicatoria,* or the "Spanish fly," and taken orally for its purported qualities as an aphrodisiac. Cantharidin can be extremely toxic to humans, but as recently as the early 1900s cantharidin was used to treat venereal disease, disorders of the bladder, bed-wetting, and warts.

Cantharidin is also poisonous to grazing animals that might accidentally consume blister beetles resting or feeding on plants. Some animals may learn to avoid large numbers of those species that are brightly marked with contrasting colors of black, red, yellow, and orange. Cantharidin sometimes attracts antlike flower beetles (Anthicidae), especially *Notoxus.* These small beetles, with their antlike heads and distinctive horn projecting from their pronotum, are sometimes found swarming over dead blister beetles.

Some adult blister beetles do not feed, but most are plant feeders, consuming leaves and flowers. The feeding activities of some *(Epicauta)* can be ruinous to alfalfa *(Medicago sativa)* and other field crops. Species of *Epicauta* and *Lytta* are often an important and conspicuous element of the desert fauna, where they are sometimes found swarming over the ground and their food plants by the hundreds or thousands. All of the known larvae are predators.

Blister beetles exhibit exceptionally diverse courtship behavior, using chemical, tactile, and visual cues. Eggs are typically laid on the ground or on plants. Larvae develop by a special type of metamorphosis called hypermetamorphosis, a process characterized by two or more distinct larval forms. The first-instar larva, or triungulin, usually looks more like a silverfish than a typical beetle larva. It has long legs and is adapted for locating host insects. Species that attack grasshopper egg masses, or are associated with the nests of solitary bees burrow into the soil to search for their hosts. Other solitary bee predators (e.g., *Meloe* and *Nemognatha*) climb up on flowers and attach themselves to visiting bees to be transported back to the nest. There they will feed on pollen, nectar, and bee larvae. Upon finding the proper host, the triungulin molts into a less-active larva with short, thick legs and begins to feed. This grublike form persists for the next few molts, until the emergence of the coarctate, a fat, legless C-shaped larva. The coarctate rides out the winter. In spring another molt produces a more-active short-legged grub that spends most if its time preparing a pupal chamber.

IDENTIFICATION: Adult California blister beetles are black, metallic blue or green, or a combination of yellow, orange, or red with black markings. They range in length from 3.0 to over 30.0 mm. Their conspicuous, antlike head is hypognathous and attached to the body by a distinct neck. The antennae are usually thread- or beadlike, consisting of eight to 11 segments. The pronotum is without margins on the sides and usually narrower than both the head and the base of the elytra. The scutellum is visible, except in *Meloe*. The long, soft, leathery elytra are typically rolled over the sides of the abdomen. In some species the elytra do not meet in a straight line over the back or are slightly to considerably shorter than the abdomen. The tarsal formula is 5-5-4. The abdomen has six segments visible from below.

The highly mobile first-instar larvae, or triungulins, are tough bodied and silverfish shaped. The head is prognathous, with three-segmented antennae and one or two pairs of simple eyes. The following four instars are C-shaped feeding grubs, with hypognathous heads. The next stage, the coarctate larva, is virtually immobile and lacks legs and mouthparts. The last instar has short legs and most closely resembles the feeding grub.

SIMILAR CALIFORNIA FAMILIES:
- soldier beetles (Cantharidae)—tarsi 5-5-5; elytra usually flattened, not rolled over abdomen
- false blister beetles (Oedemeridae)—head lacks neck

CALIFORNIA FAUNA: Approximately 120 species in 17 genera.

Pyrota palpalis (6.0 to 17.0 mm) (pl. 210) is a distinctive species found in the Mojave and Colorado Deserts. The body and legs are orange, but the elytra are usually ivory or cream colored with three broad black bands. The pronotum is also marked with two black spots. The adult is found feeding on various species in the sunflower family. It ranges from southern California and southern Utah south to northern Mexico, and east to New Mexico and western Texas. It is the only species of *Pyrota* known to occur in California.

The genus *Cordylospasta* consists of two black species, both of which occur in California. The males are fully winged, whereas the females are flightless. Their elytra are short and do not cover the bloated abdomen, and the flying wings are greatly reduced or absent. Females resemble *Meloe*, but the elytra do not overlap at their bases. *Cordylospasta fulleri* (males 7.0 to 12.0 mm, females 7.0 to 17 mm) is unique among North American blister beetles in that it has only eight to 10 antennal segments. Its range is almost entirely within the Great Basin, including Mono County, in the central eastern portion of the state. It is active April through October. *Cordylospasta opaca* (males 6.0 to 12.0 mm, females 8.0 to 19.0 mm) (pl. 211) has 11-segmented antennae and occurs throughout central and southern California from sea level to 6,500 ft. In spring the adult is found crawling over the ground or feeding on various plants. It also occurs in southern Nevada and northern Arizona.

The Inflated Beetle *(Cysteodemus armatus)* (7.0 to 18.0 mm) (pl. 212) is found in the Mojave and Colorado Deserts in spring, actively crawling on flowers or the ground, especially during the warmer parts of the day. This unusual, flightless black beetle is distinguished from all other blister beetles by its fused, inflated, almost spherical elytra. The coarsely pitted elytra are often covered with a white or yellowish nitrogenous secretion. The dead-air space beneath the elytra may serve as insulation, helping this desert beetle to regulate its body temperature. The Inflated Beetle feeds on a variety of plants, including creosote bush *(Larrea tridentata)*, goldenhead *(Acamptopappus sphaerocephalum)*, pincushion *(Chaenactis)*, desert-sunflower *(Geraea canescens)*, palafoxia *(Palafoxia arida)*, gilia *(Gilia)*, langloisia *(Langloisia matthewsii)*, and tiquilia *(Tiquilia)*. The larva has been found in the subterranean nests of solitary bees. The species also occurs in Arizona and Nevada.

The Elegant Blister Beetle *(Eupompha elegans)* (7.0 to 13.0 mm) (pl. 213) is a common and widespread species distributed south of the San Francisco Bay region. The head and thorax are black, whereas the elytra are variably marked with black and orange; there are also entirely dark blue populations. Although recorded to feed upon 20 genera of plants in eight families, the adult prefers members of the sunflower family, especially desert-sunflower. The larva is thought to feed on the larvae and provisions of ground-nesting solitary bees.

Phodaga alticeps (8.0 to 25.0 mm) (pl. 214) has a distinctly pointed head and is dull black above, with scattered gray setae on the sides of the head, bases of legs, and rear margins of the abdominal plates. It is primarily active in spring in all of California's deserts, where it feeds on the flowers and foliage of tiquilia *(Tiquilia* [formerly *Coldenia] palmeri* and *T. plicata)*. This beetle runs quickly over the ground and, when threatened, stops and spreads its elytra to expose its red abdomen, remaining motionless for brief periods. *Phodaga marmorata* (6.0 to 16.0 mm) is dull black with small patches of grayish setae scattered over the pronotum and elytra. It is most common in late summer and early fall. It feeds on the flowers and foliage of chinchweed *(Pectis papposa)*, kallostroemia *(Kallostroemia grandiflora)*, and puncture vine *(Tribulus terrestris)*.

Only one species of *Pleuropasta* occurs in the state. *Pleuropasta mirabilis* (6.0 to 13.0 mm) (pl. 215) is found throughout California's deserts. It is active in the Colorado Desert before May and during the months of June and July in the Great Basin and Mojave Desert. Adults are commonly found feeding on species of tiquilia, as well as *Cryptantha angustifolia* and *C. barbigera,* all in the family Boraginaceae. The elytra are a faded yellow with reddish brown transverse markings and have distinctive elytral ridges connected by faint cross ribs.

Tegrodera (13.0 to 29.0 mm) is a genus of large and conspicuous beetles active primarily during the spring months in Arizona, California, and adjacent Mexico. All three species occur in California. *Tegrodera latecincta* (15.0 to 26.0 mm) is associated with the sagebrush scrub plant community of the Owens and Antelope Valleys. It has a reddish head, an orange and black to completely black prothorax, and rough, light yellow elytra with a broad black band straight across the middle. The adult feeds on eriastrum *(Eriastrum eremicum)*, sagebrush *(Artemesia)*, and bassia *(Bassia hyssopifolia)*. *Tegrodera erosa* (13.0 to 29.0 mm) (pl. 216) is restricted to the chaparral and coastal sage scrub communities of the southern foothills of the Transverse Ranges to northwestern Baja California, where it feeds on eriastrum *(E. sapphirinum)* and alfalfa. Scattered dark markings and a relatively faint and angled black band on the elytra and the brown to black pronotum distinguish this species. *Tegrodera aloga* (14.0 to 26.0 mm) occurs along the Colorado River and in the Imperial Valley. The head and pronotum are reddish or orangish, the pronotum is marked with black, and the elytra have both scattered black markings and a straight black band.

Of the 24 species of *Epicauta* in California, the Punctate Blister Beetle *(E. puncticollis)* (9.0 to 12.0 mm) (pl. 217) is by far the most common. It is also one of the most frequently encountered blister beetles in the state. The adult is entirely black and covered with short, black hairs. It is found in a variety of habitats throughout the state from sea level to 8,000 ft, except in the deserts. It is typically found in summer on plants in the pea and sunflower families, feeding on alfalfa, tarweed *(Hemizonia)*, and goldenrod *(Solidago)*. The larva attacks grasshopper eggs. *Epicauta alphonsii* (6.0 to 11.0 mm) (previously known as *E. californica*) and *E. corybantica* (6.0 to 12.0 mm) are very similar to *E. puncticollis* in being all black and of similar size but are distinguished by always having some grayish hairs on the head and on the legs, especially the front coxae. *Epicauta alphonsii* is widespread in cismontane California from the foothills of the Sierra Nevada west to the coast and south to San Diego County, whereas *E. corybantica* occurs in the southeastern Mojave Desert in fall on rabbitbrush *(Chrysothamnus)*.

The genus *Linsleya* is restricted to western North America, with three species and one subspecies found in California. *Linsleya* is similar to the smaller, metallic species of *Lytta*, but the hind tibial spurs are slender, and the first antennal segment is longer than the third. *Linsleya sphaericollis* (7.5 to 12.0 mm) occurs throughout the state. It is brassy green to metallic blue, rarely black. The appendages are black, but the femora and tibiae are sometimes metallic. *Linsleya californica* (9.0 to 13.0 mm) is found along the coast, from just south of San Francisco Bay to San Diego and inland to the vicinity of Julian. It is black, except for an orange spot on the head and the metallic blue or green luster of the femora and tibiae. *Linsleya compressicornis compressicornis* (9.0 to 11.0 mm) is metallic green and distributed in the Sierra Nevada and the White Mountains in Inyo County. The subspecies *L. c. neglecta* (9.0 to 11.0 mm) (pl. 218) is metallic blue and is known from Utah and the Providence Mountains in the eastern Mojave Desert.

Of the 69 species of *Lytta* found in North America, 31 are found in California. *Lytta magister* (16.5 to 33.0 mm) (pl. 219) is a large and striking beetle with reddish legs, head, and prothorax and rough, black elytra. It is found singly or in large aggregations in the Colorado and Mojave Deserts in the spring, where it feeds on various plants including brittlebush *(Encelia farinosa)*. The smaller *L. vulnerata* (12.0 to 20.0 mm) is similar in color but has black legs and smooth elytra. This normally late-summer species is commonly found along sandy or gravelly washes and stream terraces in the Coast and Transverse Ranges. It feeds on scale broom *(Lepidospartum squamatum)*, sagebrush, broom *(Baccharis)*, rabbitbrush, and snakeweed *(Gutierrezia)*. *Lytta auriculata* (6.0 to 19.0 mm) (pl. 220) is found in southern California. Although the color of the top of the head can range from entirely orange to entirely black, it is typically marked with orange only on the hind corners and has a small spot in the middle between the eyes. The elytra are rough and black, with a faint metallic blue or green luster. One of several metallic green blister beetles in California, *L. chloris* (7.0 to 14.0 mm) is bright brassy green, with the head and pronotum finely wrinkled, appearing satiny. The hair on its body is either clear (Kern and Tulare Counties) or black (central western California). It is commonly found in spring feeding on members of the pea family, especially lupine *(Lupinus)*, in the valleys and foothills of the southern Coast Ranges and western foothills of the southern Sierra Nevada. *Lytta stygica* (7.0 to 15.0 mm) (pl. 221) is very similar to *L. chloris*, but the head and pronotum are smooth. It feeds on spring flowers in most parts of the state. Though sometimes all black, most individuals are a striking deep metallic green or blue, but not brassy. The small black blister beetle *L. nigripilis* (7.0 to 14.0 mm) is active in late spring and summer in the northern Coast Ranges and along the western foothills of the Sierra Nevada. The black *L. sublaevis* (11.0 to 20.0 mm) is found along the western foothills of the Great Central Valley from Alameda and San Joaquin Counties southward, then east along the Tehachapi Mountains and north along the eastern edge of the Great Central Valley to Tulare County. It is the only wingless species of *Lytta* and is superficially similar to those of the genus *Meloe*. A pale spot on the head and a visible scutellum easily distinguishes it. The larvae of *Lytta* parasitize ground-nesting solitary bees, especially those visiting the flowers of the food plants of the adult beetles.

All seven California species of *Meloe* are large, black beetles with short, overlapping elytra, exposing a somewhat bloated abdomen. These sluggish, flightless beetles are found walking on the ground or on food plants. The larvae lie in wait on flowers, latch onto visiting bees, and hitchhike back to the bee's nest. Montane species are usually active in spring, whereas valley and coast species emerge in fall and are active in winter. *Meloe angusticollis* (9.0 to 19.0 mm) ranges from having slightly metallic blue antennae and legs to being distinctly metallic blue throughout. It ranges from northern California north and east to southern Canada and across to northeastern United States. *Meloe barbarus* (5.0 to 14.0 mm) (pl. 222) is dull to shiny black. It is known primarily from the Coast Ranges, especially in the San Francisco Bay region, and the Sierra Nevada. *Meloe strigulosus* (9.0 to 17.0 mm) is dull to weakly shiny and is widely distributed in California, primarily along the coast, but also in other relatively temperate areas.

Of the 15 species of *Nemognatha* in California, many have bodies that are largely yellowish or orange. Adults are frequently encountered on flowers in late spring and summer. They have long mouthparts drawn out into a sucking tube that is sometimes longer than the entire body. *Nemognatha lurida apicalis* (7.0 to 15.0 mm) (pl. 223) is found throughout the western United States and most of California. The adult is found feeding on various species of the sunflower family.

COLLECTING METHODS: Blister beetles are particularly abundant during spring and summer in the chaparral and desert regions. They are commonly found mating and feeding on flowers, where they can be carefully gathered by hand. Be sure not to touch your face or other parts of your body with your hands after touching blister beetles. Sweeping and beating low vegetation can also be effective. Some species, particularly in the deserts, are attracted to lights at night.

ANTLIKE FLOWER BEETLES — Anthicidae

Of the 231 species of antlike flower beetles recorded in North America, nearly half belong to the genera *Anthicus* and *Notoxus*. They reach their greatest diversity in the arid regions of the southwest, especially in California, where nearly 100 species are found.

Some species of *Ischyropalpus* are found on flowers as adults, but most other antlike flower beetles are typically found on vegetation during the day. Many are found in association with decaying vegetation on the ground. Others are found crawling on the ground over areas of exposed soil with nearby ground-hugging vegetation, bits of debris, stones, or clumps of litter under which to hide. Adults of *Amblyderus* and *Mecynotarsus* are usually found in sandy soils along the coast or the margins of freshwater lakes and rivers, while *Tanarthrus* scurry about on salty mudflats like miniature tiger beetles (Carabidae).

Adult antlike flower beetles are opportunistic scavengers and predators of small insects and other arthropods. They also feed on pollen, nectar, sap, fungal hyphae, and spores. Species of *Anthicus* are scavengers of dead insects. A few species living in agricultural fields prey on the eggs and larvae of pests and have been considered as possible biological control agents. The omnivorous larvae live on the ground, where they have feeding habits that are generally similar to those of the adults.

Species in several genera of antlike flower beetles are attracted to the blistering chemical cantharidin and are found on dead or dying blister beetles (Meloidae). Cantharidin is a deterrent to potential predators and is collected primarily by the males for protection and to increase mating success. Some males of the genus *Notoxus* are able to concentrate the chemical in special glands at the tips of their elytra, where small amounts are released to attract females, and larger amounts are passed on to the females in reproductive fluids during mating. The females then pass the chemical along to their eggs and larvae, protecting them as well. Other antlike flower beetles have thoracic glands that produce chemicals that are particularly distasteful to ants, the primary predators of ground-dwelling insects.

IDENTIFICATION: As their common name implies, antlike flower beetles are antlike in appearance, with a distinct head and neck, and the prothorax is constricted or narrowed at the base. The head is slightly hypognathous, with 11-segmented antennae that are usually threadlike, serrate, or weakly clubbed. The prothoracic base is narrower than the base of the elytra (except in *Ischalia*). The sides are not keeled or margined (except in *Ischalia*). In some genera (*Mecynotarsus* and *Notoxus*) the pronotum has a single, prominent horn in the middle that projects out over the head. The scutellum is very small, but visible. The elytra are covered with short hairs and nearly or completely cover the abdomen. The tarsal formula is 5-5-4. The abdomen has five segments visible from below.

Mature larvae are pale yellowish white, straight bodied, and almost cylindrical or slightly flattened. The head and abdominal projections are darker than the rest of the body. The distinctive head is prognathous. The antennae are three-segmented, with a single pair of simple eyes. The prothorax is slightly longer than the following thoracic segments, which are nearly equal in size to each other. The legs are five-segmented, including the single claw. The 10-segmented abdomen is straight. The ninth segment has a pair of dark projections in many genera, obscuring the last segment from above.

SIMILAR CALIFORNIA FAMILIES:

- antlike stone beetles (Scydmaenidae)—antennal club loosely formed with three to four segments; tarsi 5-5-5
- spider beetles (Anobiidae)—antennae clubbed; tarsi 5-5-5
- checkered beetles (Cleridae)—antennae clubbed; tarsi 5-5-5
- fire-colored beetles (Pedilinae, Pyrochroidae)—pronotum not constricted or narrowed at base
- antlike leaf beetles (Aderidae)—eyes notched next to antennal insertion; first two abdominal segments fused
- orsodacnid leaf beetles (Orsodacnidae)—head distinctly hypognathous; tarsi appear 4-4-4 but are 5-5-5
- leaf beetles (Chrysomelidae)—head distinctly hypognathous; tarsi appear 4-4-4 but are 5-5-5

CALIFORNIA FAUNA: Approximately 96 species in 18 genera.

Members of the genus *Tanarthrus* are found almost exclusively on saline mudflats of inland lakes and ocean shores of the southwestern United States and northwestern Mexico. Nine of the 15 known species are found in California, six of which live in

Death Valley. These small, active beetles prey on tiny invertebrates. The last antennal segment is constricted in the middle, appearing as if it were two segments. *Tanarthrus alutaceus* (2.3 to 3.8 mm) (pl. 224) is dark brown to black, with the head and pronotum sometimes darker than the elytra. The head has minute punctures, and its basal margin is straight. This widespread species is found around coastal salt marshes of central and southern California, on saline flats in Death Valley, along the shores of the Salton Sea and Colorado River, and south into Baja California. It is also known from central Mexico. Individuals of *Tanarthrus* are a favorite food of tiger beetles *(Cicindela)*.

Of all the California antlike flower beetles, only *Mecynotarsus* and *Notoxus* have a distinctive horn on the pronotum that projects over their head. The head of *Mecynotarsus* lacks bristly hairs on the sides of the head, whereas *Notoxus,* the most commonly encountered of the two genera in the state, has a row of conspicuous hairs on each side. Of the 47 species of *Notoxus* in North America, 21 occur in California. *Notoxus calcaratus* (2.65 to 4.15 mm) (pl. 225) is distributed across all of the southern United States and is widespread throughout California. It is recorded to feed on damaged or drying fruit. It is also found feeding in large numbers in cotton fields on the sap beneath cotton bolls and leaves, but it does not harm the plants. In fact, this beetle may be beneficial because it also feeds on the eggs and small larvae of pestiferous moths.

COLLECTING METHODS: Some antlike flower beetles are found along coastal beaches, in desert sand dunes, or on alkali flats. Desert species are most numerous on flowers (e.g., *Ischyropalpus* and some *Notoxus*) and vegetation along streams. Look for wet or humid spots in otherwise dry areas, especially under loose rocks, cow pies, sprawling vegetation, algal crust, and debris on beaches. Beating and sweeping vegetation, especially oaks *(Quercus),* willows *(Salix),* and grasses are the most effective methods for collecting *Notoxus* adults. Also check leaf litter under shrubs. Some California genera *(Anthicus, Formicilla, Mecynotarsus, Notoxus,* and *Vacusus)* are attracted to cantharidin traps. The simplest cantharidin trap consists of a box of recently killed and pinned blister beetles (Meloidae). Simply leave the box out in a shaded area with the lid lifted up only slightly to protect the specimens. Another method is to set out a small container of alcohol in which blister beetles have been soaked. These traps are apparently most effective in the morning and the late afternoon. The best method for attracting other species is with a black light.

LONGHORN BEETLES Cerambycidae

Longhorns are among the most spectacular of California's beetles. They are particularly conspicuous in forested areas, where they are attracted to freshly painted surfaces, cut wood, flowers, or lights at night. Longhorns have long been popular with collectors because of their bright colors, large size, and interesting habits.

Most California longhorns are nocturnal and are dull black or brown. They spend their days hiding beneath logs and bark, emerging at dusk or in the evening to search for food and mates. A few species are cryptically marked, allowing them to hide out in the open during the day, camouflaged against the bark of trees. By contrast, most day-active, or diurnal species, are flower visitors and are brightly colored. Some sport metallic blues or greens, whereas others are yellow or orange and seem to match the colors of their favorite flowers. A few species are bright red, and in the case of the milkweed borers *(Tetraopes),* this color probably serves as a warning to birds and other predators that they taste bad. Wasp and bee mimics (e.g., the Lion Beetle *[Ulochaetes leoninus]*) are boldly marked with black and yellow bands. The slender bodies and wasplike flight of this and other similarly patterned longhorns probably dupes potential predators into thinking that they can sting.

Adult longhorns feed on wood, leaves, and flowers. Wood-feeding species consume twigs, bark, bast (the fibrous inner bark), branches, trunks, and roots. Leaf feeders consume foliage, stems, needles, cones, fruits, and sap, whereas flower visitors devour pollen, stamens, and nectar.

The usually plump, cylindrical, plant-feeding larvae are sometimes called roundheaded borers. They almost always feed internally, attacking dead and decaying wood or living trees and shrubs. Roundheaded borers chew their way into branches, trunks, stems, roots, and cones. Whereas several species girdle smaller twigs, the feeding activities of other stem feeders may produce galls and other abnormal tissue growths. A few species prefer the stems of herbaceous plants. Root feeders usually tunnel inside, but some occasionally burrow in the soil and feed on the roots from the outside.

The excavation patterns of roundheaded borers are often used as diagnostic characters for their identification. Some species tunnel between the bark and wood and, depending on species, either pupate there or tunnel their way into the sapwood. Other species attack only the heartwood and leave the outer, living sapwood intact. Trees hollowed and weakened by successive generations of borers are easily toppled under their own weight or by windstorms.

The type of boring dust, or frass, that is pushed out of the tunnels during excavation is also characteristic of the resident species. As the larvae feed they produce powdery frass, flaky chips, or long curly fibers resembling the excelsior used as packing material. Depending on the species, the tunnels are either tightly plugged with frass behind the larva or kept open. The newly emerged adult usually chews the exit holes. In a few species, the larvae make the hole and plug it with frass before pupating. Other insects, including leafcutter bees, frequently occupy empty emergence holes.

Food preferences vary among longhorn beetle larvae. Those with catholic tastes generally prefer wood that is either dead or decomposing, but will restrict their choice to either gymnosperms or angiosperms. Among those species preferring gymnosperms as hosts, many exhibit a decided preference for pines *(Pinus)* and firs *(Abies)* or cypresses *(Cupressus)* and junipers *(Juniperus)*. The few species of longhorns developing in living wood are usually quite selective and are restricted to a particular species or genus of plant.

Female longhorns use their long ovipositors to place their eggs in or under bark or in wounds or cracks in the wood. Others gnaw a slit in the wood to receive the eggs. Twig feeders place their eggs on leaf nodes, whereas root feeders choose the base of the plant near or below ground level. Because the eggs of all species require some degree of protection and moisture, few species deposit their eggs on exposed tree trunks without bark.

Some longhorns purposely create a suitable food source by girdling and killing stems. In California, girdling is accomplished by the larvae that cut off the branch's food and water supply by tunneling spirally inside the stem. The dead, brown branch tips, known as flags, are especially conspicuous on evergreens such as coast live oak *(Quercus agrifolia)*. The girdle eventually weakens the branch, causing it to break and fall to the ground, where the larva continues to feed inside, undisturbed.

California's longhorn beetles play a beneficial role in forests by recycling dead and dying trees, reducing them to humus. However, they can become serious pests when managed timber supplies are weakened or killed by the ravages of storms, fires, or severe infestations of other insects. Much of the damage to timber is the result of an accumulation of roundheaded borers attacking the heartwood.

Stressed or injured shade and ornamental trees are particularly susceptible to attack by larval longhorns. With the exceptions of the California Prionus *(Prionus californicus)*, Giant Mesquite Borer *(Derobrachus geminatus)*, Gooseberry Borer *(Xylocrius agassizi)*, and eucalyptus borers *(Phoracantha)*, the vast majority of the state's longhorns are not considered important horticultural or fruit tree pests.

IDENTIFICATION: Adult California longhorns are extremely variable in shape and are typically brown or black, but a few are metallic blue or green, mottled black, brown, or gray, or banded black and yellow or white, or with bright red or yellow elytra. Most are elongate, cylindrical, or flattened, whereas others are more robust and broad shouldered. They range in length from about 5 mm to well over 60 mm. The head is prognathous or slightly to strongly hypognathous and usually bears kidney-shaped eyes. The antennae usually consist of 11 segments, occasionally more, and are sometimes inserted in the eye notches. The antennae are nearly always at least half the length of the body or more, especially in the males. The pronotum is almost square, oval, or cylindrical; the margin, if present, is spiny or toothed. The scutellum is visible. The elytra are parallel sided and are smooth, densely punctured, or sparsely hairy, and almost always completely conceal the abdomen. Most species are winged, but a few species are flightless. The tarsal formula appears 4-4-4 but is actually 5-5-5. The fourth segment is tucked between the lobes of the heart-shaped third segment. The claws are equal in size and usually simple. The abdomen has five or six visible segments.

The mature larvae are white or yellowish white with dark heads, and are typically elongate and cylindrical, but a few are distinctly flattened. Their bodies are usually straight or only slightly curved. The broad head is usually partially retracted within the thorax but is exposed in the lepturines (e.g., *Centrodera, Lepturobosca* [formerly *Cosmosalia*], *Leptura, Judolia, Ortholeptura, Rhagium, Stenostrophia,* and *Desmocerus*). The head is prognathous, and the small antennae are three-segmented. The one to five pairs of simple eyes are either present, indistinct, or absent. The thoracic segments are usually wider than the head and the abdomen. When present, the legs are very small, with five or fewer segments, including the claw. The abdomen is 10-segmented, with telescoping segments fitted with fleshy lobes to help them move through their tunnels. The abdomen usually lacks projections.

SIMILAR CALIFORNIA FAMILIES:

- some click beetles (Elateridae) — mandibles enlarged; hind angles of pronotum extended backward
- soldier beetles (Cantharidae) — tarsi distinctly 5-5-5; body and elytra soft
- some false blister beetles (Oedemeridae) — the pronotum is always broadest in front; tarsi 5-5-4
- false longhorn beetles (Stenotrachelidae) — claws comblike; tarsi 5-5-4
- leaf beetles (Chrysomelidae) — in some, the pronotum is always broadest in front; in others the antenna and body are short; in still others the elytra are broadest at the end

CALIFORNIA FAUNA: 316 species in 143 genera.

The genus *Parandra* is represented by one species in California and looks more like a stag beetle (Lucanidae) than a longhorn. *Parandra marginicollis* (14.0 to 22.0 mm) (pl. 226) is shiny reddish brown and is active June through September. It is commonly attracted to porch lights in the coastal foothill areas of southern California. The adult is found beneath the bark of old stumps of western sycamore *(Platanus racemosa)* and white alder *(Alnus rhombifolia)*. The larva bores into the tree's heartwood, leaving the outer, living sapwood intact. In the San Fernando Valley, *Parandra* larvae were found in large colonies within the hollowed trunks of cultivated English walnut trees *(Juglans regia)*.

The Willow Root Borer *(Archodontes melanopus aridus)* (21.0 to 47.0 mm, without mandibles) (pl. 227) is a large, brown beetle with toothless, protruding mandibles that project slightly downward. It is restricted to the Colorado River drainage in western Arizona and southeastern California. *Archodontes* is active from June through September. It is found at dusk flying around broom

(Baccharis) and its larval host plant, willow *(Salix)*. It is commonly found at lights. This is the only species of *Archodontes* found in the state.

Another large, brown longhorn is the Southwestern Stump Borer *(Nothopleurus* [formerly *Stenodontes*] *lobigenis)* (21.0 to 40.0 mm, without mandibles). It is found in the Colorado Desert. The adult male has long, nearly horizontal mandibles that are longer than the head. Each mandible has two small teeth on the inner margin. The mandibles of the female are shorter than the head. The pronotal margin is lined with at least five small spines of equal size. The larva is primarily a root feeder of honey mesquite *(Prosopis glandulosa)* but sometimes will work its way up into the trunks. In Palm Springs, *Nothopleurus* larvae attack shade and other ornamental trees, including the Peruvian pepper tree *(Schinus molle)*. Generations of larvae developing in the heartwood eventually weaken trees, resulting in their collapse during desert windstorms. The adult is active at dusk and in the evening and is readily attracted to lights. Only one species of *Nothopleurus* is found in California.

The genus *Ergates* is found in Europe, Asia, and western North America. Two species occur in California. The Pine Sawyer *(E. spiculatus)* (40.0 to 65.0 mm) (pl. 228), also known as the Ponderous Pine-borer or Spiny Wood-borer, is one of the largest beetles in California. It is found in the Cascades and Sierra Nevada, as well as the Klamath Mountains and the Coast, Transverse, and Peninsular Ranges. The Pine Sawyer is uniformly brown, but the head, prothorax, and legs are sometimes darker. The antennae of the male Pine Sawyer is about two-thirds the length of the body, whereas those of the female are about one-half. The pronotum of both sexes is margined with numerous sharp spines and has three raised bumps or calluses. The head, pronotum, and elytra are rough in texture. The female lays its eggs in the crevices of recently killed or felled trees or fallen logs and stumps of pine, fir, Douglas-fir *(Pseudotsuga menziesii)*, and giant sequoia *(Sequoia sempervirens)*. It also infests wooden power and telephone poles. The larva excavates broad channels in the heartwood and packs them with coarse, woody frass (excelsior). Mature larvae can reach nearly 70 mm in length. The adult *Ergates* is active in midsummer through early fall and both sexes are usually found flying at dusk or crawling at night on infested stumps. This beetle's emergence holes are large and distinctly elliptical in shape. A second smaller and more slender species, *E. pauper* (45.0 to 50.0 mm), lacks the calluses on the pronotum and has pale brown elytra, giving it a two-tone appearance. This species occurs in the southern Sierra Nevada and the Coast and Transverse Ranges, where it attacks coast live, canyon live *(Q. chrysolepis)*, and interior live *(Q. wislezenii)* oaks. Female *Ergates* are more likely to be attracted to lights than males.

Of the two species of *Derobrachus* known in California, the Giant Mesquite Borer *(D. geminatus)* (32.0 to 70.0 mm) (pl. 229) is the most common. It is one of the largest beetles in the state and is found throughout the southwestern United States and northern Mexico. In California it lives in the Mojave and Colorado Deserts, where the larva feeds on cottonwood *(Populus)*, oak, mesquite, and other trees. The male flies at dusk and is attracted to lights during summer. The female flies to lights less frequently and is usually found hiding beneath loose bark or partially buried in the litter at the base of a tree. The antennae are 11-segmented and threadlike.

The robust, dark reddish brown California Prionus *(Prionus californicus)* (24.0 to 55.0 mm) (pl. 230) is widespread throughout western North America. In California it is found nearly everywhere but the deserts. The adult is distinguished from *Ergates* by having only three sharp spines on either side of the pronotum and saw-toothed antennae with 12 segments. The antennae of the male are more than two-thirds the length of the body and are distinctly sawlike. The antennae of the female are only about half as long as the body and are slender. A large female may carry up to 1,200 eggs. The root-feeding larva is sometimes called the Giant Root Borer because it can reach 80 mm and be as thick as a man's finger. The larva feeds primarily on living deciduous trees, injuring oaks, Pacific madrone *(Arbutus menziesii)*, and cottonwood, but will also attack fruit trees growing in light, well-drained soils, including apples *(Malus)*, and cherries and peaches *(Prunus)*. The Giant Root Borer has also been recorded in the roots of vines (e.g., hops *[Humulus lupulus]*) and grasses, as well as in dead or decomposing hardwoods and conifers. The larva may leave the root system to construct a pupal tunnel two or three feet in length, pupating at the end of the tunnel. The entire life cycle may require three to five years. The adult is active in summer through early fall, flies at dusk or in the evening, and is commonly attracted to lights. The only other species of *Prionus* in California is LeConte's Prionus *(P. lecontei)* (52.0 to 60.0 mm). LeConte's Prionus is often collected with the California Prionus and is distinguished by having 13 antennal segments. Also, the female of *P. lecontei* is larger, plumper, and flightless, while that of *P. californicus* is more slender and flies to lights. Its larva attacks living roots of oak trees in the Sierra Nevada and the Transverse and Peninsular Ranges. The adult emerges from the bases of infested trees in summer, and the male is commonly attracted to lights. Both *Prionus* and *Ergates* often carry one or more pseudoscorpions beneath their elytra. Pseudoscorpions are small scorpionlike arachnids without a stinging tail. The relationship between these beetles and pseudoscorpions is one of phoresy. The predatory arachnids depend on the beetles for transportation to new mite-infested stumps.

The Hairy Pine Borer *(Tragosoma depsarius)* (18.0 to 36.0 mm) (pl. 231) is widely distributed in coniferous forests of western and northeastern North America and Europe. In California, the adult is found primarily above 5,000 ft in the high Sierra Nevada, the Klamath Mountains, and the Cascade and Transverse Ranges, as well as the Panamint Mountains of eastern California during summer and early fall. This species resembles a smaller, hairier Pine Sawyer. The larva feeds on the sapwood of rotten pine logs. A similar species, *T. pilosicornis* (24.0 to 28.0 mm), occurs in the drier mountains of Shasta and Siskiyou Counties and dry inner Coastal Ranges and Sierra foothills south to San Diego County. The head and pronotum are nearly bald, and the elytra are more coarsely and deeply punctured than those of the Hairy Pine Borer. The adult male and occasionally the female are taken at lights in summer. The larva has been found in sound or rotten dead trunks and fallen limbs of the gray pine *(Pinus sabiniana)* and ponderosa pine *(P. ponderosa)*.

The black and coarsely punctured *Spondylis upiformis* (8.0 to 20.0 mm) is the only member of *Spondylis* in the United States and is quite common in the coniferous forests of western North America. In California it is found primarily in the Sierra Nevada and the Cascade and Transverse Ranges. The adult is distinguished by distinctly protruding mandibles and relatively short antennae that barely reach the base of the elytra. It is often encountered flying at dusk or on bright sunny days in spring and is frequently attracted to shiny objects such as cars. The larva bores into the roots of living firs and probably pine.

Arhopalus asperatus (17.0 to 31.0 mm) (pl. 232) is found in mountainous regions throughout California, except in the northern Coast Ranges and deserts. Outside of the state it is found throughout western North America and south to Honduras. The adult is usually captured at lights from July through October. The pronotum is angular at the sides. The larva attacks standing dead and dying firs, Douglas-fir, and spruce *(Picea)*. The New House Borer *(A. productus)* (20.0 to 30.0 mm) is black and found throughout western North America and attacks dead or dying firs and pines, especially in recently burned areas. It will emerge from lumber cut from these trees even after the wood has been used for building homes. Egg laying occurs only on wood covered with bark, so this beetle will not infest structural timbers. The pronotum is rounded at the sides.

Asemum nitidum (15.0 to 20.0 mm) (pl. 233) is a black species found in the northern Coast Ranges, Sierra Nevada, and the Transverse and Peninsular Ranges. The larva bores into sapwood and heartwood of pines and requires one to two years for development. *Asemum caseyi* (12.0 to 16.0 mm) is less robust and smaller than the preceding species and is pale brown to black. Its distribution is similar to that of *A. nitidum*, but it also occurs in the southern Coast Ranges. One other species, *A. striatum* (12.0 to 17.0 mm), is known from the state. It is found primarily in northern California and is distinguished from the other species by having a duller, less shiny body and more robust antennal segments.

Seven species and two subspecies of *Centrodera* occur in California. Both the yellowish brown *C. autumnata* (14.0 to 22.5 mm) and the dark brown *C. oculata* (13.0 to 22.0 mm) are found in the coastal mountain ranges. *Centrodera autumnata* is found from Humboldt County to San Diego County, while *C. o. oculata* occurs in coastal central California and the Sierra Nevada. *Centrodera o. blaisdelli* occurs from San Luis Obispo County southward. The robust, broad-shouldered *C. spurca* (20.0 to 30.0 mm) (pl. 234) is a shiny pale brownish yellow beetle. Its typically larger size and depressed, recurved pubescence of the elytra distinguish it from the similar *C. autumnata*. It is found in montane California, including the Sierra Nevada and the Transverse and Peninsular Ranges. The larva wanders freely through the soil, feeding in rotting stumps and roots of oak, madrone, service-berry *(Amelanchier)*, and sheperdia *(Sheperdia)*. The adult hides in living trees during the day and is commonly attracted to lights, especially in August and September.

Leptura obliterata soror (9.0 to 17.0 mm) (pl. 235) occurs in the Sierra Nevada, where it is found on various mountain meadow flowers, including yarrow *(Achillea)*, California-lilac *(Ceanothus)*, buckwheat *(Eriogonum)*, and lupine *(Lupinus)*. The larva feeds in old stumps of pines and firs. Seven other species of *Leptura* occur in the state.

The Yellow Velvet Beetle *(Lepturobosca* [formerly *Cosmosalia*] *chrysocoma)* (9.0 to 20.0 mm) (pl. 236) is found throughout the high mountainous regions of western North America and eastward across Canada and northeastern United States. This velvety, yellow beetle is a common visitor to a wide variety of meadow flowers and staminate catkins of Jeffrey pines *(Pinus jeffreyi)* from June through September. It is most common in the Cascade Ranges and Sierra Nevada but has also been taken in the Transverse Ranges in southern California. The larva is recorded from pine, spruce, alder, and black cottonwood *(Populus balsamifera trichocarpa)*.

Seven species of *Judolia* occur in California. *Judolia* (formerly *Anoplodera*) *instabilis* (6.0 to 15.0 mm) (pl. 237) is distributed in the Sierra Nevada and Transverse and Peninsular Ranges. The adult is active from May through September and is encountered most frequently on the flowers of lupine and buckwheat. It has also been found on carrot *(Daucus)*, lilies *(Calochortus)*, coneflower *(Rudbeckia)*, California-lilac, yarrow, sneezeweed *(Helenium)*, and cow-parsnip *(Heracleum)*. It is one of the most variable of all California longhorn beetles in terms of color and size. The elytral patterns vary from all black to pale yellow with only a few black spots. More than 60 percent of the individuals in southern California have elytra entirely black or nearly so. The number of black individuals decreases northward into the Sierra Nevada. The larva has been reared from the roots of milkvetch *(Astragalus)* and lupine.

Two species of *Ortholeptura* occur in California. *Ortholeptura insignis* (21.0 to 27.0 mm) occurs along the coast, from Mendocino County south to the Monterey Bay region. It is pale reddish brown with stripes on the elytra and only slightly pointed, if at all, elytral tips. The larva feeds in decaying hardwoods and pines.

Ortholeptura valida (17.0 to 23.0 mm) (pl. 238) is found mostly in the Cascades and Sierra Nevada but also occurs in the Transverse and northern Peninsular Ranges. It has yellowish brown elytra with dark spots and spined tips. The larva feeds in decaying conifers.

The Ribbed Pine Borer *(Rhagium inquisitor)* (9.0 to 21.0 mm) (pl. 239) is found throughout the forested regions of Europe and North America. With short, thick antennae and ribbed elytra, this mottled black beetle hardly looks like a longhorn beetle. In California it is found in the montane regions of the Klamath Mountains, Sierra Nevada, and Transverse and Peninsular Ranges. The adult is active February through July. The broad, flat larva is reported to feed mainly on pines but also will use spruce, Douglas-fir, and fir, tunneling between the bark and wood and pupating there within a ring of chewed wood. The oval pupal chamber, where the adult also overwinters, lies between the bark and the wood and is characteristically lined with coarse woody fibers to guard against attacks by marauding ants and other predators lurking beneath the bark.

All three species of *Stenostrophia* are found in California, each with distinctive black and yellow bands across the elytra. *Stenostrophia tribalteata tribalteata* (7.0 to 11.0 mm) is found on the east side of the Sierra Nevada, while *S. t. sierrae* (pl. 240) is known primarily from the Cascade Range, Klamath Mountains, western slopes of the Sierra Nevada, and Transverse and Peninsular Ranges. The adult is common on flowers, particularly ceanothus *(Ceanothus)* and buckthorn *(Rhamnus)*, but the larval habits are unknown.

Elderberry borers of the genus *Desmocerus* are distinctive, broad-shouldered beetles that typically begin to emerge in spring. Two subspecies of the Golden-winged Elderberry Borer *(D. auripennis)* are found in California. *Desmocerus a. auripennis* (23.0 to 30.0 mm) (pl. 241) is found in the Sierra Nevada and Klamath Mountains, where the larva mines stems of the blue elderberry *(Sambucus mexicana)*. The adults are variously marked with yellow and metallic green. *Desmocerus a. cribripennis* (11.5 to 21.0 mm) is found from the central Coast Ranges of California to British Columbia. The adult is much smaller, darker, and hairier than its Sierran counterpart. The larva of this subspecies feeds on red elderberry *(S. racemosa)*. The California Elderberry Longhorn Borer *(D. californicus californicus)* (13.0 to 25.0 mm) is found in the Coast Ranges, from Mendocino County southward to Orange County, where its larva feeds on blue elderberry. The male is especially active, flying about the host tree. In the afternoon both the male and female are found resting upon the foliage or limbs. The elytra of the male is mostly blue black with a border of red orange, turning dull yellow in dead specimens. The Valley Elderberry Longhorn Beetle *(D. c. dimorphus)* (13.0 to 21.0 mm) was listed as threatened by the U.S. Fish and Wildlife Service in 1980. The elytra of the male are mostly red with two pairs of dark, oblong spots; those of the female are dark bluish with reddish margins. The larva bores in the branches of blue and red elderberry growing along streams and rivers from the Sacramento River delta region south to Stockton.

Of the six species of *Necydalis* in California, *N. cavipennis* (13.0 to 24.0 mm) (pl. 242) is the most common and widespread. It is active from May to July along the coast. Most males are black, while most females are reddish brown. This slender beetle, with very short elytra, is wasplike in both appearance and behavior. It breeds in a variety of deciduous trees but seems to prefer various species of oak. The larva bores into the heartwood and is frequently found in stumps or at the bases of standing trees.

The Lion Beetle *(Ulochaetes leoninus)* (17.0 to 32.0 mm) (pl. 243) has short, yellow-tipped elytra that reveal most of the abdomen. In flight this hairy black and yellow beetle resembles a bumblebee. When disturbed, a Lion Beetle raises its abdomen forward over its back while flapping its wings, reinforcing its beelike appearance. In California *Ulochaetes* is distributed in the Sierra Nevada and Transverse and Peninsular Ranges. The female lays its eggs at the base of standing dead trees or stumps of a variety of conifers, especially ponderosa pine and Douglas-fir. The larva prefers to work the roots but will move into the lower part of the trunk, working with the grain in the sapwood and heartwood. Just before pupation it will turn at a right angle to the grain and tunnel its way out to the bark, plugging the emergence hole with frass. The larva then constructs its pupal chamber in the bark or sapwood. The adult is active June through August. The male is usually found on the host tree, whereas the female is often found flying at midday.

The sole species of *Paranoplium, P. gracile* (12.0 to 24.0 mm) (pl. 244), whose larva feeds on oak and other hardwoods, is active in summer and is divided into two subspecies of dubious validity.

Paranoplium g. gracile is found along the coast from Monterey County to San Diego County, while *P. g. laticolle* lives in the southern Sierra Nevada.

Brothylus gemmulatus (12.0 to 22.0 mm) (pl. 245) is found from Oregon and Utah south to Mazatlán, Mexico. The female has a bump on each side of the pronotum, whereas the male's pronotum is rounded on the sides. The larva bores into dry, dead branches of oaks. A second, less-widespread species, *B. conspersus* (12.0 to 22.0 mm), is darker and clothed in whitish pubescence, with two wavy bands running across the elytra. It also breeds in dry, dead oak limbs.

Two species of the Australian genus *Phoracantha,* known as eucalyptus longhorn beetles, are established in California. *Phoracantha semipunctata* (12.0 to 30.0 mm) was first found in 1984 on dying eucalyptus trees *(Eucalyptus)* in Orange County and has since spread throughout California. In June of 1995, an even more destructive eucalyptus longhorn borer, *P. recurva* (14.0 to 30.0 mm) (pl. 246), was discovered at the University of California at Riverside. This new arrival spread quickly throughout southern California and soon replaced *P. semipunctata* as the most common species of eucalyptus longhorn beetle. Both species are similar in size, shape, and color. The elytra of both beetles are dark brown, with cream to yellowish patches in clearly defined patterns. The elytra of *P. semipunctata* are mostly dark brown with a central cream-colored band divided by a dark zigzag line. The elytra of *P. recurva* are largely cream colored. The dark brown areas are mostly limited to a narrow strip along the base and apical third of the elytra. The antennae of both beetles are longer than the body. The antennae of *P. recurva* are densely clothed with golden hairs, whereas those of *P. semipunctata* are bare. Adults of both species are attracted to fallen branches and injured or water-stressed trees. The larvae of *Phoracantha* tunnel between the bark and wood but construct their pupal chambers in the heartwood.

The genus *Lampropterus* is restricted in North America to the Pacific coast, and both species occur in California. These small, slender longhorns have bluish or greenish elytra, are covered with erect hairs, and are often abundant on the blossoms of ceanothus and other shrubs. *Lampropterus cyanipennis* (5.0 to 8.0 mm) has yellowish brown legs with the tips of both the femora and tibiae black. The male prothorax is black, while that of the female is yellowish brown. It is active April through June and ranges from Oregon to southern California. Both sexes of *L. ruficollis* (4.5 to 8.0 mm) (pl. 247) are similar in color. They are black, with the prothorax partly (male) or entirely (female) reddish. Both species breed in dead oak twigs.

Rosalia occurs in Europe and southeast Asia, with one species represented in North America. The striking Banded Alder Borer *(R. funebris)* (23.0 to 40.0 mm) (pl. 248), also known as the California Laurel Borer, is distinctly marked with transverse, pale bluish white and black bands. It is distributed in mountains from the Rocky Mountains south to Arizona and New Mexico and from Alaska to southern California. Adults have been collected in large numbers emerging from alder or at drying paint. The larva also feeds on dead maple *(Acer)*, ash *(Fraxinus)*, sycamore, coast live oak, willow, eucalyptus, and California laurel *(Umbellularia californica)*.

The violet, blue, or rarely, greenish blue *Callidium antennatum* (9.0 to 14.0 mm) (pl. 249) is covered with erect, black hairs. The antennae of both the male and female are shorter than the body, and all of the femora appear swollen. It is known primarily from the northern Coast Ranges, Cascades, Sierra Nevada, and Transverse Ranges. The larva works under the bark of pine, scarring both the trunk and inner bark, and expels its woody waste, or frass, through holes in the bark. Eleven other species are known in the state.

Both the male and female Mesquite Borer *(Megacyllene antennata)* (12.0 to 25.0 mm) (pl. 250) have short antennae and are brown or reddish brown with gray or yellowish pubescence. The unmarked pronotum is clothed in gray pubescence, while the elytra are marked with broad, transverse bands of gray pubescence just before the middle and at the apices. The female lays its eggs beneath loose bark or in crevices on dead branches of mesquite and acacia *(Acacia)*. The larva is brown headed with a cream-colored body and can rapidly reduce mesquite logs and recently cut and cured mesquite posts to yellow dust. It works the sapwood before entering the hardwood to continue feeding and pupate. The Mesquite Borer occurs from Texas to southern California and is commonly attracted to lights in summer in the Colorado Desert.

The relatively short antennae and numerous broken ridges across the pronotum distinguish *Neoclytus* from other longhorns. These somewhat leggy and wasplike beetles often have contrasting bands of color across their bodies. The larvae of most species attack deciduous trees, boring through the heartwood and feeding under the bark only briefly. In some species, such as *N. balteatus* (8.0 to 16.0 mm) (pl. 251), the female is yellow and black, while the male is reddish brown. This species is widely distributed from British Columbia south to the San Gabriel Mountains. The adult has been beaten from ceanothus flowers. Thirteen other *Neoclytus* species are known in the state.

California has eight species of *Xylotrechus*. *Xylotrechus albonotatus* (10.0 to 20.0 mm) ranges from the Sierra Nevada to the Transverse and Peninsular Ranges and is active June through August. The body is black with a pattern of white, grayish, or yellowish hair on the pronotum and elytra. The elytra have a crescent-shaped mark behind each shoulder, followed by a wavy middle band, another complete or broken band at the apical third, and pale tips. The larval plant hosts include white fir *(Abies concolor)* and California red fir *(A. magnifica)*. The larva of *Xylotrechus insignis* (12.0 to 18.0 mm) works willows, gaining access to living and dying trunks through dead, injured, or diseased branches. The adult is active March through August. The female is black with yellow markings. The elytra have yellow bands at the base and apex, with a median band that curves forward along the median suture. The male is reddish brown with reduced yellow markings. It is found in all mountainous regions of the state except the deserts. The adult of the Nautical Borer *(X. nauticus)* (8.0 to 16.0 mm) (pl. 252), also known as the Oak Cordwood Borer, is grayish brown to black, with three irregular gray bands crossing the elytra. The outer angles of the elytra are armed with short spines. The antennae of both the male and female are short, less than half the length of the body. The Nautical Borer is often found in homes, where it most likely emerged from stores of firewood. It does not cause damage to the structural timbers of homes and outbuildings. The larva feeds in a wide variety of dead or dying deciduous trees, including oak, walnut, eucalyptus, madrone, pittosporum *(Pittosporum)*, and avocado *(Persea)*.

Nine species and 19 subspecies of *Crossidius* are recorded from California. Adults are black with reddish, brownish, or yellowish markings. They are active from mid- to late summer and early fall in the valleys and deserts, feeding during the day on the flowers of Great Basin sagebrush *(Artemisia tridentata)*, rabbitbrush

(Chrysothamnus), and goldenbush *(Isocoma).* The larvae feed on the roots of Great Basin sagebrush and rabbitbrush. They are found in the vicinity of the Great Central Valley and the Great Basin, western Mojave (Antelope Valley, Owens Valley), and Colorado Deserts. *Crossidius ater* (9.0 to 17.0 mm) is wholly black and is a common Great Basin species found on goldenbush and rabbitbrush August through October. The larva breeds in roots of Great Basin sagebrush. *Crossidius coralinus* (11.0 to 21.0 mm) is black with a red or orangish prothorax, elytra, and abdomen. The elytra are variously marked with black or bluish black markings. Seven subspecies are found in southern, central, and eastern California. The larva breeds in roots of rabbitbrush or goldenbush. *Crossidius c. ruficollis* is known from the southern San Joaquin Valley, where it is found on some species of goldenbush *(I. acradenia* var. *bracteosa* and *I. menziesii* var. *vernonoides).* The elytra are orangish brown. The male is without markings or with a black stripe along the elytral suture, and the female has a black marking on the apical third or two-thirds of the elytra. A similar subspecies, *C. c. ascendens,* is much redder. The elytra have black shoulders and base. Both it and *C. mojavensis* (10.0 to 18.0 mm) (pl. 253), with its dull golden brown, nearly unmarked elytra, are found in the western portion of the Antelope Valley on rabbitbrush in September and October. The larva of *C. mojavensis* breeds in Great Basin sagebrush. *Crossidius punctatus* (11.0 to 16.0 mm), from northeastern California, has a partially or wholly reddish brown pronotum. The yellowish brown elytra are variably marked, usually with the base and shoulders of the elytra black with a narrow sutural stripe running along the apical three-quarters. It breeds in rabbitbrush. Two subspecies of *C. suturalis* (14.0 to 19.0 mm) are known from southern California. *Crossidius s. minutivestis* (pl. 254) is known from the western Mojave Desert on rabbitbrush and goldenbush. *Crossidius s. pubescens* is found in the western Colorado Desert on a species of goldenbush *(I. a.* var. *eremophila),* in which it breeds.

Both species of *Plionoma* are active during the day, have a white stripe along the elytral suture, and occur in the deserts of southern California. *Plionoma rubens* (10.0 to 12.0 mm) is pale reddish with more widely separated eyes, and longer, denser hairs on the body. The adult is active from May to September. The larva feeds on catclaw *(Acacia greggii),* honey mesquite *(Prosopis glandulosa* var. *torreyana),* and palo verde *(Cercidium). Plionoma su-*

turalis (10.0 to 16.0 mm) (pl. 255) is black or reddish black, with eyes closer together, and shorter, scattered hairs. The larva bores into mesquite.

Schizax senax (12.0 to 18.0 mm) occurs from western Texas to southern California. The head is black with eyes completely divided by the insertion of the antennae. The pronotum and legs are somewhat reddish, while the elytra are black and bordered by reddish yellow pubescence and clothed overall in short white hairs. This diurnal species is active in spring and is found on mesquite, creosote bush *(Larrea tridentata)*, and blue palo verde *(C. floridum)*, breeding in the latter.

Trachyderes (formerly *Dendrobias*) *mandibularis reductus* (17.0 to 32.0 mm) (pl. 256) is a handsome beetle with shiny yellow elytra that are dark at the base and sometimes bear a narrow, darkened suture. The adult is active July through September along the Colorado River. The enlarged, horizontal mandibles distinguish the male from the female. The male also lacks marks on the elytra, except for the occasional spot on the suture, but the female has a narrow, black band that crosses the elytra midway. The larva feeds on willow. In southeastern Arizona, the subspecies *T. m. mandibularis* is often encountered on sapping broom plants and is attracted to watermelon rinds and other rotting fruit.

Members of the genus *Tragidion* sometimes have striking red elytra, somewhat resembling a large tarantula wasp *(Pepsis)* in flight. The Yucca Borer *(T. armatum armatum)* (20.0 to 30.0 mm) (pl. 257) has smooth elytra and is active in May through August. It is found on or flying near chaparral yucca *(Yucca whipplei)*. The next two species have ribbed elytra. *Tragidion peninsulare* (28.0 to 34.0 mm) is found in the Colorado Desert south to Baja California from July to October. Its larval host plants are mesquite and palo verde. *Tragidion californicum* (17.0 to 21.0 mm) is found in the Sierra Nevada and the Transverse and Peninsular Ranges. The adult is active in July and August. The larva feeds in oak, chamise *(Adenostoma fasciculatum)*, sugar bush *(Rhus ovata)*, lemonadeberry *(R. integrifolia)*, and toyon *(Heteromeles arbutifolia)*. The adult is attracted to brush fires, and charred stems of the host plants are often teeming with larvae.

The Hairy Borer *(Ipochus fasciatus)* (4.5 to 10.0 mm) (pl. 258) is a small, wingless beetle with a strongly oval and convex body that strongly resembles that of a spider. It is reddish brown or dark brown to black and clothed in long, erect hairs. The adult is common under bark of willow and oak, but it has been taken on a variety of other plants including lemonadeberry, laurel sumac *(Malosma laurina)*, toyon, milk thistle *(Silybum marianum)*, walnut, black cottonwood, mustard *(Brassica)*, castor bean *(Ricinus)*, apple, and pear *(Pyrus)*. In California this species occurs in the Coast Ranges from Monterey Bay southward to the Santa Monica Mountains and coastal foothills of the Transverse and Peninsular Ranges, as well as the northern California Channel Islands and Santa Catalina Island. The adult is collected year-round.

The Cactus Beetle *(Moneilema semipunctatum)* (15.0 to 30.0 mm) (pl. 259) is found in the Great Basin of southern Idaho to northern Baja California. In California it is found in the Owens Valley and along the interior foothills of the Transverse and Peninsular Ranges. It is also known from areas of desert scrub in the coastal plain of southern California. This species is active from April through November, reaching peak activity during July and August. The Cactus Beetle hides by day at the bases or in the crowns of teddy bear cholla *(Opuntia bigelovii)* and beavertail cactus *(O. basilaris)*, coming out at night to feed on the spiny columns or pads. The female lays its eggs at the base of the cactus. The larva usually feeds near the root collar, and the pupal cell is constructed partially or entirely in the soil. The boring activities of larvae above ground are conspicuous by the tarlike excrement and fluid expelled from the wounds created by their feeding activities. The wingless, black adult resembles the foul smelling and unrelated darkling beetles of the genus *Eleodes* (Tenebrionidae) in appearance and behavior. The flightless adult occasionally walks to lights.

The Oregon Fir Sawyer *(Monochamus scutellatus oregonensis)* (13.0 to 27.0 mm) (pl. 260) is a striking, shiny black beetle with a slight metallic tinge. It occurs primarily in the Klamath Mountains, Cascade Ranges, Sierra Nevada, and Transverse Ranges. It also occurs in Oregon and Washington. The male has extremely long antennae, much greater than the length of its body. The larva tunnels between the bark and wood but usually feeds and pupates in the heartwood. The Spotted Pine Sawyer *(M. clamator latus)* (14.0 to 29.0 mm) (pl. 261) is similar in appearance but is dark brown with scattered patches of dull white and black pubescence. It lives in the Sierra Nevada, northern Coast Ranges, Klamath Mountains, Cascades, and Transverse Ranges. The large, white larvae of both species of *Monochamus* cause considerable damage to recently felled, injured, burned, dying, or dead pine, fir, Douglas-fir, larch *(Larix)*, and spruce by excavating extensive tunnels in the sapwood and throughout the heartwood. Adults of both California species are active May through September. Adults are often found walking or resting on host tree trunks.

The broad and "square-shouldered" Spotted Tree Borer *(Synaphaeta guexi)* (11.0 to 27.0 mm) (pl. 262) is a very handsome and distinctive species. The elytra are cryptically marked with a black zigzag stripe and patches of gray and yellow orange pubescence. The antennae are as long as or longer than the body. The Spotted Tree Borer is broadly distributed in all regions except the deserts. It has been collected year-round but is most common from April to July. It frequently attacks weak and dead orchard and ornamental trees. The adult usually lays its eggs in wounded or injured areas or gnawed niches. The young larva mines under bark for a short distance then proceeds into the heartwood, leaving in its wake large tunnels loosely filled with fibrous shavings of wood. Larval hosts include maple, buckeye *(Aesculus)*, alder, citrus *(Citrus)*, fig *(Ficus)*, walnut, apple, cottonwood, peach, pear, oak, California coffeeberry *(Rhamnus californica)*, willow, elm, laurel, and wisteria *(Wisteria)*.

Five species of *Poliaenus* occur in California. *Poliaenus oregonus* (6.0 to 8.5 mm) lives in the mountains from British Columbia to southern California and the Rocky Mountains, from Idaho and Montana to Colorado and Utah. In California it is known from the northern Coast Ranges, Cascades, Sierra Nevada, and Transverse Ranges. The larva develops on the ground in broken branches and limbs of firs and Douglas-fir. The adult has bristly hairs scattered on its antennae and legs. The elytra have a broad, black band just behind the middle, while the tips are covered with short, white hairs. *Poliaenus californicus* (6.5 to 11.0 mm) has even hairier legs and antennae but lacks the black elytral band. It is known from the western foothills of the Sierra Nevada and the Transverse Ranges. Both the larva and adult are found on California flannelbush or fremontia *(Fremontodendron californica)*. In *Poliaenus obscurus* (6.0 to 9.5 mm) the legs and antennae have long hairs like *P. californicus* and a broad band across the elytra that is strongly narrowed in the middle. The larva of *P. obscurus* develops in pines. Three subspecies of *P. obscurus*

(6.0 to 9.5 mm) are recognized in the state, but these may prove to be geographic or host races. They all have elytra with weakly developed ribs that are distinctly less than twice as long as their basal width (the elytral ribs of *P. californicus* are twice as long as their basal width). *Poliaenus o. ponderosae* (pl. 263) has well-defined, black triangular patches just behind the middle of the elytra. It lives in montane California, where its larva breeds in ponderosa pine *(Pinus ponderosa)*. Another subspecies, *P. o. albidus*, has a pale reddish brown integument and mostly pale elytra without any pattern. It is found in the foothills surrounding the Great Central Valley. The larva feeds in gray pine *(Pinus sabiniana)*. The larva of *P. o. schaefferi* feeds on Colorado pinyon *(Pinus edulis)* in southern California. The adult has a dark brown integument with a variegated pattern on the elytra.

Their long antennae, bushy basal antennal segments, and the long egg-laying tube, or ovipositor, of the female distinguish California's three species of *Acanthocinus*. *Acanthocinus* (formerly *Neacanthocinus*) *obliquus* (8.0 to 17.0 mm, without ovipositor) is found primarily in the Sierra Nevada and is active April through September. The elytra are pitted at their bases, whereas those of *A. princeps* (13.0 to 24.0 mm, without ovipositor) (pl. 264) appear granular. Both species have similar activity periods and distributions. The larvae of all *Acanthocinus* are pine feeders, boring within the cambium and pupating in the bark.

There are two species of *Coenopoeus,* but only one occurs in the United States. In California, *C. palmeri* (15.0 to 27.0 mm) (pl. 265) is a large, black beetle with scattered patches of yellow gray hairs on the elytra. It is found in the western Colorado Desert. The adult is found during the day hidden among the branches of several cholla cactus species *(Opuntia)* from May through September. It emerges at dusk and in the evening to feed and mate on the cactus.

Most *Saperda* live in eastern United States, but three species do occur in California. The color pattern, large size, and spined elytral apices of *S. calcarata* (18.0 to 25.0 mm) (pl. 266) will distinguish it from all other species in the genus. It is distributed across the United States and southern Canada and is active June through September. It occurs in the northern Sierra Nevada. The larva works the wood of cottonwood and willow. *Saperda horni* (10.5 to 20.0 mm) is grayish and mottled with orange brown. The pronotum and elytra are sparsely covered with distinctive, craterlike pits. It bores within living willow stems.

Milkweed borers of the genus *Tetraopes* are intimately associated with milkweeds *(Asclepias)* throughout their lives. The red- and black-spotted adults are usually encountered on the flowers, whereas their larvae bore into the stems and roots. Both the larvae and adults incorporate the toxins in milkweed sap into their own defense system. Adult beetles reduce their ingestion of toxins by first cutting the midrib of a leaf with their mandibles and bleeding off the toxic, milky sap before feeding. As the name of the genus implies, the eyes of all *Tetraopes* are completed divided. Three species occur in California. *Tetraopes basalis* (8.0 to 17.0 mm) is widespread in the Coast Ranges, Great Central Valley, western foothills of the Sierra, and foothills of the Transverse and Peninsular Ranges. Individuals from northern populations tend to be paler than those found in the south. The adult is active from April through August. *Tetraopes femoratus* (8.0 to 19.0 mm) (pl. 267) occurs in the Great Basin in eastern California from May through October. *Tetraopes sublaevis* (11.5 to 18.0 mm) occurs in the southern Sierra of Kern County southward along the interior foothills of the Peninsular Ranges of San Diego County and eastward across the Colorado Desert to Arizona.

COLLECTING METHODS: Some species of longhorns are readily collected on meadow flowers and the blooms of buckwheat, milkweed, buckbrush or California-lilac, and other plants during the day. Many nocturnal species are attracted to lights. Black lights are effective for attracting several species, and mercury vapor lights draw beetles from longer distances. Placing these lights in open areas surrounded by trees will attract specimens into the clearing. Be sure to check the shadowy areas just beyond the lights, because many species will not come in all the way to the light. Beating and sweeping the branches and leaves of host plants day and night will produce small numbers of specimens. Inspect downed logs and freshly cut wood throughout the day and evening. Cover freshly cut stumps with green branches from the same species of tree. Visiting longhorns attracted to the scent of cut wood will take cover among the foliage. Species that feed on broad-leaved trees and shrubs are often attracted to fermenting molasses or watermelon rinds, whereas conifer-feeding species are sometimes attracted to fresh paint and solvents, such as turpentine. Pitfall trapping will produce some cactus longhorns in areas with plenty of native cacti. Another productive method is to collect infested wood and place it in a plastic container as a rearing box. Carefully seal the box so that recently emerged beetles cannot escape. Cut a hole at one end of the box and insert a small jar or vial with the opening directed inside. Emerging beetles and other insects will be drawn to the light and enter the jar or vial. Add water occasionally to the container so that the wood does not dry out.

LEAF BEETLES and SEED BEETLES — Chrysomelidae

Leaf beetles comprise one of the largest families of beetles in the world, second only to the weevils or snout beetles (Curculionidae). Some species are brightly colored or marked to indicate their distastefulness to predators. Both adults and larvae attack numerous kinds of plants, eating the bark, stems, leaves, flowers, seeds, and roots. Most species feed on flowering plants, but a few prefer conifers, ferns, and their allies. The majority of leaf beetles are specialists, feeding only on a single species of plant or groups of closely related plants. Some leaf beetles even eat aquatic plants. The adults graze on leaves and other vegetative structures above the water surface, while larvae feed underwater.

Seed beetles (formerly known as seed weevils [Bruchidae]) were once considered related to leaf-rolling weevils (Attelabidae) and weevils, but current research suggests that they are actually related to leaf beetles. The vast majority of seed beetles use legumes as food. They are especially common in the southwest, where they attack the seeds of acacia *(Acacia)*, milkvetch *(Astragalus)*, palo verde *(Cercidium)*, mesquite *(Prosopis)*, and mimosa *(Mimosa)*.

Most female leaf beetles lay their eggs directly on their host plants, either singly or in small groups. Using intricate rectal plates, some species carefully apply a protective coating of their own feces on the eggs. Most larvae require living plant food to develop and spend much of their time grazing out in the open on the surfaces of leaves and other plant structures. They feed singly or in small, tightly packed groups. Group feeders eventually disperse as they mature. Leaf mining species tunnel between the upper and lower surfaces of living leaves, leaving discolored blotches, blisters, or meandering tunnels trailing in their wake. Many larvae feed on dried plant materials, such as dead leaves or bark on old twigs. Others feed on ant eggs and waste materials and may even scavenge the bodies of dead ants.

Female seed beetles lay their eggs on the seedpod or directly on the seed surface. Several larvae may develop inside a single seed, but in some species cannibalism results in only one larva developing per seed. Larvae typically feed within the seed or just inside the seed jacket and pupate within the excavation created by their feeding activities. Most species are specialists, restricting their feeding activities to just one species or genus of plant, but a few are more generalized and will attack several plant genera. They avoid the plant's defensive toxins in the pods and seed coats by only feeding inside the seed.

Leaf beetles generally pupate in the soil, sometimes in a "cocoon," and usually in a cavity or special chamber dug by the larva. A few species anchor themselves to their host plants. The two California species of *Ophraella* pupate on their host plants inside a meshlike cocoon. Seed beetles pupate inside seeds.

Leaf beetles employ various strategies to defend themselves against predators. Larvae grazing on the surfaces of leaves have evolved glands that shoot out of their bodies like party favors. The tissues of these glands are impregnated with noxious chemicals derived either directly from the host plant or produced by the larvae themselves. The larvae of tortoise beetles carry fecal material and cast larval skins over their backs like an umbrella, whereas other leaf beetles and seed beetles construct cases from their waste that cover their entire bodies as protection from potential predators.

Adult flea beetles *(Altica* and *Disonycha)* have enlarged hind legs capable of propelling them out of harm's way. Warty leaf beetles *(Exema* and *Neochlamisus)* escape detection by resembling waste material. These small and short, chunky, dark beetles resemble caterpillar feces and are no doubt overlooked by predators and collectors alike. Other leaf beetles sequester the defensive chemicals of their host plants and incorporate them into their own defense system. Some females cover their eggs with a protective coating of their own feces. When alarmed, seed beetles feign death by tucking their head, antennae, and legs tightly against their bodies.

Many species of leaf beetles and seed beetles are of economic importance. Garden and crop pests damage plants directly through their feeding activities such as defoliation, leaf mining, and root boring. Others inadvertently inoculate plants with injurious diseases. Some seed beetles, many introduced from Europe, attack leguminous crops such as alfalfa, beans, lentils, and peas.

Those requiring fresh pods for egg-laying produce just one generation per year, whereas those capable of using stored, dried seeds breed continually and are generally considered to be more serious economic pests.

Although several species originating outside of North America have become established in California as pests, not all foreign beetles are unwelcome. Several species have been intentionally introduced to control invasive weeds. The Klamathweed *(Hypericum perforatum)*, also called St. John's wort, is European in origin and infests hundreds of thousands of acres in the Pacific Northwest, including northern California. Displacing more desirable range plants, the toxic Klamathweed causes mouth sores and intestinal distress in livestock. Based on research conducted in Australia, researchers at the University of California began testing two small, shiny leaf beetles, *Chrysolina hyperici* and *C. quadrigemina*, and a root borer of the family Buprestidae, *Agrilus hyperici*, all of European origin. The first importation of the beetles into California arrived from Australia in 1944. Released in several localities in northern California, both species of *Chrysolina* were well established by 1948 and were quite successful in reducing or eliminating the pest weed throughout its range. *Chrysolina hyperici* appears to be restricted to the coastal mountains, whereas *C. quadrigemina* has become widespread throughout the state. Some seed beetles have been used as biological control agents to combat other weed species, such as Scotch broom *(Cytisus scoparius)* in California and elsewhere.

IDENTIFICATION: California's leaf and seed beetles are extremely variable in shape. They are long and cylindrical, square or oval, and convex or flattened, ranging in length up to 11.0 mm. Many species are uniformly black or brown, mottled with yellowish or reddish hues, or iridescent blue, green or bronze, or are two-toned with distinctive patterns. The head is usually hypognathous but is sometimes prognathous. The antennae are usually 11-segmented and threadlike, saw-toothed, almost leaflike, feathery, fan shaped, or occasionally almost threadlike or gradually clubbed. The pronotum is broader than the head, triangular or rectangular, and usually distinctly margined and is sometimes broadly expanded and flattened. The scutellum is visible. The elytra usually conceal the abdomen (leaf beetles), or leave the last segment exposed (seed beetles). The elytral margins are usually parallel sided or broadly rounded but are sometimes broadly expanded and flattened. The elytral surface is rounded or flattened and may be smooth and shining, or with distinct grooves or lines of pits, or scattered hairs. The tarsi appear 4-4-4 but are actually 5-5-5, with the fourth segment small and hidden between the lobes of the third. The hind legs of flea beetles (e.g., *Altica* and *Disonycha*) are enlarged for jumping. Claws are usually equal in size and simple or with a single broad tooth. The abdomen has five segments visible from below.

The larvae are extremely variable and are humpbacked, caterpillarlike, spindle shaped, long and cylindrical, or somewhat C shaped. They vary in color from whitish to cream in subterranean forms to metallic or patterned in the exposed leaf feeders. The body, which may be covered with tiny spines, fleshy lobes, or just plain, may be covered or encased in fecal material or bits of plant detritus. The head is either distinct or partially withdrawn into the thorax and is usually hypognathous and has a pair of small antennae with one to three segments. There are zero to six pairs of simple eyes. The legs, if present, are four- or five-segmented, including the claw. The abdomen may appear to have eight segments visible from above; the ninth and tenth segments are usually hidden beneath. The tenth segment may have a false foot, like a caterpillar. In species that cover themselves with feces, segment eight has special forks that hold the feces over the back.

SIMILAR CALIFORNIA FAMILIES:

- some checkered beetles (Cleridae)—antennae gradually or distinctly clubbed; tarsi distinctly 5-5-5
- soft-winged flower beetles (Melyridae)—antennae sawtoothed; tarsi distinctly 5-5-5
- some pleasing fungus beetles (Erotylidae)—antennae with distinct, flat clubs
- some handsome fungus beetles (Endomychidae)—tarsi appear 3-3-3 but are 4-4-4
- some lady beetles (Coccinellidae)—tarsi appear 3-3-3 but are 4-4-4
- some darkling beetles (Tenebrionidae)—tarsi 5-5-4
- antlike flower beetles (Anthicidae)—head with neck
- some longhorn beetles (Cerambycidae)—pronotum usually not parallel sided, widest at middle or behind
- leaf-rolling weevils (Attelabidae)—antennae with three-segmented club
- weevils or snout beetles (Curculionidae)—snout usually longer; antennae elbowed with distinct, three-segmented club

CALIFORNIA FAUNA: Approximately 534 species in 109 genera.

Two species of *Crioceris* have been introduced into North America from Europe. The Asparagus Beetle *(C. asparagi)* (4.7 to 7.0 mm) (pl. 268) is metallic blue black with an orange or reddish orange pronotum. The elytra are dark with variable yellow markings. The black may extend broadly along the elytral suture and meet with black patches on the shoulders. The sutural stripe usually extends to the sides at the middle and near the tip of the elytra. Both the adult and larva feed on asparagus and are particularly abundant in the asparagus-growing districts in the northern parts of the state and along the coast of southern California. The female lays its eggs on the young shoots of asparagus. The mature larva constructs a pupal chamber in the soil. The entire life cycle takes about a month, producing two or three generations annually. The adult overwinters in plant debris left in the fields and is capable of stridulation. This species has become widespread in the Pacific Northwest and northeastern United States. The head and elytra of the Spotted Asparagus Beetle *(C. duodecimpunctata)* (4.9 to 6.6 mm) are red or reddish orange. Each elytron usually has six black spots. This species also feeds exclusively on asparagus.

The head of the oval and somewhat flattened Eggplant Tortoise Beetle *(Gratiana pallidula)* (5.0 mm) (pl. 269), also called the Green Tortoise Beetle, is hidden from view. This dull green beetle is coarsely pitted, with each puncture grayish or dark. The female emerges from hibernation in spring to lay eggs singly or in small groups on plants in the Solanaceae family, such as eggplants, tomatoes, and potatoes. It is especially common on the native wild nightshade *(Solanum)*. The flat, pale greenish larva are perfectly camouflaged on the leaves of their host plants and are difficult to see. Both the adult and larva chew holes in the foliage and may kill all the leaves on the plant. The sole American species of this genus is found across the southern United States and is common in southern California.

The Golden Tortoise Beetle *(Charidotella* [formerly *Metriona*] *sexpunctata bicolor)* (5.0 to 6.0 mm) is the only member of the genus in California. It is broadly oval in outline and humpbacked, like a tortoise. The head is hidden from above, and the upper surface is a shiny iridescent gold, with three small black spots on each elytron. The color changes slightly while the beetle is feeding, but the metallic luster is lost permanently upon its death. The dull brown larva covers itself in a protective shell of its own feces. The Golden Tortoise Beetle is distributed throughout much of California, where it feeds on members of the morning glory family and is well established across the continent and to South America.

The Eucalyptus Beetle *(Trachymela sloanei)* (6.0 to 7.0 mm) (pl. 270) resembles a mottled brown lady beetle (Coccinellidae). Although originally from Australia, it is currently found throughout California, where it feeds on both red gum *(Eucalyptus camaldulensis)* and blue gum *(E. globulus)*. Its feet are broad and thick and covered below with thick brushy pads. The caterpillar-like larva is grayish brown with thick bumps of different sizes scattered on the body. The Eucalyptus Beetle lays small batches of eggs in cracks and crevices on or beneath bark. The entire life cycle takes two or three months. Both the adult and larva are voracious leaf feeders, but it is not clear whether they pose a serious threat to the tree.

Only one species of *Gastrophysa* is known from the state. The Green Dock Beetle *(G. cyanea)* (4.0 to 5.0 mm) (pl. 271) is common throughout much of California and Oregon in spring and summer. The head, legs, and underside are black, sometimes with a hint of iridescence, while the upper side is entirely shiny metallic green or blue. Clusters of long, orange eggs are laid on the leaves of dock *(Rumex)*.

The genus *Trirhabda* is conspicuous throughout the southwest. Adults are usually pale yellow with a single black or brown spot between the eyes and usually three spots on the pronotum. The elytra are variable and are either striped or entirely green or blue with a narrow, pale band along the outer margins. Egg laying and pupation take place in the soil. The larva sometimes resembles a dark, metallic blue green caterpillar. Both the adult and larva are active in late spring and summer, feeding on a wide variety of plants, especially members of the sunflower family. Sixteen species are known from California. *Trirhabda luteocincta* (8.5 to 10.5 mm) feeds on California sagebrush *(Artemisia californica)* in southern California and also occurs in Arizona, New Mexico, and Baja California. The spot on the body has a metallic luster, and the pronotum is shiny. The elytra are dark green, blue, or violet, sometimes with a pale stripe running partially down the middle. The body underneath is dark with a metallic luster. *Trirhabda nitidicollis* (6.0 to 9.0 mm) is primarily a Great Basin species. The markings on the head and pronotum, as well as the elytra, may be metallic or not. The pronotum is large, polished, and without any depressions. The stripes on the elytra are blue, green, or even purple. The underside of the body is pale. It feeds on rubber rabbitbrush *(Chrysothamnus nauseosus)*, matchweed *(Gutierrezia sarothrae)*, and sagebrush. *Trirhabda diducta* (6.5 to 9.0 mm) is found throughout the state, wherever its food plant, yerba santa *(Eriodictyon californicum)*, grows. The head has a broad, black spot with a metallic luster, while the spots on the pronotum do not. The elytra have blue, green, or purple marginal and sutural stripes that usually meet before the wing tips. The body underneath is pale, sometimes with a dark margin along the thorax and abdomen. *Trirhabda geminata* (5.5 to 7.0 mm) (pl. 272) occurs in dry chaparral and coastal sage scrub communities in California and Arizona and is found on brittlebush *(Encelia farinosa)*. The spots on the head and pronotum are dark brown or black, without any metallic luster. The marginal and sutural stripes of the elytra are usually distinct, while there is only a trace of a middle stripe. *Trirhabda confusa* (7.0 to 8.5 mm) (pl. 273) occurs in southern California, the Great Central and Owens Valleys, and northward along the coast to Humboldt County. The broad spot on the head is metallic, but those on the pronotum may be faintly or not at all metallic. The elytra are entirely metallic green, except along the margin, or striped with a pale stripe down the middle, and covered with dense, silky hairs. The abdomen is mostly pale. This species feeds on sagebrush.

The Elm Leaf Beetle *(Xanthogaleruca luteola)* (6.0 to 6.5 mm) (pl. 274) was introduced from Europe and is now established throughout much of North America and is a serious pest of elm trees *(Ulmus)*. The adult is yellowish to olive green, with two spots and an hourglass mark on the pronotum, and broad, black stripes following the edge of each elyton. The mature larva is yellowish with a pair of dark stripes running down the length of the body. The larva damages the leaves by grazing on their surface, leaving behind only a skeletonlike network of veins. Pupation takes place in the soil near the base of the tree.

Two species of *Diabrotica* are found in California. The most common and widespread species is the Western Spotted Cucumber Beetle *(D. undecimpunctata)* (6.0 to 7.5 mm) (pl. 275). It has greenish yellow pronotum and elytra. Each elytron bears six large, black spots. The remaining body is black. The larva feeds on the roots of grasses and legumes in wooded foothills and canyons, completing its growth in about 60 days. Pupation takes place in an earthen pupal chamber in the soil. This beetle usually passes the winter in the adult stage, emerging January through March to lay its eggs and die. It is common in home gardens and flowerbeds, where it feeds on cucumber, corn, and other plants and is sometimes a serious, yet localized, pest of fruit trees.

The Western Striped Cucumber Beetle *(Acalymma trivittatum* [often misspelled as *trivittata*]) (4.0 to 6.0 mm) (pl. 276) is the only member of the genus in California and occurs throughout the state. The head is black, with a greenish yellow pronotum and black and white (or yellow) striped elytra. The overwintering female emerges in early spring and lays eggs at the base of various host plants, including almond, apple, apricot, bean, beet, cucumber, melons, pea, pumpkin, squash, and watermelon. The larva attacks roots while the adult feeds on all structures of the plant above ground. Three generations are produced annually.

Flea beetles are so named because of their enlarged hind legs and tremendous jumping capability. The Alder Flea Beetle *(Altica ambiens)* (5.0 to 6.0 mm) (pl. 277) is dark shiny blue and occurs from Alaska to California and New Mexico. The larva skeletonizes leaves of alders *(Alnus)*, while the adult chews holes in them. The mature larva is brown to black above and yellowish below, with a shining black head, thorax, and legs. Pupation occurs in late summer in the litter below the host plant. The adult emerges within 10 days, feeds until the end of the season, then hibernates through winter. It emerges the following spring to mate and lay eggs. *Altica biguttata* is also metallic blue but is less shiny and feeds on willows *(Salix)*. A metallic green species, *A. prasina*, also feeds on willow. Sixteen additional species of *Altica* live in California.

The larvae of *Chaetocnema* feed in the roots of dichondra lawns and can cause serious damage when present in large numbers. Because of their greatly enlarged hind legs and ability to jump, the adults are often called dichondra flea beetles. Several species in the east are important pests of corn and other crops. Nineteen species occur in California. The Dichondra Flea Beetle *(C. repens)* (1.5 mm) (pl. 278) is a small, reddish bronzy black beetle that has distinct rows of pits running the length of the elytra. The adult overwinters down in the thatch of the lawn and emerges on warm days to feed on the leaf blades of dichondra *(Dichondra repens).* It ranges from southern California to Florida.

Nine species of *Disonycha* occur in California. *Disonycha alternata* (6.8 to 7.7 mm) (pl. 279) is widespread throughout North America, where it feeds on willow growing along streams and moist canyon bottoms. The head is red with black eyes and antennae. The pronotum is red with a narrow, yellow border all around and four black spots arranged in an arc. The underside is mostly red, and the legs are red with mostly black tibiae and tarsi on the female, whereas the hind legs of the male are totally black. The elytra are pale yellow with five longitudinal black stripes.

The large and beautiful Blue Milkweed Beetle *(Chrysochus cobaltinus)* (6.5 to 11.5 mm) (pl. 280) is a robust and oblong species that may be metallic green or a deep metallic blue with or without a greenish tinge. Groups of beetles are often seen feeding and mating on the leaves of milkweeds *(Asclepias),* while the larva probably feeds on the roots. The adult usually spends most of its life in a small area and lives for about a month. A disturbed beetle will tuck its legs in and fall to the ground, meanwhile producing a foul-smelling and foul-tasting fluid. The milky sap of milkweed is brimming with bitter toxins that discourage most herbivores, but the Blue Milkweed Beetle is undeterred and feeds exclusively on the plant. Unlike other insects that feed on milkweed, it apparently sequesters very little if any of the toxic cardenolides (cardio-active steroids) found in milkweed tissue and sap. Before feeding it will bite through the mid-rib of a leaf, bleeding it of much of its toxic sap. On the other hand, the female applies a protective coating of feces to its eggs that contains high concentrations of the toxin.

The genus *Glyptoscelis* ranges from Argentina and Chile to southern Canada, with most species occurring north of Mexico. They are robust, reddish brown, black, or bronzy beetles covered with coarse white, brown, or yellow hairs that lie flat on the body. Some have faint spots or stripes on the elytra, created by hairy patches with different densities. *Glyptoscelis* feed on several species of woody trees and shrubs, including conifers, but are not considered to be important pests. Fifteen species are found in California. The Grape Bud Beetle *(G. squamulata)* (6.5 to 10.0 mm) is covered in broad, whitish or pale brownish scales, giving it a snowy appearance. It is widespread in California and has also been found in Arizona, Nevada, and Utah. It is found on California buckwheat *(Eriogonum fasciculatum)*, parish mallow *(Sphaeralcea ambigua* var. *rosacea)*, and willow, as well as on commercially grown grapes, oranges, and peaches. *Glyptoscelis albida* (6.0 to 10.5 mm) (pl. 281) is grayish white and is known from the Sierra Nevada and the Transverse, Peninsular, and southern Coast Ranges. It has been found on California buckwheat and chamise *(Adenostoma fasciculatum)*.

Cryptocephalus beetles are small, robust, and almost perfectly cylindrical, with threadlike antennae. Little is known of their biology, but most larvae probably live in plant litter and are covered with a case made with bits of leaves and their own fecal material. Adults feed on a wide variety of plants, including some crop and garden plants, but none are known to be important pests. Twelve species occur in California. The Red and Yellow Leaf Beetle *(C. castaneus)* (3.9 to 5.3 mm) (pl. 282) is common in the interior valleys, where it feeds on strawberry, blackberry *(Rubus)*, willow, and alfalfa *(Medicago sativa)*. Its markings vary, with the pronotum ranging from red to dull orange with four black to red to dark orange bright stripes. The elytra are creamy yellow to orange with variable markings of black, dark red, or orange.

Members of the genus *Saxinis* are metallic blue or green and typically bear a bright red spot on the shoulder of each elytron. The small, yet chunky, metallic blue Red-shouldered Leaf Beetle *(S. saucia)* (4.0 to 5.0 mm) (pl. 283) and its subspecies are common in late spring and summer in California and range throughout western North America. They feed on wild buckwheat *(Eriogonum)*, toyon *(Heteromeles arbutifolia)*, and ceanothus *(Ceanothus)* and occasionally attack the young buds of apricots, almonds, plums, and other nursery stock. The delicate yellow eggs are laid on old flower buds and are covered with wax, resembling brownish seeds. The case-bearing larvae may inhabit the nest of ants. Two additional species of *Saxinis* occur in California.

The three species of *Bruchus* in California occur throughout the United States and were all accidentally introduced from Europe. The Pea Weevil *(B. pisorum)* (4.0 to 5.0 mm) (pl. 284) is black or brownish with a distinct spot of white scales on the pronotum. The elytra have small, scattered, white and gray patches and a zigzag band just past the middle. The exposed abdominal segment is white with two black spots. The lower margin of the hind femora has a single, large tooth. The adult mostly overwinters with peas in storage, but also in the field. It emerges in spring as the peas come into bloom and feeds on pollen and nectar. Although several eggs are laid on each pea seed, only one matures to reach adulthood. The Broadbean Weevil *(B. rufimanus)* (3.0 to 4.0 mm) is similar to the Pea Weevil but is smaller and the exposed abdominal segment is black. Both species produce only one generation per year. The Vetch Bruchid *(B. brachialis)* (2.5 to 3.0 mm) is black with a mottled white pattern. The male's antennae are entirely yellow, but only the first five and last segment are so in the female.

Acanthoscelides is the largest genus of sand beetles in California with nine species. Each hind femur has a large tooth on the margin underneath, followed by three to four smaller teeth. The Bean Weevil *(A. obtectus)* (2.5 to 3.5 mm) (pl. 285) is originally from the New World and has become a major pest of leguminous crops worldwide. The elytra are velvety gray or brown with pale stripes. The tips of their antennae are red, while the remaining segments are dark brown. Most of the abdomen, tips of the elytra, and legs are reddish yellow. Two or three small teeth follow the large tooth on the underside of the hind femur. The female lays its eggs on the seedpods of beans, lentils, and peas in the field. The larva is cannibalistic: only one adult emerges from each infested seed. Storage of leguminous products under warm, damp conditions encourages reinfestation, and under the right conditions, Bean Weevils will reproduce continually.

COLLECTING METHODS: Some adult leaf beetles may be collected by hand as they feed on leaves and flowers of weeds and bushes, but sweeping and beating is more efficient for collecting specimens. They are more abundant in open and weedy areas than in forested habitats. To avoid damaging host plants, specimens feeding in agricultural fields and gardens should be collected by hand. Adult seed beetles are sometimes common on vegetation and flowers. Beating and sweeping is the most efficient means of collecting these species. Others are reared in captivity by collecting seeds and seedpods and placing them in a secure container until the adults emerge.

LEAF-ROLLING WEEVILS   **Attelabidae**

Leaf-rolling weevils are so named because of the egg-laying habits of some females. First they lay one or more eggs on a leaf. Then they bite and cut the leaf with their mandibles and use their enlarged front legs to roll up the leaf around the eggs. The rolled-up leaf resembles a barrel and serves as both food and shelter for the developing larvae.

Although the common name "leaf roller" is applied to the entire family, none of the California species are known to engage in this activity. In fact, all of the state's leaf-rolling weevils belong to a group known as "tooth-nosed snout beetles," a reference to the toothed inner and outer margins of the mandibles.

Females of some species lay their eggs on oaks *(Quercus)*. Depending on the species, they lay their eggs on the terminal buds or on new leaf growth of oaks. Some deposit their eggs in sections of leaves they have cut with their mandibles. Another species develops in the fruit of desert peach *(Prunus andersonii)*. The leaf-cutting and egg-laying activities of most California leaf-rolling weevils rarely harm the plant, and the beetles are seldom of any economic importance. However, rose curculios *(Merhynchites)* sometimes damage cultivated blackberries, raspberries, and roses.

Of the 44 species of leaf-rolling weevils currently recognized in the United States and Canada, the larval development of only 12 species has been documented in the literature.

IDENTIFICATION: California leaf-rolling weevils are small (5.5 mm or less) and more or less elongate and covered with short, stiff hairs, but not scales. They are quite variable in color, ranging from reddish, green or metallic blue, to black with metallic reflections. The head is usually wider at the base, almost triangular in shape. The 11-segmented antennae are not elbowed and have a loose, three-segmented club. The mouthparts are located on the tip of a long, nearly straight snout. The pronotum is distinctly narrower than the base of the elytra. The minute scutellum is visible. The elytra together are nearly as wide as long and cover the entire abdomen. The legs are all equally developed, with a tarsal formula that appears 4-4-4 but is 5-5-5; the fourth segment is small and partially hidden within the heart-shaped third segment. Claws are equal in size with tooth- or clawlike processes at their bases. Abdominal segments are progressively smaller toward the rear, with sutures distinct except between the first and second segments, which are fused together.

The mature larvae of *Merhynchites* are whitish, stout, C shaped, and grublike. The head is longer than wide and partially withdrawn into the thorax. It has two-segmented antennae and one pair of simple eyes. The legs are absent. The abdomen is 10-segmented, without projections.

SIMILAR CALIFORNIA FAMILIES:

- pine flower snout beetles (Nemonychidae)—rare, associated with male pollen-bearing flowers of pines; antennae straight with three-segmented club; beak with a distinct labrum
- straight-snouted weevils (Brentidae)—antennae straight with three-segmented club; body outline usually pear shaped; long, cylindrical trochanter with femur attached at tip is unique
- fungus weevils (Anthribidae)—beak broad; antennae not elbowed, with club faint or absent in species with long antennae; abdomen exposed beyond tips of elytra
- weevils or snout beetles (Curculionidae)—antennae elbowed with a compact one- or three-segmented club

CALIFORNIA FAUNA: 18 species in six genera.

Of the four species of *Merhynchites* known in North America, two occur in California. The Eastern Rose Curculio *(M. bicolor)* (4.5 to 5.5 mm) and Western Rose Curculio *(M. wickhami)* (4.5 to 5.5 mm) (pl. 286) are known as rose weevils. Both species are found throughout the state, especially in the mountains, but not in the deserts. They are very similar in appearance. The pronotum and elytra of both species are red. The head of *M. bicolor* is usually red, whereas that of *M. wickhami* is usually black, but this characteristic is not always reliable. In dorsal view, the eyes of *M. bicolor* are bulging, the front surface almost forming a right angle with the snout. By contrast, the eyes of *M. wickhami* are less bulging, forming less than a 45 degree angle with the snout. The elytra of *M. bicolor* are moderately grooved and smooth between grooves, but those of *M. wickhami* are weakly grooved and appear faintly rough between grooves. They also differ in their biology. The larva of *M. bicolor* develops in rose hips, whereas that of *M. wickhami* prefers the flower buds. Although both species are found most frequently on wild roses *(Rosa)*, they are sometimes considered minor pests when their feeding activities extend to cultivated roses. Adults puncture rose buds, preventing the blooms from developing, while the larvae develop in the hips or flower buds. They also attack the fruit of blackberries and raspberries *(Rubus)*, also members of the rose family.

COLLECTING METHODS: Look for bright red rose curculios on the flowers of blackberries, raspberries, and roses in summer. In other genera *(Deporaus, Haplorhynchites,* and *Rhynchites),* the first generation of adults emerges in spring and may be collected by inspecting sunflowers, or beating and sweeping the flowers or developing leaves and fruits of oaks, manzanitas *(Arctostaphylos),* and desert peach.

WEEVILS or SNOUT BEETLES — Curculionidae

The weevils, also called snout beetles, compose the largest family of animals, with over 60,000 species known worldwide. Most weevils are immediately recognizable by their short, broad beaks or their long, outstretched snouts and elbowed antennae. Either way, weevils are well equipped to chew their way through the leathery skins of seedpods or the tough shells of nuts. In contrast, the mouthparts of the wood-boring ambrosia beetles and bark beetles (Scolytinae, Curculionidae) are not extended at all. Weevils are generally hard bodied, and although most are good fliers, a number of species are wingless or nearly so and are incapable of flight. A few groups are capable of reproducing without mating, a cloning process called parthenogenesis. In many populations of parthenogenetic weevils, males are entirely unknown.

Most weevils feed on living plants and are associated with virtually every kind of aquatic or terrestrial plant and their various parts. Many larvae burrow into the stems of plants, whereas others feed on sick or healthy trees. A few species feed on leaves, living or dead, mining inner tissues or grazing on their outer surfaces. The activities of a few root-feeding species may produce galls, characteristically shaped swellings on the roots. Some adults attack fruits and nuts, but most feed on leaves, pollen, flowers, or fungi or burrow into wood. Adults may or may not feed on the same host as the larvae. The feeding preferences of many species, such as the Rice Weevil *(Sitophilus oryzae),* Granary Weevil *(S. granarius),* Boll Weevil *(Anthonomus grandis),* and numerous bark beetles (e.g., species of *Dendroctonus* and *Ips),* make them important pantry, garden, agricultural, and forest pests.

The majority of weevils are associated with flowering plants, but some are associated with conifers, especially pines *(Pinus).* Adult and larval generalists consume a wide variety of plants, whereas specialists feed on a single genus or family of plants.

Most larval generalists burrow through the soil and feed externally on roots, while the adults eat leaves. Pupation takes place in the soil. Specialist larvae usually feed internally in the stems, roots, leaves, or reproductive structures of one or more closely related species of plants. However, the larvae of the Alfalfa Weevil *(Hypera postica)* and its relatives prefer to feed externally on the leaves, flowers, or seedpods of their host. Specialists usually pupate inside the tissues of the host plant or in the soil near the base of the plant, but *Hypera* constructs loosely woven cocoons attached to the host plant.

A great number of weevils associated with ornamental plants were introduced into North America from Europe in imported plants and stored products, or in ballast brought by ships at the turn of the nineteenth century. Weevils infesting agricultural products are routinely intercepted at ports of entry into the United States and Canada. Others were deliberately introduced for biological control purposes, especially for the control of invasive weeds in western North American grasslands and southeastern wetlands. Exotic weevils have been introduced into California to control numerous invasive plants, including Spanish broom *(Cytisus multiflorus)*, purple loosestrife *(Lythrum salicaria)*, gorse *(Ulex europaea)*, knapweed and starthistle *(Centaurea)*, and thistle *(Cirsium)*.

IDENTIFICATION: Adult weevils are broadly oval, long, and cylindrical, to strongly humpbacked, ranging from a few millimeters to 25.0 mm in length. Most species are scaled, with varied patterns on a black, brown, or gray background, but a few species may be whitish, gray, green, red, or with metallic inflections. The head is hypognathous, often with the mouthparts extended into a short, broad beak or stretched into a long, thin, and curved snout. The eyes are present, reduced, or very rarely absent. The 11-segmented antennae are elbowed with a compact, three-segmented club. The pronotum is slightly wider than the head. The scutellum is small or hidden. The elytra are rounded or parallel sided, bare or scaled, and almost or completely conceal the abdomen. The tarsal formula is usually 5-5-5, rarely 4-4-4 or appearing so, with segment four very small and hidden within the lobes of the larger heart-shaped third segment. The claws are nearly always equal in size and simple. The abdomen has five segments visible from below.

The mature larvae are plump, whitish, yellowish, greenish, or gray and are nearly cylindrical, slightly curved, or C-shaped grubs. They are usually covered with fine hairs. The usually distinct head is hypognathous and is seldom retracted inside the prothorax. The antennae are one- or two-segmented. The simple eyes are usually absent. The legs are absent. The 10-segmented abdomen usually has three or four distinct folds or wrinkles on the back and lacks projections.

SIMILAR CALIFORNIA FAMILIES:

- a narrow-waisted bark beetle (*Rhinosimus viridiaeneus*, Salpingidae) — rare, associated with dead wood; antennae straight; body with a distinct green metallic sheen
- pine flower snout beetles (Nemonychidae) — rare, associated with male pollen-bearing flowers of pines; antennae straight (not elbowed) with three-segmented club; beak with a distinct labrum
- straight-snouted weevils (Brentidae) — antennae straight with three-segmented club; body outline usually pear shaped; long, cylindrical trochanter with femur attached at tip is unique
- some leaf-rolling weevils (Attelabidae) — antennae straight with loose three-segmented club; elytra nearly as wide as long, covering the abdomen; body never covered with scales; claws lobed or toothed at base
- fungus weevils (Anthribidae) — beak broad; antennae not elbowed, with club faint or absent in species with long antennae; abdomen exposed beyond tips of elytra

CALIFORNIA FAUNA: Approximately 570 species in 154 genera.

The black and somewhat flattened *Yuccaborus frontalis* (10.0 to 13.0 mm) breeds in Joshua trees *(Yucca brevifolia)* in the Mojave Desert. The wing covers are deeply grooved and slightly shorter than the abdomen. The larva feeds in galleries chewed between the bark and the interior of the tree. The adult is sometimes attracted to lights. It is the only species in the genus in California.

The Rice Weevil *(Sitophilus oryzae)* (pl. 287) and its close relative *S. zeamais* are similar in size (2.0 to 3.5 mm) and are nearly cosmopolitan. They are pests of stored cereals, especially rice. They have a snout of moderate length, a dull, reddish brown body that is coarse in texture, and are capable of flight. The elytra often have four reddish yellow spots. The flightless Granary Weevil *(S. granarius)*, is similar in size and appearance but is usually dark brown or black overall, has oval rather than round pits on the prothorax, and also infests wheat and barley products. These species apparently originated in Southeast Asia.

Both the adult and larva of the Cactus Weevil *(Metamasius* [previously known as *Cactophagus*] *spinolae validus)* (15.0 to 25.0 mm) (pl. 288) are associated with beavertail cactus *(Opuntia basilaris)* in the coastal plains of southern California, especially in San Diego County. It may also occur among the few patches of saguaros *(Carnegiea gigantea)* and other cacti growing in the southeastern Colorado Desert along the Colorado River. This large black weevil with distinctly grooved wing covers is also found in Arizona, Baja California, Sinaloa, and Sonora. It is the largest weevil in California and the only member of the genus in the state.

Two species of *Rhodobaenus* occur in the United States, but only one is represented in California. The red and black *R. tredecimpunctatus* (6.0 to 10.0 mm), sometimes known as the Cocklebur Billbug, is broadly distributed in eastern Canada, across the United States, and in northern Mexico. The pronotum is red with a round, black spot in the middle and two similar spots on either side. The elytra are red with a total of eight black spots. The adult is found on the blooms of sunflowers, while the larva mines the stems. The adult emerges from its pupa in late summer and overwinters.

The genus *Scyphophorus* (8.0 to 24.0 mm) includes only two species, both of which occur in California. The Yucca Weevil *(S. yuccae)* (pl. 289) is a large, black weevil with flattened and distinctly grooved wing covers. It is commonly found clambering among the spines and flower stalks of blooming yuccas, especially chaparral yucca *(Yucca whipplei)*. It also occurs in Arizona, Texas, and Baja California. A second species, the Sisal Weevil *(S. acupunctatus),* is very similar in appearance. It attacks native and exotic agaves *(Agave)*. It is widespread in the Western Hemisphere and also occurs in portions of Hawai'i, Borneo, Java, Australia, Kenya, and Tanzania, where it feeds on introduced species of agave. Adult *Scyphophorus* feed on the sap of living plants, while the larvae bore into the pithy centers of the flower stalks and crowns.

The large and distinctive Tule Billbug *(Sphenophorus aequalis)* (9.0 to 21.0 mm) is black, with the pronotum and elytra distinctively striped, to varying degrees, with a shining white, yellow, or grayish enamel-like coating. The underside is mostly whitish or tan. It appears in early spring and uses its long snout or "bill" to sever the heads of growing stalks of tule *(Scirpus)* at ground level, killing the top. Three subspecies live in stands of tule growing along permanent rivers, streams, and other wetlands of the northern coast and Sacramento Valley *(S. a. discolor)*, the southern coastal plains and Colorado Desert *(S. a. picta)* (pl. 290), and select springs in the Great Basin and Mojave Desert *(S. a. ochrea)*. The larvae feed on *Scirpus* roots, but the adults attack grasses and sometimes become pests in adjacent grain fields. Many other species of *Sphenophorus* occur in California.

The genus *Curculio* contains three species in California, all of which are strongly rounded and have a long and curved snout, nearly twice the length of the body. The California Acorn Weevil *(C. occidentis* [formerly known as *C. uniformis])* (5.0 to 10.0 mm) (pl. 291), also called the California Filbert Weevil, is a dark reddish brown beetle clothed with yellow scales and patches of yellowish brown scales on the elytra. This species is widespread in the state and is found on live and deciduous oaks *(Quercus)* in the mountains and foothills. In the Pacific Northwest it is a pest of filberts.

Both the adult and the larva of the Eucalyptus Snout Beetle *(Gonipterus scutellatus)* (6.0 to 10.0 mm) cause extensive damage to blue gum *(Eucalyptus globulus)* by defoliating entire sections of trees planted as windbreaks. The adult feeds along the leaf margin, leaving in its wake a distinctive series of irregular notches. The larva is a leaf miner. This species was first detected in Ventura County in 1994. The adult is dark to orange brown and has a short, broad snout. When viewed from above it has rounded projections on either side located just below the base of the wing cover. The larva is sluglike and slimy in appearance and produces chains of stringy frass that cling to its body. The young larva is yellowish with small black dots, whereas a mature individual is yellow green with numerous small black dots and a pair of dark stripes running the length of the body. This beetle's distribution appears limited to Ventura, Los Angeles, and Santa Barbara Counties.

The genus *Listroderes* is native to South America, where populations include both males and females. In the United States, accidentally introduced populations of *Listroderes* consist entirely of females that reproduce parthenogenetically. The Vegetable

Weevil *(L. costirostris)* (6.4 to 10.0 mm) (pl. 292) was inadvertently imported from Brazil in the 1920s and has become established in many other parts of the world. Both the adult and larva feed on the foliage of numerous garden plants and agricultural crops. The eggs are laid singly on the stems or crowns of food plants, or in the soil. The plump, pale greenish larva is fond of tuberous roots, such as potatoes, and feeds at night. It is sometimes a pest on dichondra *(Dichondra repens)* lawns. The brown adult is also active at night during summer. It is easily distinguished from other California weevils by a V-shaped marking across the lower portion of the elytra. The adult is active throughout the year and is common in southern California, Nevada, and Arizona. The closely related *L. difficilis* is sometimes confused with *L. costirostris*. However, it is found throughout southeastern United States to Arizona and does not occur in California. It has numerous round, metallic scales mixed with the coarse setae on the underside of its thorax (metasternum), whereas *L. costirostris* lacks metallic scales.

Of the 64 species of *Trigonoscuta* described from western United States and adjacent southern Canada, 60 occur in California. Some entomologists question whether these are all valid species. They are associated with coastal and desert dune habitats, where they feed on various plants. Coastal species are sometimes common among the roots of beach grasses. *Trigonoscuta* are black or dark red, with a slightly paler underside and antennae. They are clothed above with brownish and whitish, oval scales intermixed with white, brown, or blackish scales. *Trigonoscuta morroensis* (6.0 to 7.6 mm) (pl. 293) was described from the coastal dunes in and around Morro Bay.

Fuller's Rose Weevil *(Naupactus* [formerly known as *Asynonychus] godmanni)* (5.0 to 9.0 mm) (pl. 294) is originally from South America. This nocturnal species is active from late spring through early winter throughout most of California. It attacks numerous weeds, cultivated flowers, and vegetables and is occasionally a pest in citrus groves. Eggs are glued in small batches to the food plants, but the legless grub feeds underground on roots. This short-nosed, gray brown weevil is often found resting on the walls of buildings or hiding beneath bark or leaves during the day. An introduced Argentinean species, the White Fringed Beetle *(N.* [formerly *Graphognathus] leucoloma)* (11.0 to 13.0 mm), is widespread throughout the southeastern United States.

A small, isolated population of this conspicuous gray beetle with broad, white stripes on either side of its body was discovered in the Chino Hills area of San Bernardino County in 1984. The male is unknown in North America, but the female is capable of reproducing parthenogenetically.

Ophryastes is a genus of large, whitish to pale gray weevils with brownish or blackish stripes that live in dry habitats throughout the western United States and adjacent southern Canada, with six species in California. Adults are flightless and feign death when disturbed. *Ophryastes desertus* (11.0 to 23.0 mm) (pl. 295) is known from the Colorado and Mojave Deserts, where it has been found on saltbush *(Atriplex)* and pluchea *(Pluchea)* in spring and summer. It is sometimes seen crawling on the ground in the early morning hours. Another species, *O. geminatus* (5.5 to 19.20 mm), is found primarily in the Great Basin and Mojave Desert on creosote bush *(Larrea tridentata)*, dalea *(Dalea)*, buckwheat *(Eriogonum)*, rabbitbrush *(Chrysothamnus)*, and snakeweed *(Gutierrezia)* in late winter.

Five species of *Otiorhyncus* (formerly known as *Brachyrhinus*) occur in California, and all were inadvertently introduced from Europe. They are generalists and feed on many garden and crop plant species. Their larvae feed externally in the soil on roots, whereas the adults tend to feed on fresh foliage, flowers, or buds. They have short, broad snouts that are notched in the middle and lobed at the sides. *Otiorhyncus cribricollis* (7.0 to 8.0 mm) was first found in Los Angeles County in 1927. It now occurs throughout coastal California and the Great Central Valley, as well as Arizona, Nevada, New Mexico, Texas, and Baja California. It is distinguished from all other *Otiorhynchus* in the state by having the tips of its front tibiae expanded all the way around, rather than rounded on one side. The following two species lack teeth on their femora. *Otiorhynchus meridionalis* (7.0 to 10.0 mm) (pl. 296) was first reported in San Jose in 1931 and has since become widespread in the coastal and central regions of the state. Its snout is almost completely without scales and has a ridge running down the middle. *Otiorhynchus rugostriatus* (6.0 to 8.0 mm) is found on both the east and west coasts. Its distribution in the state is similar to that of the other species in the genus. It lacks the ridge on its snout and has double rows of curled, reddish hairs on the elytra. The first femora of the remaining two species are toothed. The Black Vine Weevil *(O. sulcatus)* (6.0 to 10.0 mm) is covered with patches of yellow scales and sometimes has reddish legs. It is a pest of gardens, crops, and citrus groves. The white or pinkish larva feeds on roots, causing most of the damage. The adult is found at the base of strawberry, blackberry, azaleas, and other ornamentals. It occurs primarily in the Pacific Northwest and northeastern United States. In California it is common in the San Francisco Bay region and the southern coastal plains. The tooth of the front femur is simple, not notched. The Strawberry Root Weevil *(O. ovatus)* (5.0 to 6.2 mm) is reddish brown with a distinctively grooved pronotum, and the front femoral teeth are notched. The larva attacks strawberry plants underground, consuming the root and crown. The flightless adult feeds on the leaves and attacks numerous plants, including strawberries, raspberries, and beets. The male is unknown in North America, and the female reproduces by parthenogenesis. This weevil is found throughout southern Canada, the northern United States, and along coastal California. It seems to frequently turn up in houses, usually in the sink or bathtub.

Of the 17 species of *Hypera* found in North America, six are introduced from the Old World. Both adults and larvae feed on the foliage of various members of the pea and buckwheat families. Three species occur in California. The reddish black, brown, brownish black, or black Alfalfa Weevil *(H. postica)* (3.0 to 5.1 mm) is the most common weevil on alfalfa *(Medicago sativa)* and burclover (e.g., *M. arabica*, *M. polymorpha*). The legs and antennae are always paler than the rest of the body. The body is covered with notched scales that range from ashy gray to brown. This weevil reduces crop yield and curbs seed production by damaging young alfalfa plants early in the season. The Clover Leaf Weevil *(H. punctata)* (5.0 to 10.0 mm) has a brown, scaly body and a short snout. It has long been established in clover *(Melilotus)* and alfalfa throughout the United States and Canada. Both the adult and the larva devour the leaves of their host plants. Pupation takes place in a loosely woven cocoon attached to the food plant. Both species were accidentally introduced from Europe.

The species of *Apleurus* are long and slender to somewhat robust, black weevils covered with dense patches or stripes of scales. The prothorax has a distinct bump on each side toward the front. Four species are known to occur in California. *Apleurus albovestitus* (11.8 to 21.4 mm) (pl. 297) is widespread in the arid regions of central, southern, and eastern California. It is found on a variety of plants, especially the asters. *Apleurus jacobinus* (5.8 to 14.3 mm) occurs from the San Francisco Bay regions south along the central coast, the arid regions of the Great Central Valley, and the deserts of southern California and Baja California. The adult is found throughout the year on a variety of plants, including tarweed *(Hemizonia)*.

Members of the genus *Dendroctonus* appear to have the elytra cut off and lack any spines at the tip. They typically seek out weak, dead, or dying trees on which to lay their eggs, but a few will attack healthy trees. *Dendroctonus* beetles work in pairs to bore through bark and carve the brood chamber between the bark and wood. The egg galleries are always packed with wood dust, except for the portion where the beetles are working. Six species are known in California. The largest species is the Red Turpentine Beetle *(D. valens)* (5.7 to 9.5 mm) (pl. 298). This reddish brown beetle attacks all species of pine found within its range. The stout Mountain Pine Beetle *(D. ponderosae)* (4.0 to 7.5 mm) is dark brown to black. It is the most destructive species of bark beetle in western North America, attacking lodgepole *(Pinus contorta)*, ponderosa *(P. ponderosa)*, western white *(P. monticola)*, and sugar *(P. lambertiana)* pines. The Western Pine Beetle *(D. brevicomis)* (3.0 to 5.0 mm) is dark brown and causes extensive mortality of ponderosa pine throughout western United States and Canada. The Jeffrey Pine Beetle *(D. jeffreyi)* (8.0 mm) is similar to the mountain pine beetle, but feeds only on Jeffrey pine *(P. jeffreyi)*. It can be a serious pest in old growth stands and timber-producing areas of northeastern California. It normally breeds in scattered mature trees, but it will also attack lightning-struck trees and trees recently toppled by high winds. During outbreaks this beetle can kill groups of 20 to 30 trees regardless of age or health. It is sometimes found with the California Flatheaded Borer *(Phaenops californica,* Buprestidae) or the Pine Engraver *(Ips pini,* Curculionidae). The Douglas-fir Beetle *(D. pseudotsugae)* (4.4 to 7.0 mm) is a hairy, dark brown to black species with reddish elytra.

Twelve species of *Scolytus* are recorded in California. Their abdomen appears "sawed off." The Fir Engraver *(S. ventralis)* (4.0 mm) (pl. 299) is one of the largest species in the genus. The second abdominal plate of the male has a sharp tubercle that is absent in the female. It is a major pest of firs *(Abies)*, regularly killing hundreds of thousands of trees annually in California, with periodic outbreaks killing many more. Trees infected with root-rot fungus or defoliated by the Douglas-fir Tussock Moth *(Orgyia pseudotsugata)* are likely to be attacked. This beetle inoculates its brood chambers with a brown-staining fungus to assure successful development of the brood, which feeds on fungus-infested wood. The egg galleries are excavated in the inner bark and are cut deeply and transversely across the grain of the wood up to 15 cm away from the central chamber. One or two generations are produced each year.

Engraver beetles of the genus *Ips* attack mostly pines and spruces *(Picea)*. The elytra appear cut off and are tipped with spines around the edges. The reddish brown Pine Engraver *(I. pini)* (3.5 to 4.2 mm) (pl. 300) is found across North America. One of the most common of the bark beetles, this species can become a serious pest in almost any species of pine in California, especially ponderosa, Jeffrey, and lodgepole pines. It normally attacks downed wood or limbs already killed by *Dendroctonus,* but it frequently develops large populations that attack healthy, living trees. The California Five-spined Ips *(I. paraconfusus)* (4.0 to 4.5 mm) attacks all species of pines in its range, especially ponderosa and sugar pines. This reddish brown to black species prefers smaller trees but will infest the tops of larger individuals. The Emarginate Ips *(I. emarginatus)* (5.5 to 7.0 mm) is commonly associated with the Mountain Pine Beetle and Jeffrey Pine Beetle, attacking ponderosa, lodgepole, and western white pines. The shiny dark brown adult engraves long, parallel egg galleries that connect at various points. The two- to four-foot long galleries run up and down the length of the tree. The state has 15 species of *Ips.*

The California Oak Ambrosia Beetle *(Monarthrum scutellare)* (3.5 to 4.1 mm) is small, slender bodied, cylindrical, and dark brown and attacks coast live oaks *(Quercus agrifolia)*. The male takes about three to four weeks to completely excavate the main gallery and nuptial chamber under the bark. The female bores up to four secondary galleries leading away from the nuptial chamber, sometimes with the assistance of the male. Mating occurs when the galleries have been completed. The female then carves niches along the sides of the galleries into which she lays 50 or more eggs. Three species of *Monarthrum* occur in the state.

COLLECTING METHODS: Weevil diversity is greatest in areas where there is plenty of moisture. Desert species are often active during the late winter and early spring months. Developing foliage, especially plants in bloom or bearing fruit, are especially attractive to many weevils. Beating and sweeping a wide variety of plants during the day and especially after dark will increase the diversity of your catch. A few species come to lights. Adults often play dead when disturbed and may go unnoticed. Ambrosia and bark beetles are often collected with a net, by hand, or at lights after they have emerged from the host tree in spring and summer and are in search of a new host. Look for piles of fine sawdust in bark crevices or on the ground under a dead or dying tree, and peel back the bark to search for adults. Rearing ambrosia and bark beetles from infested wood is the most productive method of collection, especially those species developing in cones, twigs, and branches. Freshly killed trees are more attractive to beetles than those that have been dead for a long time.

CHECKLIST OF NORTH AMERICAN BEETLE FAMILIES

Families with * are not known to be established in California.

Suborder Archostemmata

- [ ] Reticulated beetles (Cupedidae)
- [ ] Telephone-pole beetles (Micromalthidae)*

Suborder Myxophaga

- [ ] Minute bog beetles (Microsporidae)
- [ ] Skiff beetles (Hydroscaphidae)

Suborder Adephaga

- [ ] Wrinkled bark beetles (Rhysodidae)
- [ ] Ground beetles and tiger beetles (Carabidae)
- [ ] Whirligig beetles (Gyrinidae)
- [ ] Crawling water beetles (Haliplidae)
- [ ] False ground beetles (Trachypachidae)
- [ ] Burrowing water beetles (Noteridae)
- [ ] Trout-stream beetles (Amphizoidae)
- [ ] Predaceous diving beetles (Dytiscidae)

Suborder Polyphaga
Series Staphyliniformia
Superfamily Hydrophiloidea

- [ ] Water scavenger beetles (Hydrophilidae)
- [ ] False clown beetles (Sphaeritidae)
- [ ] Clown beetles (Histeridae)

Superfamily Staphylinoidea

- ☐ Minute moss beetles (Hydraenidae)
- ☐ Featherwing beetles (Ptiliidae)
- ☐ Primitive carrion beetles (Agyrtidae)
- ☐ Round fungus beetles (Leiodidae)
- ☐ Antlike stone beetles (Scydmaenidae)
- ☐ Carrion beetles (Silphidae)
- ☐ Rove beetles (Staphylinidae)

Series Scarabaeiformia
Superfamily Scarabaeoidea

- ☐ Stag beetles (Lucanidae)
- ☐ False stag beetles (Diphyllostomatidae)
- ☐ Bess beetles (Passalidae)*
- ☐ Enigmatic scarab beetles (Glaresidae)
- ☐ Hide beetles (Trogidae)
- ☐ Rain beetles (Pleocomidae)
- ☐ Earth-boring scarab beetles (Geotrupidae)
- ☐ Sand-loving scarab beetles (Ochodaeidae)
- ☐ Scavenger scarab beetles (Hybosoridae)
- ☐ Pill scarab beetles (Ceratocanthidae)*
- ☐ Bumblebee scarab beetles (Glaphyridae)
- ☐ Scarab beetles (Scarabaeidae)

Series Elateriformia
Superfamily Scirtoidea

- ☐ Plate-thigh beetles (Eucinetidae)
- ☐ Minute beetles (Clambidae)
- ☐ Marsh beetles (Scirtidae)

Superfamily Dascilloidea

- ☐ Soft-bodied plant beetles (Dascillidae)
- ☐ Cedar beetles or cicada parasite beetles (Rhipiceridae)

Superfamily Buprestoidea

- ☐ Schizopodid beetles or false jewel beetles (Schizopodidae)
- ☐ Metallic wood-boring beetles or jewel beetles (Buprestidae)

Superfamily Byrroidea

- ☐ Pill beetles or moss beetles (Byrrhidae)
- ☐ Riffle beetles (Elmidae)
- ☐ Long-toed water beetles (Dryopidae)
- ☐ Travertine beetles (Lutrochidae)*
- ☐ Minute marsh-loving beetles (Limnichidae)
- ☐ Variegated mud-loving beetles (Heteroceridae)
- ☐ Water penny beetles (Psephenidae)
- ☐ Ptilodactylid beetles (Ptilodactylidae)
- ☐ Chelonariid beetles (Chelonariidae)*
- ☐ Forest stream beetles (Eulichadidae)
- ☐ Callirhipid beetles (Callirhipidae)*

Superfamily Elateroidea

- ☐ Artematopodid beetles (Artematopodidae)
- ☐ Texas beetles (Brachypsectridae)
- ☐ Rare click beetles (Cerophytidae)
- ☐ False click beetles (Eucnemidae)
- ☐ Throscid beetles (Throscidae)
- ☐ Click beetles (Elateridae)
- ☐ Net-winged beetles (Lycidae)
- ☐ Long-lipped beetles (Telegeusidae)*
- ☐ Glowworms (Phengodidae)
- ☐ Fireflies, lightning bugs, and glowworms (Lampyridae)
- ☐ False soldier beetles and false firefly beetles (Omethidae)
- ☐ Soldier beetles (Cantharidae)

Series Bostrichioformia
Superfamily Derodontoidea

- ☐ Jacobsoniid beetle (Jacobsoniidae)*
- ☐ Toothed-neck fungus beetles (Derodontidae)

Superfamily Bostrichoidea (includes Dermestoidea)

- ☐ Wounded-tree beetles (Nosodendridae)
- ☐ Skin beetles (Dermestidae)
- ☐ Bostrichid beetles (Bostrichidae)
- ☐ Deathwatch beetles and spider beetles (Anobiidae)

Series Cucujiformia
Superfamily Lymexyloniodea

- ☐ Ship-timber beetles (Lymexylidae)

Superfamily Cleroidea

- ☐ Bark-gnawing beetles (Trogossitidae)
- ☐ Checkered beetles (Cleridae)
- ☐ Soft-winged flower beetles (Melyridae)

Superfamily Cucujoidea (formerly Clavicornia)

- ☐ Cryptic slime mold beetles (Sphindidae)
- ☐ Short-winged flower beetles (Kateretidae)
- ☐ Sap beetles (Nitidulidae)
- ☐ Palmetto beetles (Smicripidae)
- ☐ Root-eating beetles (Monotomidae)
- ☐ Silvanid flat bark beetles (Silvanidae)
- ☐ Parasitic flat bark beetles (Passandridae)*
- ☐ Flat bark beetles (Cucujidae)
- ☐ Lined flat bark beetles (Laemophloeidae)
- ☐ Shining flower beetles and shining mold beetles (Phalacridae)
- ☐ Silken fungus beetles (Cryptophagidae)
- ☐ Pleasing fungus beetles (Erotylidae)
- ☐ Fruitworms (Byturidae)
- ☐ False skin beetles (Biphyllidae)
- ☐ Bothriderid beetles (Bothrideridae)
- ☐ Minute bark beetles (Cerylonidae)
- ☐ Handsome fungus beetles (Endomychidae)
- ☐ Lady beetles (Coccinellidae)
- ☐ Minute hooded beetles, minute fungus beetles, and hooded beetles (Corylophidae)
- ☐ Minute brown scavenger beetles (Latridiidae)

Superfamily Tenebrionoidea

- ☐ Hairy fungus beetles (Mycetophagidae)
- ☐ Archaeocryptic beetles (Archeocrypticidae)*
- ☐ Minute tree-fungus beetles (Ciidae)
- ☐ Polypore fungus beetles (Tetratomidae)

- ☐ False darkling beetles (Melandryidae)
- ☐ Tumbling flower beetles (Mordellidae)
- ☐ Ripiphorid beetles (Ripiphoridae)
- ☐ Zopherid beetles (Zopheridae)
- ☐ Darkling beetles (Tenebrionidae)
- ☐ Jugular-horned beetles (Prostomidae)
- ☐ Synchroa bark beetles (Synchroidae)*
- ☐ False blister beetles (Oedemeridae)
- ☐ False longhorn beetles (Stenotrachelidae)
- ☐ Blister beetles (Meloidae)
- ☐ Palm beetles and flower beetles (Mycteridae)
- ☐ Conifer bark beetles (Boridae)
- ☐ Dead-log beetles (Pythidae)
- ☐ Fire-colored beetles (Pyrochroidae)
- ☐ Narrow-waisted bark beetles (Salpingidae)
- ☐ Antlike flower beetles (Anthicidae)
- ☐ Antlike leaf beetles (Aderidae)
- ☐ False flower beetles (Scraptiidae)

Superfamily Chrysomeloidea

- ☐ Longhorn beetles (Cerambycidae)
- ☐ Megalopodid leaf beetles (Megalopodidae)
- ☐ Orsodacnid leaf beetles (Orsodacnidae)
- ☐ Leaf beetles and seed beetles (Chrysomelidae)

Superfamily Curculionoidea

- ☐ Pine flower snout beetles (Nemonychidae)
- ☐ Fungus weevils (Anthribidae)
- ☐ Cycad weevils (Belidae)*
- ☐ Leaf-rolling weevils, thief weevils, and tooth-nosed snout beetles (Attelabidae)
- ☐ Straight-snouted weevils and pear-shaped weevils (Brentidae)
- ☐ New York weevils (Ithyceridae)*
- ☐ Weevils or snout beetles (Curculionidae)

CALIFORNIA'S SENSITIVE BEETLES

The following four species are listed on the Federal Endangered Species list:

Carabidae

Delta Green Ground Beetle *(Elaphrus viridis)*, threatened
Ohlone Tiger Beetle *(Cicindela ohlone)*, endangered

Scarabaeidae

Mount Hermon June Beetle *(Polyphylla barbata)*, endangered

Cerambycidae

Valley Elderberry Longhorn Beetle *(Desmocerus californicus dimorphus)*, threatened

The following species live in sensitive habitats, and many of them have been proposed for listing on the federal endangered species list as threatened or endangered species:

Carabidae

Greenest Tiger Beetle *(Cicindela tranquebarica viridissima)*
Humboldt Ground Beetle *(Scaphinotus longiceps)*
Oblivious Tiger Beetle *(Cicindela latesignata obliviosa)*, extinct?
Sacramento Valley Tiger Beetle *(Cicindela hirticollis abrupta)*
San Joaquin Tiger Beetle (*Cicindela tranquebarica* undescribed subspecies)
Sandy Beach Tiger Beetle *(Cicindela hirticollis gravida)*

Scaphinotus behrensi
Siskiyou Ground Beetle *(Nebria gebleri siskiyouensis)*
South Forks Ground Beetle *(Nebria darlingtoni)*
Trinity Alps Ground Beetle *(Nebria sahlberbii triad)*

Dytiscidae

Curved-footed Hygrotus Diving Beetle *(Hygrotus curvipes)*
Death Valley Agabus Diving Beetle *(Agabus rumppi)*
Leech's Skyline Diving Beetle *(Hydroporus leechi)*
Simple Hydroporus Diving Beetle *(Hydroporus simplex)*
Travertine Band-thigh Diving Beetle *(Hygrotus fontinalis)*
Wooly Hydroporus Diving Beetle *(Hydroporus hirsutus)*

Hydrophilidae

Leech's Chaetarthrian Water Scavenger Beetle *(Chaetarthria leechi)*
Ricksecker's Water Scavenger Beetle *(Hydrochara rickseckeri)*

Hydraenidae

Wilber Springs Minute Moss Beetle *(Ochthebius reticulatus)*
Wing-shoulder Minute Moss Beetle *(Ochthebius crassalus)*

Glaresidae

Kelso Dune Glaresis Scarab *(Glaresis arenata)*

Pleocomidae

Santa Cruz Rain Beetle *(Pleocoma conjugens conjugens)*

Glaphyridae

Bumblebee Scarab *(Lichnanthe ursina)*
White Sand Bear Scarab *(Lichnanthe albopilosa)*

Scarabaeidae

Andrews' Dune Scarab *(Pseudocotalpa andrewsi)*
Atascadero June Beetle *(Polyphylla nubila)*
Carlson's Dune Beetle *(Anomala carlsoni)*
Ciervo Aegialian Scarab *(Aegialia concinna)*
Death Valley June Beetle *(Polyphylla erratica)*
Delta June Beetle *(Polyphylla stellata)*

Hardy's Dune Beetle *(Anomala hardyorum)*
Saline Valley Snow-front June Beetle *(Polyphylla anteronivea)*
San Clemente Island Coenonycha Beetle *(Coenonycha clementina)*

Elmidae

Brownish Dubiraphian Riffle Beetle *(Dubiraphia brunnescens)*
Giuliani's Dubiraphian Riffle Beetle *(Dubiraphia giulianii)*
Microcylloepus formcoideus
Microcylloepus similis
Pinnacles Optioservus Riffle Beetle *(Optioservus canus)*
Wawona Riffle Beetle *(Atractelmis wawona)*

Eucnemidae

Dohrn's Elegant Eucnemid Beetle *(Paleoxenus dohrni)*

Tenebrionidae

Channel Islands Dune Beetle *(Coelus pacificus)*
Globose Dune Beetle *(Coelus globosus)*
San Joaquin Dune Beetle *(Coelus gracilis)*

Meloidae

Hopping's Blister Beetle *(Lytta hoppingi)*
Moestan Blister Beetle *(Lytta moesta)*
Mojave Desert Blister Beetle *(Lytta insperata)*
Molestan Blister Beetle *(Lytta molesta)*
Morrison's Blister Beetle *(Lytta morrisoni)*

Anthicidae

Antioch Dunes Anthicid *(Anthicus antiochensis)*
Sacramento Anthicid *(Anthicus sacramento)*

Cerambycidae

Rude's Longhorn Beetle *(Necydalis rudei)*

Curculionidae

Blaisdell's Trigonoscuta Weevil *(Trigonoscuta blaisdelli)*
Brown-tassel Trigonoscuta Weevil *(Trigonoscuta brunnotasselata)*

Dorothy's El Segundo Dune Weevil *(Trigonoscuta dorothea dorothea)*
Doyen's Trigonoscuta Dune Weevil *(Trigonoscuta doyeni)*
Lange's El Segundo Dune Weevil *(Onychobaris langei)*
Nelson's Miloderes Weevil *(Miloderes nelsoni)*
Santa Catalina Island Trigonoscuta Weevil *(Trigonoscuta catalina)*
Santa Cruz Island Shore Weevil *(Trigonoscuta stantoni)*

COLLECTIONS, SOCIETIES, AND OTHER RESOURCES

Selected Beetle Collections in California

The collections are listed alphabetically by city.

Essig Museum of Entomology, University of California, Berkeley
    http://essig.berkeley.edu/
Bohart Museum of Entomology, University of California, Davis
    http://bohart.ucdavis.edu/
Entomology Section, Natural History Museum, Los Angeles
    http://www.nhm.org/research/entomology/
Entomology Research Museum, University of California, Riverside
    http://www.entmuseum.ucr.edu/
Entomology Laboratory, California Department of Food and Agriculture, Plant Pest Diagnostics Center, Sacramento
    http://www.cdfa.ca.gov/phpps/ppd/Entomology/Entomology.htm
Entomology Department, San Diego Natural History Museum, San Diego
    http://www.sdnhm.org/research/entomology/
Department of Entomology, California Academy of Sciences, San Francisco
    http://www.calacademy.org/research/entomology/
J. Gordon Edwards Entomology Museum, San Jose State University, San Jose
    http://www.biology.sjsu.edu/entmuseum/
Department of Invertebrate Zoology, Santa Barbara Museum of Natural History, Santa Barbara
    http://www.sbnature.org/collections/invert/entom/cbphomepage.htm

Societies and Websites Promoting the Study of Beetles

The California Beetle Project
    http://www.sbnature.org/collections/invert/entom/cbphome
    page.htm
The Coleopterists' Society
    http://www.coleopsoc.org/
The Lorquin Entomological Society
    http://www.nhm.org/research/entomology/links.html
The Pacific Coast Entomological Society
    http://www.cdfa.ca.gov/phpps/ppd/Entomology/PCES/
The Young Entomologists Society
    http://members.aol.com/YESbugs/bugclub.html

California Sources for Books and Collecting Equipment

Acorn Naturalists
(800) 422-8886
Acorn@aol.com
www.acornnaturalists.com

BioQuip Products
(310) 667-8800
bioquip@aol.com
www.bioquip.com

GLOSSARY

Antenna A jointed, paired, sensory appendage attached to the head but not associated with the mouth. The plural is *antennae*.

Arthropoda A phylum of organisms characterized by a segmented body, jointed legs and appendages, and an exoskeleton.

Berlese funnel A device, usually consisting of a funnel filled with a freshly collected sample of leaf litter or similar material, suspended above a receptacle. Animals in the sample are driven into the receptacle as the upper layers dry out.

Biodiversity The variety of living organisms in a given place and time, including their genetic diversity and the communities they form.

Biological control The control of pests by use of their natural predators, parasites, competitors, or pathogens.

Canthus A sclerotized process of the beetle head that wholly or partially divides the eye into upper and lower regions.

Club An enlarged portion on the tip of the antenna, composed of one to many segments (antennomeres).

Clypeus The sclerite or exoskeletal plate of the head that covers the mouthparts in beetles.

Coarctate The immobile larval stage in the hypermetamorphic development of blister beetles (Meloidae).

Coleoptera Beetles. Insects distinguished by having chewing mouthparts, leathery or sheathlike forewings that usually meet in a straight line down their back, and development by complete metamorphosis (e.g., egg, larva, pupa, adult).

Common name The popular or local name given to an organism. This name often varies from region to region.

Compound eyes Eyes with multiple facets, or lenses.

Coxa The basal segment of the insect leg. The plural is *coxae*.

Crepuscular Pertaining to organisms that are active at dusk or dawn.

Elytron The hardened forewing of beetles. The plural is *elytra*.

Frons The anterior, upper portion of the head capsule between the clypeus and the vertex.

Gin trap A defensive pinching device on opposable sclerites on the dorsal surface of the abdomen of some beetle pupae.

Habitat The physical environment occupied by an organism.

Hypognathous Having the head vertically oriented such that the mouthparts are directed downward.

Indicator species A species whose presence, absence, or abundance is used to gauge the health or condition of a particular habitat or ecosystem.

Insectivorous Pertaining to organisms that eat insects.

Instar The stage between molts of an immature insect.

Larva The young, preadult insect. Includes nymphs, naiads, caterpillars, grubs, and maggots. The plural is *larvae*.

Mandibles The first pair of jaws in beetles; used for chewing, digging, burrowing, mating, and defense.

Maxillae The second pair of jaws in beetles, located just behind the mandibles; used to manipulate food.

Morphology The study of the structural form and function of an organism's body or its parts.

Ocellus The simple eye found in most larvae and some adult insects; in adults, located on the top of the head between the compound eyes. The plural is *ocelli*.

Palp A jointed, paired, fingerlike appendage, associated with the mouthparts; used to help manipulate food. The plural is *palpi* or *palps*.

Parasite An organism that lives upon, or in, another organism and derives its existence at the expense of its host. Parasites usually do not kill the host.

Parthenogenesis The development of organisms from unfertilized eggs.

Plastron A bubble or film of air on the outside of the body, used by many aquatic insects to breathe under water.

Polymorphic Having different forms.

Population A group of individuals of a given species that inhabits the same area at the same time.

Prognathous Having the head horizontally oriented such that the mouthparts are directed forward.

Pronotum The upper surface of the prothorax, the beetle's midsection. The remaining two sections of the beetle thorax are tightly joined together and are hidden under the elytra.

Pupa The intermediate stage between the larva and adult in insects that develop by complete metamorphosis (e.g., egg, larva, pupa, adult). The plural is *pupae*.

Rugose Wrinkled.

Sclerite Any piece or plate of insect exoskeleton surrounded by membranes or sutures.

Scutellum The small, usually triangular sclerite at the base of and between the elytra.

Species The basic unit of biological classification. For sexually reproducing beetles, a species constitutes a group of individuals that can potentially reproduce with one another and is reproductively isolated from other species.

Specimen A collected and properly preserved individual organism and its collection data that is stored and used in a teaching or research collection.

Sternite A portion or sclerite of the sternum.

Sternum The underside of a body segment of an insect.

Stria In beetles, an impressed line or groove running the length of the elytra or pronotum, often with rows of pits or punctures. The plural is *striae* or *striations*.

Stridulation The act of producing sound by rubbing one body surface against another, usually filelike spines, teeth, or bumps over a ridgelike surface.

Subspecies A taxonomically distinct geographical or host-specific population, or group of populations, of a particular species.

Suture A groove or seam that marks the line of contact between two sclerites.

Systematics The study of biological classification and diversity that establishes the natural relationships of organisms based on their shared evolutionary history.

Tarsal formula The number of tarsal segments on the foreleg, middle leg, hind leg, respectively (e.g., 4-4-4, 5-5-4, or 5-5-5).

Tarsus The foot of an insect formed by a series of articulated segments, or tarsomeres, attached to the tip of the tibia and often bearing claws. The plural is *tarsi*.

Taxonomy The theory and practice of classifying organisms and arranging them into a system.

Tergite A plate or sclerite of the tergum.

Tergum The upper surface of any body segment of an insect.

Tracheae The elongate, tubular network of invaginations of the cuticle through which the cells of insects obtain oxygen.

Truncate Pertaining to an abrupt ending, as if cut off.

Urogomphus A paired fixed or articulating process attached to the posterior end of the ninth abdominal segment of some beetle larvae. The plural is *urogomphi*.

Vermiform Resembling a worm.

Vertex The top of the head between the eyes.

SELECTED GENERAL REFERENCES

Because of limited space only a few of the hundreds of papers consulted for this field guide are included here. For a complete listing of all of the taxonomic and natural history papers pertinent to the taxa mentioned in the family accounts, please visit www.sbnature.org/calbeetles/fieldguide/suppl.htm.

Andrews, F. G., A. R. Hardy, and D. Guiliani. 1979. *A report. The coleopterous fauna of selected California sand dunes.* In fulfillment of Bureau of Land Management Contract CA-960–1285-DEOO. Sacramento: California Department of Food and Agriculture.

Arnett, R. H., Jr., and M. C. Thomas, eds. 2001. *American beetles. Vol. 1. Archostemata, Myxophaga, Adephaga, Polyphaga: Staphyliniformia.* Boca Raton, FL: CRC Press.

Arnett, R. H., Jr., M. C. Thomas, P. E. Skelley, and J. H. Frank, eds. 2002. *American beetles. Vol. 2. Polyphaga: Scarabaeoidea through Curculionidae.* Boca Raton, FL: CRC Press.

Booth, R. G., M. L. Cox, and R. B. Madge. 1990. *IIE guides to insects of importance to man. 3. Coleoptera.* Oxon, UK: International Institute of Entomology (an institute of CAB International), Natural History Museum.

Dajoz, R. 1998. Some Coleoptera of the sand dunes of the Coachella Valley Preserve (southern California) and description of a new species of the genus *Edrotes* (Coleoptera, Tenebrionidae). *Nouvelle Revue Entomologique* (N.S.) 15(4):317–328.

Ebeling, W. 1978. *Urban entomology.* Berkeley: University of California Division of Agricultural Sciences.

Essig, E. O. 1915. Injurious and beneficial insects of California. *Supplement to the Monthly Bulletin, State Commission of Horticulture* 4(4):1–541.

Essig, E. O. 1926. *Insects of western North America. A manual and textbook for students in colleges and universities and a handbook for county, state, and federal entomologists and agriculturalists, as well as foresters, farmers, gardeners, travelers, and lovers of nature.* New York: MacMillan.

Evans, A. V., and J. N. Hogue. 2004. *Introduction to California beetles.* Berkeley and Los Angeles: University of California Press.

Furniss, R. L., and V. M. Carolin. 1977. *Western forest insects.* Miscellaneous Publication No. 1339. Washington, DC: U.S. Department of Agriculture, Forest Service.

Gorham, J. R., ed. 1991. *Insect and mite pests in food.* 2 vols. Agriculture Handbook 655. Washington, DC: United States Department of Agriculture.

Hatch, M. H. 1953. *The beetles of the Pacific Northwest. Part I: Introduction and Adephaga.* University of Washington Publications in Biology, vol. 16. Seattle: University of Washington.

Hatch, M. H. 1957. *The beetles of the Pacific Northwest. Part II. Staphyliniformia.* University of Washington Publications in Biology, vol. 16. Seattle: University of Washington.

Hatch, M. H. 1962. *The beetles of the Pacific Northwest. Part III. Pselaphidae and Diversicornia I.* University of Washington Publications in Biology, vol. 16. Seattle: University of Washington.

Hatch, M. H. 1965. *The beetles of the Pacific Northwest. Part IV. Macrodactyles, Palpicornes, and Heteromera.* University of Washington Publications in Biology, vol. 16. Seattle: University of Washington.

Hatch, M. H. 1971. *The beetles of the Pacific Northwest. Part V. Rhipiceroidea, Sternoxi, Phytophaga, Rhynchophora, and Lamellicornia.* University of Washington Publications in Biology, vol. 16. Seattle: University of Washington.

Hogue, C. L. 1993. *Insects of the Los Angeles Basin.* Los Angeles: Natural History Museum of Los Angeles County.

Papp, C. S. 1984. *Introduction to North American beetles.* Sacramento: Entomography Publications.

Pearson, D. L., C. B. Knisley, and C. J. Kazilek. 2005. *A field guide to the tiger beetles of the United States and Canada. Identification, natural history, and distribution of the Cicindelidae.* New York: Oxford University Press.

Powell, J. A., and C. L. Hogue. 1979. *California insects.* California Natural History Guides 44. Berkeley and Los Angeles: University of California Press.

Ritcher, P. O. 1966. *White grubs and their allies: A study of North American scarabaeoid larvae.* Corvalis: Oregon State University Press.

Solomon, J. D. 1995. *Guide to the insect borers of North American broadleaf trees and shrubs.* Agricultural Handbook 706. Washington, DC: U.S. Department of Agriculture, Forest Service.

Stehr, F. W., ed. 1991. *Immature insects.* Vol. 2. Dubuque, IA: Kendall/Hunt Publishing.

Usinger, R. L., ed. 1956. *Aquatic insects of California.* Berkeley and Los Angeles: University of California Press.

White, R. E. 1983. *A field guide to the beetles of North America.* Peterson Field Guide Series. Boston: Houghton Mifflin.

ART CREDITS

Figures (except maps)
JAMES N. HOGUE

Plates
Except as noted below, all of the photographs are our own.

GREG BALLMER 1, 8, 9, 42, 48, 55, 64, 79, 84, 86, 89, 96, 100, 183, 211, 217, 222, 259

CHARLES BELLAMY 87

PETER BRYANT 28, 175

MICHAEL CATERINO 27, 130

ROSSER GARRISON 112

CHARLES HOGUE 4, 76, 90, 128, 177, 252, 260, 264, 269, 274, 278, 298

FRANK HOVORE 254

STEVE PRCHAL 30, 121, 229, 236, 250

JACQUES RIFKIND 158

ROBYN WAAYERS 290

INDEX

Page references in **boldface type** refer to main discussions of families.

abdomen, morphology, 3, 4
Abies, 18, 114, 128, 149, 176, 218, 235, 275
   concolor, 52, 94, 176, 247
   magnifica, 247
Abronia, 57, 217
   villosa, 115
Acacia, 120, 246, 254
   greggii, 120, 127, 248
acacia, 120, 246, 254
Acalymma trivittatum, 261, pl. 276
Acamptopappus sphaerocephalum, 226
Acanthocinus, 252
   (formerly *Neacanthocinus*)
      obliquus, 252
   princeps, 252, pl. 264
Acanthoscelides, 264
   obtectus, 264, pl. 285
Acer, 127, 167, 246
   glabrum, 93
   macrophyllum, 93, 210
Achillea, 242
   millefolium, 105
Aclypea, 86
Acmaeodera, 127
   connexa, 127, pl. 88
   cribricollis, 127
   gibbula, 43 (fig.), 127
Acmaeoderopsis, 128
   guttifera, 128
   hualpaiana, 128
   jaguarina, 128

Acorn Weevil, California, 271, pl. 291
acrobat beetles. *See Eleodes*
Adalia bipunctata, 203, pl. 177
Adenostoma fasciculatum, 18, 103, 113, 180, 249, 263
Aesculus, 251
   californica, 111
Aethina, 184
Agabus, 72–73
   disintegratus, 73
   lutosus, 72–73
   regularis, 73, pl. 26
   strigulosus, 72
Agave, 270
agave, 270
Agrilus, 125, 133–134
   angelicus, 133
   hyperici, 256
   politus, 133–134
   walsinghami, 134, pl. 100
Agrilus, Common Willow, 133–134
Agriolimax agrestis, 58
Agyrtes, 85
Agyrtidae, **83–85**, pl. 40
Ahasverus, 188
Alaus, 146
   melanops, 149–150, pl. 120
alder, 93, 129, 242, 251
   red alder, 93
   white alder, 130, 237
Alder Borer, Banded, 246, pl. 248
Alder Flea Beetle, 261, pl. 277

Aleochara, 89
alfalfa, 184, 192, 194, 224, 228, 255, 263, 274
Alfalfa Leafcutting Bee, 165
Alfalfa Weevil, 268, 274
alkali mallow, 208
Allenrolfea occidentalis, 129, 174
Allium, 179, 218
Alnus
 rhombifolia, 129, 130, 237
 rubra, 93
Altica, 255, 257, 261
 ambiens, 261, pl. 277
 biguttata, 261
 prasina, 261
Amblonoxia palpalis, 6 (fig.), 31 (figs.), 35 (figs.), 106, 111–112, pl. 65
Amblycheila, 56, 58, 63
 schwarzi, 58, pl. 8
Amblyderus, 231
ambrosia beetles, 195, 196, 277
 California Oak Ambrosia beetle, 276
Ambrosia chamissonis, 217
Amelanchier, 242
American elm, 121
American Spider Beetle, 171–172
Ammophila arenaria, 83
Ampedus, 148–149
 cordifer, 41 (fig.), 148–149, pl. 115
 occidentalis (formerly *bimaculatus*), 149
Amphicerus, 168
 cornutus, 168, pl. 141
Amphizoa, 68
 insolens, 69, pl. 24
Amphizoidae, **68–69**, pl. 24
Amphotis, 184
Anaspis sp., 45 (fig.), 47 (fig.)
Anchomenus funebris, 61, pl. 16
Anchomma, 214
Aneflus calvatus, 33 (fig.), 43 (fig.)
Anelastes, 146
 californicus, 146
 druryi, 145, pl. 109
Anobiidae, **170–174**, pls. 145–150
 parasites of, 194
Anobium punctatum, 170, 173

Anorus, 119, 120, 121
 parvicollis, 120
 piceus, 45 (fig.), 120, pl. 83
ant lions, 217
antennae
 beetle identification and, 32–35, 33 (figs.), 35 (figs.), 40, 41 (figs.)
 morphology, 5, 6 (fig.)
antennal segments, 5
antennomeres, 5
Anthaxia, 132, pl. 97
 aenescens, 132
 inornata (formerly *expansa*), 132
 prasina, 132
Anthicidae, **231–234**, pls. 224–225
 body, 45 (fig.)
 cantharidin as attracting, 224
 head, 45 (fig.)
 sensitive species, 287
Anthicus, 231, 233
Anthonomus grandis, 267
Anthrax, 55
Anthrenus, 163, 165, 166
 flavipes, 165
 lepidus, 165
 museorum, 165
 verbasci, 165, pl. 137
antlike flower beetles, 224, **231–234**
ants
 harvester ants, 118, 213
 mound-building ants, 118
 scarabs eating, 117–118
Anulocaulis annulata, 123
Apatides fortis, 37 (fig.), 168–169, pl. 142
Aphelosternus, 83
aphids, 159, 162, 203, 204
Aphodius, 106, 108, 119
 fimetarius, 108, pl. 57
 fossor, 108
 granarius, 108
 hamatus, 108
 lividus, 108, pl. 58
 pardalis, 108
 pseudolividus, 108
 vittatus, 108
Aphorista morosa, 41 (fig.), 198, pl. 172

Aplastus, 147, 148
　molestus, 148
　optatus, 148
　speratus, 148
Apleurus, 274–275
　albovestitus, 274–275, pl. 297
　jacobinus, 275
apple, 98, 120, 132, 169, 239, 250, 251
Appletree Borer, Flatheaded, 132–133
Apteraliplus, 66, 67, 68
　parvulus, 67
Apteroloma, 85
aquatic habitats, 19
Arbutus menziesii, 93, 127, 167, 239
Archodontes, 237–238
　melanopus aridus, 237–238, pl. 227
Arctostaphylos, 102, 114, 167, 193, 267
Argemone munita, 194
Argoporis bicolor, 219, pl. 202
Arhopalus
　asperatus, 241, pl. 232
　productus, 241
Armored Darkling Beetle, 219
arrow weed, 128, 133
Artemisia, 182, 227
　californica, 114, 259
　tridentata, 113, 122, 247
Asbolus, 215
　laevis, 215, pl. 190
　verrucosus, 215, pl. 189
Asclepias, 207, 253, 262
Asclera excavata, 33 (fig.), 41 (fig.), 45 (fig.)
Asemum
　caseyi, 241
　nitidum, 241, pl. 233
　striatum, 196, 241
ash, 52, 93, 246
Ashy Gray Lady Beetle, 200, 204, pl. 181
Asian Lady Beetle, Multicolored, 199, 203–204, pl. 179
Asidina confluens, 217
asparagus beetles
　Asparagus Beetle, 258, pl. 268
　Spotted Asparagus Beetle, 258
aspen, 129
aspirators, 22

asters, 207, 275
Astragalus, 194, 242, 254
Ataenius, 106, 108–109
　platensis, 109, pl. 59
　spretulus, 108–109
Ataenius, Shining Black Turfgrass, 108–109
Atholus bimaculatus, 82–83, pl. 39
Athous, 150
　axillaris, 150, pl. 122
Atriplex, 273
　lentiformis, 127
　leucophylla, 217
Attagenus, 162–163, 164–165
　megatoma, 165
Attelabidae, **264–267**, pl. 286
　mouthparts, 5
　seed beetles and, 254
Auduoin's Night-stalking Tiger Beetle, 58
Aulicus, 178, 180
　bicinctus, 180
　terrestris, 180
avocado, 247
Axion, 204
　plagiatum, 201
azaleas, 274

Baccharis, 229, 238
　emoryi, 208
　pilularis, 99, 169, 179
baccharis, 208
back crawlers, 117
bacteria
　as digestive aid, 166, 170
　F.B.I. (fungi, bacteria, insects), 10
Banded Alder Borer, 246, pl. 248
Banded Glowworm, Western, pl. 127
bark beetles
　collecting, 277
　flat bark beetles, **190–191**
　Green Bark Beetle, 176, pl. 153
　as pine pest, 276
　predators of, 81, 89, 176, 177, 179, 180, 181, 184, 196

INDEX 303

bark beetles (cont.)
  Red Flat Bark Beetle, 190–191, pl. 166
  silvanid flat bark beetles, **188–189**
  Wrinkled Bark Beetle, 53, pl. 3
  wrinkled bark beetles, **52–54**
bark-gnawing beetles, **174–177**
Basidiomycetes, 176
bassia, 228
Bassia hyssopifolia, 228
beach-bur, 217
beach grass, European, 83
Bean Weevil, 264, pl. 285
Bear Scarab, White Sand, 104–105
beavertail cactus, 250, 270
bee flies, 55
Bee Scarab, 105, pl. 56
beech, 52
Beechey Ground Squirrel, 109
bees, 177, 179, 206, 208
  Alfalfa Leafcutting Bee, 165
  solitary bees, 165, 179, 225, 226
beetle collections, 2, 289
  See also collecting beetles
Behren's Rain Beetle, 99
beloperone, 130
Berlese funnels, 19, 80, 137
Berosus, 76, 78
  punctatissimus, 78, pl. 31
  striatus, 78
Betula occidentalis, 93
Big-headed Ground Beetle, 60, pl. 14
Big-headed Tiger Beetle, Pan-American, 58–59, pl. 10
big-leaf maple, 93, 210
big saltbush, 127, 133
bigcone spruce, 146
billbugs
  Cocklebur Billbug, 270
  Tule Billbug, 270–271, pl. 290
binoculars, 23
biological controls
  antlike flower beetles as, 231
  for Brown Garden Snail, 90
  for Horn Fly, 81, 83
  for introduced weeds, 15, 256, 263, 268

  lady beetles as, 199, 201, 203–204
  soldier beetles as, 159
bioluminescence, 153–154, 156–157
birch, water, 93
Bird Nest Carpet Beetle, 165
Black and White Click Beetle, 150, pl. 121
Black Burying Beetle, 88
Black California Cedar Beetle, 122
Black Calosoma, 57, pl. 5
Black Carpet Beetle, 164–165
black cottonwood, 242, 250
black oak, California, 93
Black Rain Beetle, 99
Black Vine Weevil, 273–274
Black Water Beetle, Giant, 79, pl. 34
Black-bellied Clerid, 181
blackberries, 263, 265, 266, 274
blister beetles, **224–230**
  for attracting antlike flower beetles, 231, 233–234
  collecting, 18
  Elegant Blister Beetle, 227, pl. 213
  elytra, 7
  false blister beetles, **221–224**
  Punctate Blister Beetle, 228, pl. 217
blue death feigner. *See* Desert Ironclad Beetle
blue elderberry, 243, 244
blue gum, 93, 127, 259, 271
Blue Milkweed Beetle, 262, pl. 280
blue palo verde, 116, 117, 128, 133, 249
body, beetle identification and, 30, 31 (figs.), 36–39, 37 (figs.), 39 (figs.), 46, 47 (figs.)
body regions, morphology, 3, 4 (fig.)
Bolbelasmus, 101, 102, 103
  hornii, 102
Bolboceras, 101, 102, 103
  obesus, 102, pl. 53
Bolbocerastes, 101, 102, 103
  imperialis, 102
  regalis, 102, pl. 54
Boll Weevil, 267
bombardier beetles, 61–62
  False Bombardier Beetle, 61, pl. 18

Boraginaceae, 227
Boreal Long-lipped Tiger Beetle, 60
borers
    Banded Alder Borer, 246, pl. 248
    branch borers. *See* bostrichid
        beetles
    California Elderberry Longhorn
        Borer, 243–244
    California Flatheaded Borer, 132,
        275, pl. 96
    California Laurel Borer. *See* Banded
        Alder Borer
    Clover Stem Borer, 192, 194, pl. 168
    eucalyptus borers, 236
    Flatheaded Appletree Borer,
        132–133
    Flatheaded Cedar Borer, 133
    Flatheaded Pine Borer, 132
    Giant Mesquite Borer, 236, 239,
        pl. 229
    Giant Palm Borer, 178, pl. 140
    Giant Root Borer, 239
    Golden-winged Elderberry Borer,
        243, pl. 241
    Gooseberry Borer, 236
    Hairy Borer, 249–250, pl. 258
    Hairy Pine Borer, 240, pl. 231
    Lead Cable Borer, 169, pl. 143
    Mesquite Borer, 246, pl. 250
    milkweed borers, 234, 253
    Nautical Borer, 247, pl. 252
    New House Borer, 241
    Oak Cordwood Borer. *See* Nautical
        Borer
    Pacific Flatheaded Borer, 133
    Ponderous Pine-borer. *See* Pine
        Sawyer
    Ribbed Pine Borer, 243, pl. 239
    roundheaded borers, 234–235, 236
    Sculptured Pine Borer, 128, pl. 90
    shot-hole borers. *See* bostrichid
        beetles
    Southwestern Stump Borer, 238
    Spiny Wood-borer. *See* Pine Sawyer
    Spotted Limb Borer, 169
    Spotted Tree Borer, 251, pl. 262
    Stout's Hardwood Borer, 167–168,
        pl. 139
    twig borers, 7, 166
    Valley Elderberry Longhorn Borer,
        25, 244, 285
    Western Cedar Borer, 131
    Western Twig Borer, 168, pl. 141
    Wharf Borer, 222, pl. 207
    Willow Root Borer, 237, pl. 227
    See also wood-boring beetles
bostrichid beetles, **166–170**, 194
Bostrichidae, **166–170**, pls. 139–144
    body, 37 (fig.)
    parasites of, 194
    pronotum, 7
bothriderid beetles, **194–196**
Bothrideridae, **194–196**, pls. 169–170
Brachinus, 54, 56, 61–62
    costipennis, 62
    favicollis, 62, pl. 20
    mexicanus (formerly *fidelis*), 62
    pallidus, 62
Brachys, 125–126
bracket fungi, 193
braconid wasps, 216
branch borers. *See* bostrichid beetles
Brassica, 250
brittlebush, 18, 123, 229, 260
Broadbean Weevil, 264
Brodiaea, 124, 179
brodiaea, 124, 179
Bromus tectorum, 218
broom, 229, 237, 249
    chaparral broom, 99, 169, 179
    scale broom, 16–17, 18, 229
    Scotch broom, 256
    Spanish broom, 268
Brothylus
    conspersus, 45 (fig.), 245
    gemmulatus, 245, pl. 245
Brown Dung Beetle, 110, pl. 61
Brown Garden Snail, 87, 90
Brown Leatherwing Beetle, 161, pl. 132
Brown Spider Beetle, 172
Bruchid, Vetch, 264
Bruchidae, 254
Bruchus, 263–264
    brachialis, 264
    pisorum, 263–264, pl. 284
    rufimanus, 264

Brychius, 66, 67–68
buckbrush, 100, 253
buckeye, 251
   California buckeye, 111
buckthorn, 243
buckwheat, 18, 129, 207, 242, 253, 263, 273
   California buckwheat, 111, 113, 206, 262–263
   desert buckwheat, 128
Bud Beetle, Grape, 262–263
Buffalo Flower Beetle, 164, pl. 136
bumblebee scarab beetles, **103–105**
   Rathvon's Bumblebee Scarab, 105
Buprestidae, **125–134**, pls. 87–100
   as biological control, 256
   distribution patterns, 13
   head, 43 (fig.)
   parasites of, 194
   predators of, 177, 196
buprestids
   Golden Buprestid, 130–131, pl. 93
   Spotted Flower Buprestid, 127, pl. 88
Buprestis, 126, 130–131
   subornata, 130
   viridisuturalis, 130
burclover, 274
burr sage, 115
burying beetles, 85–86
   Black Burying Beetle, 88
   collecting, 20
   Red and Black Burying Beetle, 87, pl. 42

Cable Borer, Lead, 169, pl. 143
cactus, beavertail, 250, 270
Cactus Beetle, 250, pl. 259
Cactus Flower Beetle, 187
Cactus Weevil, 270, pl. 288
Cadelle, 175, 177
California
   beetle distribution patterns, 10–14
   beetle families, 26
   beetle species, 2
   maps, 11, 13
   seasons, 15–17
California Academy of Sciences, 2
California Acorn Weevil, 271, pl. 291

California black oak, 93
California buckeye, 111
California buckwheat, 111, 113, 206, 262–263
California Cedar Beetle, Black, 122
California coffeeberry, 251
California Elderberry Longhorn Borer, 243–244
California fan palm, 168
California Filbert Weevil. *See* California Acorn Weevil
California Five-spined Ips, 276
California Flatheaded Borer, 132, 275, pl. 96
California Glowworm, 158, pl. 129
California juniper, 113, 124, 131, 211
California Lady Beetle, 203
California laurel, 93, 167, 246
California Laurel Borer. *See* Banded Alder Borer
California Night-stalking Tiger Beetle, 58, pl. 9
California Oak Ambrosia Beetle, 276
California Prionus, 236, 239–240, pl. 230
California red fir, 247
California sagebrush, 114, 259
California Tiger Beetle, 59–60, pl. 12
California-lilac, 18, 99, 103, 110, 127, 242, 253
Calitys, 175, 176, 177
   scabra, 176, pl. 152
Callidium antennatum, 246, pl. 249
Calochortus, 242
Calochromus, 152, 161
Calocedrus decurrens, 131, 146, 176
Calosoma, 54, 56–57
   latipenne, 57
   parvicollis, 57
   prominens, 8 (fig.)
   scrutator, 57, pl. 6
   semilaeve, 4 (fig.), 5 (fig.), 31 (figs.), 57, pl. 5
Calosoma, Black, 57, pl. 5
Camissonia claviformis, 115
Camponotus, 118
Canis latrans, 98
Cannabis, 206

Cantharidae, **159–162**, pls. 132–134
    body, 45 (fig.)
    head, 45 (fig.)
cantharidin, 221, 224, 231, 233–234
Cantharis, 159, 160
Canthon, 107
    simplex, 8 (fig.), 109, pl. 60
canyon live oak, 99, 114, 239
captive beetles, 23–24
Carabidae, **54–63**, pls. 4–20
    antennae, 6 (fig.)
    body, 4 (fig.), 31 (figs.)
    collecting, 19
    endangered and threatened species, 285
    extinct species, 25
    mouthparts, 4–5, 5 (fig.)
    sensitive species, 285–286
    tarsi, 8 (fig.)
    wrinkled bark beetles as, 53
carcasses, as beetle habitat, 20
cardenolides, 262
Carnegiea gigantea, 270
carpet beetles
    Bird Nest Carpet Beetle, 165
    Black Carpet Beetle, 164–165
    Furniture Carpet Beetle, 165
    Varied Carpet Beetle, 165, pl. 137
Carpobrotus, 83
Carpophilus, 184–185, 186–187
    hemipterus, 187
    humeralis, 33 (fig.), 187, pl. 163
    pallipennis, 187
carrion beetles, **85–88**
    feeding habits, 20
    Garden Carrion Beetle, 86–87, pl. 41
    primitive carrion beetles, **83–85**
    Satin Carrion Beetle, 87
Carrion Dermestid, Common, 164, pl. 135
carrot, 116, 242
Carrot Beetle, Western, 116, pl. 75
Carthartosylvanus, 188–189
Castilleja, 111
castor bean, 250
catclaw, 120, 127, 128, 133, 248
Cathartus, 188

cattail, 73
ceanothus, 111, 114, 243, 245, 247, 263
Ceanothus, 18, 99, 103, 110, 111, 127, 242, 243, 263
    cuneatus, 100
cedar
    incense-cedar, 131, 146, 176
    western red cedar, 131, 223
cedar beetles, **121–122**
    Black California Cedar Beetle, 122
cedar borers
    Flatheaded Cedar Borer, 133
    Western Cedar Borer, 131
Cellar Beetle, Hairy, 197
Centaurea, 268
Centrodera, 237, 241–242
    autumnata, 241
    oculata, 241
    oculata blaisdelli, 241
    oculata oculata, 241
    spurca, 241–242, pl. 234
Cerambycidae, **234–254**, pls. 226–267
    antennae, 6 (fig.), 33 (fig.), 41 (fig.)
    body, 45 (fig.)
    head, 43 (fig.)
    larval tunnels, 179
    mouthparts, 5
    parasites of, 194
    predators of, 177
    pronotum, 45 (fig.)
    sensitive species, 287
    threatened species, 25, 285
Ceratophyus, 101, 103
    gopherinus, 35 (fig.), 102–103, pl. 55
Cercidium, 248, 254
    floridum, 116, 128, 133, 249
    microphyllum, 127
Cercocarpus, 18, 114, 127
Ceruchus, 92, 93–94
    punctatus, 94
    striatus, 94
Chaenactis, 226
    carphoclinia, 123
Chaetocnema, 261–262
    repens, 261–262, pl. 278

chafers
    fruit chafers, 117
    masked chafers, 106, 116
    mouthparts, 5
Chalcolepidius, 146, 147
    webbi, 150, pl. 121
Chalcophora angulicollis, 128, pl. 90
chamise, 18, 103, 113, 114, 115, 180, 249, 263
chaparral broom, 99, 169, 179
chaparral yucca, 249, 270
Charadrius vociferus, 55
Charcoal Beetle, 131–132
Charidotella (formerly *Metriona*) *sexpunctata bicolor*, 258–259
Chauliognathus, 160, 162
cheat grass, 218
checkered beetles, **177–182**
    Ornate Checkered Beetle, 179–180, pl. 157,
cherry, 93, 98, 120, 239
Chilocorus, 204
    fraternus, 201
    orbus, 201, pl. 174
chinchweed, 227
Chinese Beetle. *See* Green Fig Beetle
Chlaenius, 54, 61
    cumatilis, 61, pl. 18
    sericeus viridifrons, 61
Chnaunanthus, 119
cholla, 187, 252
    teddy bear cholla, 250
Chrysobothris, 132–133
    biramosa calida, 132
    caurina (formerly *beeri*), 132
    cupressicona, 126
    femorata, 132–133
    leechi, 132
    mali, 133
    monticola, 133, pl. 98
    nixa, 133
    octocala, 133, pl. 99
Chrysochus cobaltinus, 8 (fig.), 41 (fig.), 47 (fig.), 262, pl. 280
Chrysolina
    hyperici, 256
    quadrigemina, 256

Chrysomelidae, **254–264**, pls. 268–285
    antennae, 33 (fig.), 41 (fig.)
    body, 47 (fig.)
    head, 31 (fig.), 45 (fig.)
    mouthparts, 5
    tarsi, 8 (fig.)
Chrysophana conicola, 126
Chrysothamnus, 16, 228, 248, 273
    nauseosus, 134, 260
    nauseosus mohavensis, 124
chuparosa, 130
cicada parasite beetles. *See* cedar beetles
Cicindela, 56, 59–60
    californica, 59–60, pl. 12
    californica mojavi, 59–60
    californica pseudoerronea, 60
    collecting, 62–63
    feeding habits, 233
    latesignata obliviosa, 25
    lemniscata, 60, pl. 13
    longilabris, 60
    longilabris perviridis, 60
    ohlone, 25, 285
    oregona, 6 (fig.), 59, pl. 11
    tranquebarica, 25
Cicindelidae, 55
Cigarette Beetle, 170, 174, pl. 150
Ciliated Dune Beetle, 217
Cirsium, 268
Citrus, 251
citrus, 251, 274
Citrus Mealybug, 161
classification, 9
Cleridae, **177–182**, pls. 155–159
    body, 39 (fig.)
    collecting, 20
    pronotum, 41 (fig.), 45 (fig.)
clerids
    Black-bellied Clerid, 181
    Red-bellied Clerid, 180–181
click beetles, **146–151**
    Black and White Click Beetle, 150, pl. 121
    collecting, 18
    elytra, 7
    false click beetles, **144–146**

Grape Click Beetle. *See* Pacific Coast Wireworm
Western Eyed Click Beetle, 149–150, pl. 120
Clinidium calcaratum, 53–54
clover, 274
   burclover, 274
   owls clover, 111
   red clover, 192
Clover Leaf Weevil, 274
Clover Stem Borer, 192, 194, pl. 168
clown beetles, **80–83**
   collecting, 19, 20
   elytra, 7
clypeus, defined, 4
Coach Horse, Devil's, 90, pl. 45
coast live oak, 52, 93, 124, 176, 179, 201, 212, 236, 239, 246, 276
coast redwood, 94
Coccinella
   californica, 203
   septempunctata, 203, pl. 178
Coccinellidae, **198–204**, pls. 173–181
   defensive strategy, 197
   mouthparts, 41 (fig.)
Cocklebur Billbug, 270
cockroaches, 206
Coelocnemis, 220
   californica, 220, pl. 205
   magna, 220
Coelus, 217
   ciliatus, 217
   globosus, 217, pl. 196
Coenonycha, 113, 118
   ampla, 113
   fusca, 113
   testacea, 113, pl. 68
Coenopoeus, 252
   palmeri, 252, pl. 265
coffeeberry, California, 251
collecting beetles
   basics, 22–23
   best seasons, 15–17
   sources of supplies, 290
   to keep in captivity, 23–24
   to rear, 24

where to look, 17–21
See also beetle collections
Collops, 182, 183–184, pl. 161
   bipunctatus, 184
Colorado pinyon, 252
Common Carrion Dermestid, 164, pl. 135
Common Mealworm, 219–220, pl. 203
Common Willow Agrilus, 133–134
compost, as beetle habitat, 21
Condalia, 128
condalia, 128
coneflower, 242
Confused Flour Beetles, 14
Coniontis, 216, pl. 194
   abdominalis, 216
Conoderus, 149
   exsul, 149, pl. 119
conserving beetles, 24–25, 285–288
Convergent Lady Beetle, 199–200, 202, pl. 176
Cordwood Borer, Oak. *See* Nautical Borer
Cordylospasta, 226
   fulleri, 226
   opaca, 226, pl. 211
Coreopsis, 18
coreopsis, 18
Cotalpa flavida, 115, pl. 71
Cotinis, 106, 119
   mutabilis (formerly *texanus*), 117, 119, pl. 78
cottonwood, 82, 129, 130, 212, 239, 251, 252
   black cottonwood, 242, 250
   Fremont cottonwood, 93, 115
Cottony Cushion Scale, 201
Coulter pine, 173
cow-parsnip, 242
coyote, 96, 98
coyote brush, 169, 179
Cranberry Grub, 103
crawlers, back, 117
crawling water beetles, 7, **65–68**
Cremastocheilus, 106, 117–118
   angularis, 118
   armatus, 118

Cremastocheilus (cont.)
  schaumii, 118
  westwoodi, 118, pl. 80
Creophilus maxillosus, 90, pl. 44
creosote bush, 102, 115, 127, 226, 249, 273
Crioceris, 258
  asparagi, 258, pl. 268
  duodecimpunctata, 258
Crossidius, 247–248
  ater, 248
  coralinus, 248
  coralinus ascendens, 248
  coralinus ruficollis, 248
  mojavensis, 248, pl. 253
  punctatus, 248
  suturalis, 248
  suturalis minutivestis, 248, pl. 254
  suturalis pubescens, 248
Cryptantha
  angustifolia, 227
  barbigera, 227
Cryptocephalus, 263
  castaneus, 263, pl. 282
Cryptoglossa, 215
  (formerly *Centrioptera*) *muricata*, pl. 191
Cryptolaemus montrouzieri, 201, pl. 173
Cryptophilus integer, 192
Ctenicera, 151
  conjugens (formerly *lecontei*), 151
Cucujidae, **190–191**, pl. 166
Cucujus, 190, 191
  clavipes clavipes, 191
  clavipes puniceus, 190–191, pl. 166
cucumber beetles
  Western Spotted Cucumber Beetle, 260–261, pl. 275
  Western Striped Cucumber Beetle, 261, pl. 276
Cultellunguis, 161–162
  americana, 161–162, pl. 133
Cupedidae, **51–52**, pls. 1–2
Cupressus, 131, 179, 235
  forbesii, 133
  macrocarpa, 133
  sargentii, 131

Curculio, 271
  occidentis (formerly *uniformis*), 31 (fig.), 271, pl. 291
Curculionidae, **267–277**, pls. 287–300
  antennae, 6 (fig.)
  head, 31 (fig.)
  mouthparts, 5
  parasites of, 195
  predators of, 80, 81, 89, 176, 177, 179, 180, 182, 184, 196
  sensitive species, 287–288
curculios
  Eastern Rose Curculio, 266
  Western Rose Curculio, 266, pl. 286
Cushion Scale, Cottony, 201
Cybister, 74
  ellipticus, 74
  explanatus, 74
Cybocephalus, 184, 185
Cyclocephala, 106, 116
  hirta, 116
  longula, 116
  melanocephala, 116
  pasadenae, 116, pl. 74
Cyclodinus sp., 45 (figs.)
Cymatodera, 177, 178–179, 181
  californica, 45 (fig.), 179
  oblita, 179, pl. 155
  ovipennis, 179
  pseudotsugae, 39 (fig.), 41 (fig.)
cypress, 131, 179, 235
  Sargent's cypress, 131, 133
  tecate cypress, 133
Cypriacis (formerly *Buprestis*) *aurulenta*, 130–131, pl. 93
Cysteodemus armatus, 226, pl. 212
Cytisus
  multiflorus, 268
  scoparius, 256

Dacne, 193
  californica, 193
Dalea, 273
dalea, 273
Dampwood Termite, Pacific, 118
Danosoma brevicornis, 149, pl. 116
darkling beetles, **212–221**
  Armored Darkling Beetle, 219

collecting, 18, 20
distribution patterns, 12
Wooly Darkling Beetle, 219, pl. 200
Dascillidae, **119–121**, pls. 82–83
head, 43 (figs.)
pronotum, 45 (fig.)
Dascillus, 119–120, 121
cervinus, 120
davidsoni, 120, pl. 82
plumbeus, 43 (figs.), 120
Datura, 116, 129
Daucus, 242
Davidson's Beetle, 120
dead animals, as beetle habitat, 20
deathwatch beetles, **170–174**
Deathwatch Beetle, 173
parasites of, 194
decomposition, role of beetles, 10
defense strategies, 10
Delta Green Ground Beetle, 25, 285
Dendroctonus, 180, 196, 267, 275, 276
brevicomis, 275
jeffreyi, 275
monticolae, 180
ponderosae, 180, 275
pseudotsugae, 180, 275
valens, 275, pl. 298
Deporaus, 267
Deretaphrus, 195
oregonensis, 196, pl. 169
Dermestes, 164, 166
maculatus, 164
marmoratus, 37 (fig.), 164, pl. 135
Dermestid, Common Carrion, 164, pl. 135
Dermestidae, **162–166**, pls. 135–138
collecting, 20
head, 37 (fig.)
Derobrachus, 239
geminatus, 236, pl. 229
desert buckwheat, 128
Desert Ironclad Beetle, 215, pl. 189
desert peach, 265, 267
desert tea, 123
desert trumpet, 127
deserts, California, 12
desert-sunflower, 123, 226, 227

Desmocerus, 237, 243 244
auripennis, 243
auripennis auripennis, 243, pl. 241
auripennis cribripennis, 243
californicus californicus, 243–244
californicus dimorphus, 25, 244, 285
Destroyer, Mealybug, 201, pl. 173
Devastating Grasshopper, 87
Devil's Coach Horse, 90, pl. 45
Diabolical Ironclad Beetle, 212
Diabrotica, 260–261
undecimpunctata, 260–261, pl. 275
Dicerca, 129
hornii, 129, pl. 92
tenebrica, 129
tenebrosa, 129
Dicerca, Poplar, 129
Dichelonyx, 106, 114
backi (formerly *fulgida*), 114
pusilla, 114
truncata, 114
valida vicina, 114, pl. 69
dichondra, 262, 272
Dichondra Flea Beetle, 261–262, pl. 278
Dichondra repens, 262, 272
Dictyoptera simplicipes, 47 (fig.), 153, pl. 125
Dinacoma, 111
caseyi, 111
marginata, 111, pl. 64
Dinapate, 169
hughleechi, 168
wrighti, 168, pl. 140
Dineutus, 64, 65
solitarius, 65, pl. 21
Dinocampus
coccinellae, 199–200
terminatus, 200
Diplotaxis, 106, 113, 118
moerens moerens, 113
sierrae, 113, pl. 67
subangulata, 113
Dipodomys, 80, 116
Disonycha, 255, 257, 262
alternata, 262, pl. 279
Distichlis spicata, 218
distribution patterns, 10–14
Ditylus quadricollis, 223

diving beetles
Mono Lake Hygrotus Diving Beetle, 25
predaceous diving beetles, 13, 19, **70–75**
Sunburst Diving Beetle, 73–74, pl. 29
Yellow-spotted Diving Beetle. *See* Sunburst Diving Beetle
See also water beetles
dock, 259
Dock Beetle, Green, 259, pl. 271
Dohrn's Elegant Euchemid Beetle, 33 (figs.), 144, 146, pl. 110
dooryard peach, 133
Douglas-fir, 52, 53, 128, 130, 149, 150, 176, 221, 238, 241, 243, 244, 251
Douglas-fir Beetle, 275
Douglas-fir Tussock Moth, 276
Downy Leather-winged Beetle, 162
Dried Fruit Beetle, 187
Drugstore Beetle, 170, 174, pl. 148
Dryopidae, **137–140**, pl. 105
collecting, 19
tarsi, 8 (fig.)
dry-rot termite, subterranean, 120
Dubiraphia, 136
brunnescens, 136
giulianii, 136
dune beetles
Ciliated Dune Beetle, 217
Globose Dune Beetle, 217, pl. 196
dung, as beetle habitat, 20
dung beetles
Brown Dung Beetle, 110, pl. 61
collecting, 20, 21, 119
European Dung Beetle, 108, pl. 57
as introduced beneficial, 15
Spotted Dung Beetle, 79–80, pl. 35
dung scarabs, 106, 107
dung-rolling scarabs, 109
Dusty June Beetle, 111–112, pl. 65
Dystaxia, 124
elegans, 124
murrayi, 124
Dytiscidae, **70–75**, pls. 25–30
distribution patterns, 13
extinct species, 25

sensitive species, 286
tarsi, 8
Dytiscus, 73
marginicollis, 8 (fig.), 73, pl. 28

earth-boring scarab beetles, 6–7, **100–103**
Eastern Rose Curculio, 266
Edrotes, 217–218
ventricosus, 218, pl. 197
Eggplant Tortoise Beetle, 258, pl. 269
Elaphrus viridis, 25, 285
Elater, 148
lecontei, 148, pl. 113
pinguis, 148
Elateridae, **146–151**, pls. 111–124
antennae, 6 (fig.), 33 (fig.), 41 (figs.)
elytra, 7
head, 31 (fig.)
elderberry, 179
blue elderberry, 243, 244
red elderberry, 243, 244
elderberry borers
California Elderberry Longhorn Borer, 243–244
Golden-winged Elderberry Borer, 243, pl. 241
Valley Elderberry Longhorn Borer, 25, 244, 285
Elegant Blister Beetle, 227, pl. 213
Eleodes, 213, 218–219
armatus, 37 (fig.), 219
clavicornis, 219
gigantea, 219, pl. 199
(formerly *Amphidora*) *nigropilosa*, 219, pl. 201
(formerly *Cratidus*) *osculans*, 219, pl. 200
Eleodes, Gigantic, 219, pl. 199
Elephant Beetle, Sleeper's, 116–117, pl. 76
Ellychnia, 157, 158, 159
californica, 158, pl. 129
elm, 52, 251, 260
American elm, 121
Elm Leaf Beetle, 260, pl. 274

Elmidae, **134–137**, pl. 101–104
    collecting, 19
    sensitive species, 287
elytra
    morphology, 3, 7
    taxonomic role, 9
elytral suture, 7
Emarginate Ips, 276
Encelia farinosa, 18, 123, 229, 260
endangered and threatened species, 14, 25, 285
    See also sensitive species
Endomychidae, **196–198**, pls. 171–172
    head, 41 (fig.)
Endomychus, 198
    limbatus, 198, pl. 171
Englemann spruce, 223
English walnut, 237
engraver beetles
    Fir Engraver, 275–276, pl. 299
    Pine Engraver, 275, 276, pl. 300
Enoclerus, 177, 180–181, 181
    lecontei, 181
    quadrisignatus, 181, pl. 158
    sphegeus, 180–181
Ephedra, 128
    californica, 123
Epicauta, 224, 228
    alphonsii (formerly *californica*), 228
    corybantica, 228
    puncticollis, 228, pl. 217
Epuraea, 184
Eremosaprinus, 83
Eretes sticticus (formerly *occidentalis*), 74, pl. 30
Ergates, 238–239, 240
    pauper, 239
    spiculatus, 238–239, pl. 228
Eriastrum
    eremicum, 227
    sapphirinum, 228
eriastrum, 227, 228
Eriodictyon, 111
    californicum, 260
Eriogonum, 18, 129, 207, 242, 263, 273
    deserticola, 128
    elongatum, 129

fasciculatum, 18, 111, 113, 206, 263
inflatum, 127
nudum, 129
Ernobius, 173
    melanoventris, 173
    montanus, 173, pl. 146
Eronyxa expansus, 175
Erotylidae, **191–194**, pls. 167–168
Eryniopsis, 160
Erysimum, 132
erythraeid mite, 55
Eschscholzia, 124
Esselenia vanduzeei, 180
Eubrianax edwardsii, 47 (fig.), 143, pl. 108
Eucalyptus, 167, 204, 245
    camaldulensis, 259
    globulus, 93, 127, 259, 271
eucalyptus, 167, 169, 204, 245, 246, 247
Eucalyptus Beetle, 259, pl. 270
eucalyptus borers, 236
eucalyptus longhorn beetles, 245
Eucalyptus Psyllid, Red Gum, 204
Eucalyptus Snout Beetle, 271
Eucnemidae, **144–146**, pls. 109–110
    antennae, 33 (figs.), 35 (fig.)
    body, 33 (fig.)
    sensitive species, 287
Eumecomera, 223
    bicolor, 223
    cyanipennis, 223, pl. 208
    obscura, 223
Euoniticellus intermedius, 109
Euphoria, 106, 119
    fascifera trapezium, 117, pl. 79
Eupompha elegans, 227, pl. 213
European beach grass, 83
European Dung Beetle, 108, pl. 57
Eusattus, 216–217
    difficilis, 216
    dilatatus, 216–217, pl. 195
    muricatus muricatus, 216
Euspilotus, 83
Euthysanius, 147, 148
    lautus, 6 (fig.), 33 (fig.), 41 (fig.), 148, pls. 111–112

Euvrilleta, 174
   distans, 174, pl. 149
Exema, 255
exoskeleton, morphology, 3–4
extinct species, 25
Eyed Click Beetle, Western, 149–150, pl. 120

Fagus, 52
false blister beetles, **221–224**
False Bombardier Beetle, 61, pl. 18
false click beetles, **144–146**
false jewel beetles. *See* schizopodid beetles
false wireworms, 213
families
   checklist, 279–283
   defined, 9
   identifying, 26–27
   order of presentation, 26
fan palm, California, 168
F.B.I. (fungi, bacteria, insects), 10
Federal Endangered Species list, 25, 285–288
Ficus, 169, 251
Fiery Searcher, 57, pl. 6
fig, 169, 251
Fig Beetle, Green, 117, pl. 78
filbert, 271
Filbert Weevil, California. *See* California Acorn Weevil
Fimbriate Rain Beetle, 99–100
fir, 18, 114, 128, 130, 149, 176, 218, 235, 238, 241, 242, 243, 251, 275
   California red fir, 247
   Douglas-fir, 52, 53, 128, 130, 149, 150, 176, 221, 238, 241, 243, 244, 251
   white fir, 52, 94, 176, 247
Fir Engraver, 275–276, pl. 299
Fir Sawyer, Oregon, 250, pl. 260
Firebug. *See* Charcoal Beetle
fireflies, **156–159**
   elytra, 7
Five-spined Ips, California, 276
Five-spotted Lady Beetle, 202–203
flags, defined, 235
flannelbush, 114, 251

flat bark beetles, **190–191**
   Red Flat Bark Beetle, 190–191, pl. 166, title page
   silvanid flat bark beetles, **188–189**
flatheaded borers
   California Flatheaded Borer, 132, 275, pl. 96
   Flatheaded Appletree Borer, 132–133
   Flatheaded Cedar Borer, 133
   Flatheaded Pine Borer, 132
   Pacific Flatheaded Borer, 133
flea beetles, 255, 261–262
   Alder Flea Beetle, 261, pl. 277
   Dichondra Flea Beetle, 261–262, pl. 278
fleabane, salt marsh, 207–208
flies
   bee flies, 55
   Horn Fly, 81, 83
   Spanish fly. *See* blister beetles
   tachinid flies, 216, 217
   Therevid flies, 217
Flour Beetles, Confused, 14
flower beetles
   antlike flower beetles, 224, **231–234**
   Buffalo Flower Beetle, 164, pl. 136
   Cactus Flower Beetle, 187
   collecting, 18
   soft-winged flower beetles, 7, **182–184**
   tumbling flower beetles, 18, 182, **204–206**
Flower Buprestid, Spotted, 127, pl. 88
flowers, as beetle habitat, 18
Fomes, 176, 193, 218
forest pests, 14
   See also specific beetles
Formica, 118
Formicilla, 233
foxes, 98
Franseria, 115
Fraxinus, 52, 93, 246
Fremont cottonwood, 93, 115
fremontia, 251
Fremontodendron californicum, 114, 251
Fringed Beetle, White, 272–273
Fruit Beetle, Dried, 187

fruit chafers, 117
Fuller's Rose Weevil, 272, pl. 294
fungi, 176
    as beetle habitat, 19, 218
    bracket, 193
    F.B.I. (fungi, bacteria, insects), 10
fungus beetles
    handsome fungus beetles, **196–198**
    pleasing fungus beetles, **191–194**
Furniture Beetle, 170, 173
Furniture Carpet Beetle, 165

galls, 125, 205
    moth larva galls, 179
Garden Carrion Beetle, 86–87, pl. 41
Garden Slug, Gray, 58
Garrya, 127
Gastrophysa, 259
    cyanea, 259, pl. 271
genera, defined, 9
Geomysaprinus, 83
Geotrupidae, **100–103**, pls. 53–55
    antennae, 35 (fig.)
    pronotum, 6–7
Geraea canescens, 123, 226
ghost beetle. *See* Desert Ironclad Beetle
Giant Black Water Beetle, 79, pl. 34
Giant Green Water Beetle, 73, pl. 28
Giant Mesquite Borer, 236, 239, pl. 229
Giant Palm Borer, 178, pl. 140
Giant Root Borer, 239
giant sequoia, 131, 238
Giant Tiger Beetle, Mojave, 58, pl. 8
Gibbium psylloides, 171, 172
Gigantic Eleodes, 219, pl. 199
Gilia, 226
gilia, 226
girdlers, 125, 235–236
    Pacific Oak Twig Girdler, 133
Glaphyridae, **103–105**, pl. 56
    sensitive species, 286
Glaresidae, 286
Glischrochilus, 185
globemallow, 127
Globose Dune Beetle, 217, pl. 196
glowworms, **153–159**
    California Glowworm, 158, pl. 129
    elytra, 7

Pink Glowworm, 158, pl. 128
Western Banded Glowworm, pl. 127
Glycaspis brimblecombii, 204
Glyptoscelimorpha, 124–125
    juniperae juniperae, 124
    juniperae viridiceps, 124
    marmorata, 124–125
    viridis, 124, pl. 86
Glyptoscelis, 262–263
    albida, 263, pl. 281
    squamulata, 262–263
Golden Buprestid, 130–131, pl. 93
Golden Tortoise Beetle, 258–259
goldenbush, 248
goldenhead, 226
goldenrod, 228
Golden-winged Elderberry Borer, 243, pl. 241
Gonipterus scutellatus, 271
Gooseberry Borer, 236
Gopher Beetle, 102–103, pl. 55
gorse, 268
grain beetles
    Merchant Grain Beetle, 189
    Saw-toothed Grain Beetle, 14, 189, pl. 165
Grainary Weevil, 267, 269–270
grape, 111, 169, 263
Grape Bud Beetle, 262–263
Grape Click Beetle. *See* Pacific Coast Wireworm
Grapevine Hoplia, 111, pl. 62
grass
    cheat grass, 218
    European beach, 83
    salt grass, 218
    See also turf pests
grasshoppers, 178, 228
    Devastating Grasshopper, 87
    Lubber Grasshopper, 180
Gratiana pallidula, 45 (fig.), 258, pl. 269
Gray Garden Slug, 58
gray pine, 58, 240, 252
Great Basin sagebrush, 113, 122, 247, 248
Green Bark Beetle, 176, pl. 153
Green Dock Beetle, 259, pl. 271

Green Fig Beetle, 117, pl. 78
Green Ground Beetle, Delta, 25, 285, 285
Green Tortoise Beetle. *See* Eggplant Tortoise Beetle
Green Water Beetle, Giant, 73, pl. 28
ground beetles, **54–63**
    Big-headed Ground Beetle, 60, pl. 14
    collecting, 19, 20
    coxae, 7
    Delta Green Ground Beetle, 25, 285
    distribution patterns, 13
    mouthparts, 4–5
    Murky Ground Beetle, 60
    wrinkled bark beetles as, 53
ground squirrels, 16, 80
    California Ground Squirrel, 109
grubs, 106
    Cranberry Grub, 103
gum
    blue gum, 93, 127, 259, 271
    red gum, 259
Gutierrezia, 128, 229, 273
    sarothrae, 260
Gyascutus dianae, 128–129, pl. 91
Gymnopyge, 119
Gypsy Moth, 165
Gyretes, 65
    californicus, 65
Gyrinidae, **63–65**, pls. 21–22
    mouthparts, 5
Gyrinus, 64, 65
    plicifer, 65, pl. 22

habitats
    for collecting beetles, 17–21
    loss of, 24–25
    sensitive, 25, 67, 285–288
Haematobia irritans, 81, 83
Hairy Borer, 249–250, pl. 258
Hairy Cellar Beetle, 197
Hairy June Beetle, 114, pl. 70
Hairy Pine Borer, 240, pl. 231
Hairy Rove Beetle, 90, pl. 44
Haliplidae, **65–68**, pl. 23
    body, 31 (fig.)
    coxae, 7

Haliplus, 66, 67, 68
ham beetles
    collecting, 20
    Red-legged Ham Beetle, 181, pl. 159
    Red-shouldered Ham Beetle, 181
hand lenses, 22, 26
handsome fungus beetles, **196–198**
Haplorhynchites, 267
Hardwood Borer, Stout's, 167–168, pl. 139
Harmonia axyridis, 199, 203–204, pl. 179
Harpalus, 60
    caliginosus, 60
head
    beetle identification and, 30, 31 (figs.), 36, 37 (figs.), 42–45, 43 (figs.), 45 (figs.)
    morphology, 4, 5 (fig.)
Helenium, 242
Helichus, 138, 139
    suturalis, 138, 139
Helix aspersa, 87, 90
Hemicoelus gibbicollis, 173
Hemiopsida rubusta, 144
Hemiphileurus illatus, 117, pl. 77
Hemizonia, 228, 275
hemlock, 176
    mountain hemlock, 94, 131
hemp, 206
Heracleum, 242
Heterlimnius koebelei, 136, pl. 101
Heteroceridae, **140–142**, pl. 106–107
    antennae, 33 (fig.), 35 (fig.)
    collecting, 19
Heteromeles arbutifolia, 93, 120, 249, 263
Heterosilpha, 86, 88
    aenescens, 87
    ramosa, 37 (fig.), 86–87, pl. 41
hide beetles, **94–96**
    collecting, 20
    Hide Beetle, 164
Hippodamia, 202–203
    convergens, 41 (fig.), 199–200, 202, pl. 176
    quinquesignata, 202–203

Hister, 80, 81, 83
  abbreviatus, 83
  sellatus, 83
Hister Beetle, Striated, 83
Histeridae, **80–83**, pls. 36–39
  antennae, 33 (fig.)
  body, 33 (fig.)
  elytra, 7
Hololepta, 82
  populnea, 82
  vicina, 82
  yucateca, 82
honey mesquite, 102, 117, 127, 238, 248
Hoplia, 106, 110–111
  callipyge, 111, pl. 62
  dispar, 110
Hoplia, Grapevine, 111
Horn Fly, biological controls for, 81, 83
Horse, Devil's Coach, 90, pl. 45
House Borer, New, 241
Hydraenidae, 286
Hydrochara, 78
  lineata, 35 (figs.), 78, pl. 32
Hydrophilidae, **75–80**, pls. 31–35
  antennae, 35 (figs.)
  body, 37 (fig.), 39 (fig.)
  collecting, 19
  as larval food, 62
  sensitive species, 286
Hydrophilus triangularis, 35 (fig.), 79, pl. 34
Hygrotus artus, 25
Hygrotus Diving Beetle, Mono Lake, 25
Hyles lineata, 57
Hypera, 274
  postica, 268, 274
  punctata, 274
Hypericum perforatum, 256
hypermetamorphosis, 121, 225
Hypocaccus, 80
hypognathous mouthparts, 5, 40, 41 (figs.)
Hyporhagus gilensis, 212

ice plant, 83
Icerya purchasi, 201

identifying beetles, 3, 25 27
  key, 30–47
Iliotona cacti, 82, pl. 38
incense-cedar, 131, 146, 176
Incense-cedar Scale, 175
indigo bush, 127
Inflated Beetle, 226, pl. 212
interior live oak, 180, 239
iodine bush, 129, 174
Ipelates, 85
Iphthiminus, 220–221
  serratus, 220–221, pl. 206
Ipochus fasciatus, 179, 249–250, pl. 258
Ips, 267, 276
  emarginatus, 276
  paraconfusus, 276
  pini, 275, 276, pl. 300
  predators of, 179, 180
ips
  California Five-spined Ips, 276
  Emarginate Ips, 276
ironclad beetles
  Desert Ironclad Beetle, 215, pl. 189
  Diabolical Ironclad Beetle, 212
  Ironclad Beetle, 212, pl. 187
  Plicate Ironclad Beetle, 212, pl. 188
ironclads. *See* zopherid beetles
ironwood, 117, 128
Ischalia, 232
Ischyropalpus, 231, 233
islands, California, 13–14
Isocoma, 248
  acradenia var. *bracteosa*, 248
  acradenia var. *eremophila*, 248
  menziesii var. *vernonoides*, 248

Japanese Beetle, as misnomer. *See* Green Fig Beetle
Jeffrey pine, 149, 150, 173, 179, 196, 211, 242, 275, 276
Jeffrey Pine Beetle, 275, 276
jewel beetles. *See* metallic wood-boring beetles
jewel beetles, false. *See* schizopodid beetles
Jimson weed, 116, 129
Joshua tree, 269

Judolia, 237, 242
   (formerly *Anoplodera*) *instabilis*, 242, pl. 237
Juglans regia, 237
Juncus, 73
June beetles, 106
   Dusty June Beetle, 111–112, pl. 65
   Hairy June Beetle, 114, pl. 70
   Mount Hermon June Beetle, 25, 112–113, 285
   Ten-lined June Beetle, 112, pl. 66
   White-lined June Beetle, 112
June bugs, 106
juniper, 235
   California juniper, 113, 124, 131, 211
   Utah juniper, 124
Juniperella mirabilis, 131, pl. 94
Juniperus, 235
   californica, 113, 124, 131, 211
   osteospermae, 124
Justicia californica, 130

kallostroemia, 227
Kallostroemia grandiflora, 227
kangaroo rat, 80, 116
Kelp Beetle, 218
Killdeer, 55
Klamathweed, 256
knapweed, 268

Lacon
   rorulenta, 31 (fig.), 149, pl. 118
   sparsus, 149, pl. 117
lady beetles, **198–204**
   Ashy Gray Lady Beetle, 200, 204, pl. 181
   California Lady Beetle, 203
   Convergent Lady Beetle, 199–200, 202, pl. 176
   Five-spotted Lady Beetle, 202–203
   Multicolored Asian Lady Beetle, 199, 203–204, pl. 179
   Seven-spotted Lady Beetle, 203, pl. 178
   Two-spotted Lady Beetle, 203, pl. 177

Ladybird Beetle Parasite, 199–200
ladybird beetles. *See* lady beetles
ladybugs. *See* lady beetles
Laemostenus complanatus, 61, pl. 17
lakes, as beetle habitat, 19
Lampropterus, 245–246
   cyanipennis, 245–246
   ruficollis, 246, pl. 247
Lampyridae, **156–159**, pls. 128–131
   antennae, 6 (fig.), 33 (fig.)
   elytra, 7
langloisia, 226
Langloisia matthewsii, 226
Languria, 191, 194
   californica, 194
   convexicollis, 194
   mozardi, 192, 194, pl. 168
Languriidae, 191
Languriinae, 191
Lanius ludovicianus, 115
Lanternarius, 141
   brunneus, 141, pl. 106
   parrotus, 141
   sinuosus, 141
Lara, 137
   avara, 137, pl. 104
larch, 251
Larix, 251
Larrea tridentata, 102, 115, 127, 226, 249, 273
larvae, identifying, 3, 26
Lasioderma serricorne, 170, 174, pl. 150
laurel, 251
   California laurel, 93, 167, 246
Laurel Borer, California. *See* Banded Alder Borer
laurel sumac, 250
Lead Cable Borer, 169, pl. 143
leaf beetles, **254–264**
   as biological controls, 15
   Elm Leaf Beetle, 260, pl. 274
   mouthparts, 5
   Red and Yellow Leaf Beetle, 263, pl. 282
   Red-shouldered Leaf Beetle, 263, pl. 283
leaf litter, as beetle habitat, 20–21
Leaf Weevil, Clover, 274

Leafcutting Bee, Alfalfa, 165
leaf-rolling weevils, 254, **264–267**
Leather-winged Beetle, Downy, 162
Leatherwing Beetle, Brown, 161, pl. 132
LeConte's Prionus, 240
legs, morphology, 7–8, 8 (fig.)
lemonadeberry, 179, 249, 250
Lentinellus, 193
Lepidospartum squamatum, 17, 229
Leptura, 237, 242
    obliterata soror, 242, pl. 235
Lepturobosca (formerly *Cosmosalia*), 237
    chrysocoma, 242, pl. 236
Liatongus, 107
    californicus, 109
lichens, as beetle habitat, 19
Lichnanthe, 103, 104–105
    albipilosa, 104–105
    apina, 105, pl. 56
    rathvoni, 105
    ursina, 105
    vulpina, 103
lights, as beetle habitat, 21
lilies, 242
Limb Borer, Spotted, 169
Limonius, 150–151
    californicus, 150–151, pl. 123
    canus, 151
Linsleya, 228–229
    californica, 228–229
    compressicornis compressicornis, 229
    compressicornis neglecta, 229, pl. 218
    sphaericollis, 228
Lion Beetle, 234, 244, pl. 243
Listroderes, 271–272
    costirostris, 31 (fig.), 272, pl. 292
    difficilis, 272
Listrus (formerly *Amecocerus*), 182, 184, pl. 162
Lithocarpus densiflorus, 93
Little Bear, 115, pl. 72
live oak, 58
    canyon live oak, 99, 114, 239
    coast live oak, 52, 93, 124, 176, 179, 201, 212, 236, 239, 246, 276
    interior live oak, 180, 239

lizard beetles, 191, 192, 194
lodgepole pine, 150, 276
Loggerhead Shrike, 115
longhorn beetles, 10, **234–254**
    collecting, 18–19
    distribution patterns, 12
    elytra, 7
    eucalyptus longhorn beetles, 245
    mouthparts, 5
    parasites of, 194
    predators of, 177, 196
longhorn borers
    California Elderberry Longhorn Borer, 243–244
    Valley Elderberry Longhorn Borer, 25, 244, 285
Long-lipped Tiger Beetle, Boreal, 60
long-toed water beetles, 19, **137–140**
loosestrife, purple, 268
Lubber Grasshopper, 180
Lucanidae, **91–94**, pls. 47–49
lupine, 18, 111, 229, 242
Lupinus, 18, 111, 229, 242
Lustrous Night-stalking Tiger Beetle, 58
Lycidae, **151–153**, pls. 125–126
    body, 47 (fig.)
    mouthparts, 5
Lyctidae, 166
Lyctus, 169
    planicollis, 169, pl. 144
Lymantria dispar, 165
Lythrum salicaria, 268
Lytta, 224, 229–230
    auriculata, 229, pl. 220
    chloris, 229
    magister, 229, pl. 219
    nigripilis, 229
    stygica, 229, pl. 221
    sublaevis, 229–230
    vesicatoria, 224
    vulnerata, 229

Macrosiagon, 206, 207
    cruenta, 207–208, pl. 183
madrone, 242, 247
    Pacific madrone, 93, 127, 167, 239
ma'kech, 209

Malachius, 43 (figs.), 182, 183, pl. 160
  bipustulatus, 182
mallow
  alkali, 208
  globemallow, 127
  parish mallow, 263
Malosma laurina, 250
Malus, 98, 120, 132, 169, 239
Malvella leprosa, 208
mandibles, 4
manzanita, 102, 103, 114, 167, 193, 267
maple, 93, 127, 132, 167, 169, 246, 251
  big-leaf maple, 93, 210
  mountain maple, 93
Margarinotus sexstriatus, 83
Marmot, Yellow-bellied, 109
Marmota flaviventris, 109
masked chafers, 106, 116
matchweed, 260
maxillae, 4
May beetles, 106
mealworms, 213
  Common Mealworm, 219–220, pl. 203
  Super Mealworm, 220
  Yellow Mealworm. *See* Common Mealworm
Mealybug, Citrus, 161
Mealybug Destroyer, 201, pl. 173
Mecynotarsus, 231, 232, 233
Medicago
  arabica, 274
  polymorpha, 274
  sativa, 184, 192, 224, 263, 274
Megacephala, 58
Megachile rotundata, 165
Megacyllene antennata, 246, pl. 250
Megalodacne, 191, 192
  fasciata, 193, pl. 167
Megapenthes tartareus, 148, pl. 114
Megasoma, 119
  sleeperi, 116–117, pl. 76
Megeleates sequoiarum, 218, pl. 198
Melanophila, 132
  consputa, 131–132
  occidentalis, 132
Melanoplus devastator, 87

Melasis, 145
  rufipennis, 33 (fig.), 35 (fig.)
Meligethes, 187
  nigrescens, 187
  rufimanus, 33 (fig.), 35 (fig.), 39 (fig.), 187, pl. 164
Melilotus, 274
Meloe, 225, 230
  angusticollis, 230
  barbarus, 230, pl. 222
  strigulosus, 230
Meloidae, **224–230**, pls. 210–223
  attracting antlike flower beetles with, 231, 233–234
  elytra, 7
  sensitive species, 287
Melyridae, **182–184**, pls. 160–162
  elytra, 7
  head, 43 (figs.)
Mephitis, 98
Merchant Grain Beetle, 189
Merhynchites, 265–266
  bicolor, 266
  wickhami, 266, pl. 286
mesothorax, 6
mesquite, 18, 127, 128, 133, 168, 239, 246, 249, 254
  honey mesquite, 102, 117, 127, 238, 248
Mesquite Borer, 246, pl. 250
  Giant Mesquite Borer, 236, 239, pl. 229
Messor, 118, 213
Metaclisa marginalis, 41 (fig.)
metallic wood-boring beetles, **125–134**
  as biological controls, 15
  collecting, 18
  distribution patterns, 12
  parasites of, 194
  predators of, 177, 196
  See also borers
Metamasius (formerly *Cactophagus*) *spinolae validus*, 270, pl. 288
metamorphosis
  complete, 3, 9
  hypermetamorphosis, 121, 225
metathorax, 6

Mezium, 171–172
　affine, 172
　americanum, 172
Microphotus, 157, 159
　angustus, 33 (fig.), 158, pl. 128
milk thistle, 250
milkvetch, 194, 242, 254
milkweed, 207, 253, 262
Milkweed Beetle, Blue, 262, pl. 280
milkweed borers, 234, 253
millipedes, 155
Mimosa, 254
mimosa, 254
mites
　erythraeid mite, 55
　phoretic mite, 85, 148
Mojave Giant Tiger Beetle, 58, pl. 8
Monarthrum, 276
　scutellare, 276
Moneilema semipunctatum, 250, pl. 259
Mono Lake Hygrotus Diving Beetle, 25
Monochamus, 250–251
　clamator latus, 250–251, pl. 261
　scutellatus oregonensis, 250, pl. 260
Mordella, 205–206
　albosuturalis, 206
　grandis, 205–206
　hubbsi, 33 (fig.), 43 (fig.), 206, pl. 182
Mordellidae, **204–206**, pl. 182
　collecting, 18
　elytra, 33 (fig.)
　head, 43 (fig.)
　larval host plant, 182
Mordellistena, 182, 204, 206
Mormon tea, 128
morning glory family, 259
morphology, 3–8, 4 (fig.)
　abdomen, 3, 4
　antennae, 5, 6 (fig.)
　body regions, 3, 4 (fig.)
　elytra, 3, 7
　exoskeleton, 3–4
　head, 4, 5 (fig.)
　legs, 7–8, 8 (fig.)
　mouthparts, 3, 4–5, 5 (fig.)
　thorax, 3, 4, 5–7
　wings, 3

Morus, 212
mosquito larvae, predators of, 77, 79
mosses, as beetle habitat, 19
moths
　Douglas-fir Tussock Moth, 276
　Gypsy Moth, 165
　larva galls, 179
　White-lined Sphinx Moth, 57
Mount Hermon June Beetle, 25, 112–113, 285
mountain hemlock, 94, 131
mountain maple, 93
Mountain Pine Beetle, 275, 276
mountain-mahogany, 18, 114, 127
mountains, California, 11–12
mouthparts
　beetle identification and, 40, 41 (figs.)
　morphology, 3, 4–5, 5 (fig.)
Muck Beetle. *See* Western Carrot Beetle
mud dauber wasps, 165
mud-loving beetles, variegated, 19, **140–142**
mulberry, 212
Müllerian mimicry, 214
Multicolored Asian Lady Beetle, 199, 203–204, pl. 179
Murky Ground Beetle, 60
Museum Beetle, 165
mushrooms, as beetle habitat, 19
mustard, 250
Mycetaea subterranea, 197
Myzia, 204
　interrupta, 204
　subvittata, 204, pl. 180

Nacerdes, 223
　melaneura, 222, pl. 207
Nanularia, 129
　brunneata, 129
　monoensis, 129
　obrienorum, 129
Naupactus
　(formerly *Asynonychus*) *godmanni*, 272, pl. 294
　(formerly *Graphognathus*) *leucoloma*, 272–273

Nausibius, 188, 189
Nautical Borer, 247, pl. 252
Necrobia, 20, 178, 181, 182
   ruficollis, 181
   rufipes, 181, pl. 159
   violacea, 181
Necrophilus, 85
   hydrophiloides, 84, pl. 40
Necydalis, 244
   cavipennis, 244, pl. 242
needle, Spanish, 127
Nemognatha, 225, 230
   lurida apicalis, 230, pl. 223
Neochlamisus, 255
Neoclytus, 246–247
   balteatus, 247, pl. 251
Neoheterocerus gnatho, 33 (fig.), 35 (fig.), 141, pl. 107
Neopachylopus sulcifrons, 82
Neotoma, 80, 117
nests, as beetle habitat, 21
net-winged beetles, 5, **151–153**
New House Borer, 241
Nicrophorus, 85–86, 86, 87–88
   defodiens, 87–88
   guttula, 88, pl. 43
   marginatus, 87, 88, pl. 42
   nigrita, 6 (fig.), 35 (fig.), 88
nightshade, wild, 258
night-stalking tiger beetles
   Auduoin's Night-stalking Tiger Beetle, 58
   California Night-stalking Tiger Beetle, 58, pl. 9
   Lustrous Night-stalking Tiger Beetle, 58
Nitidulidae, **184–188**, pls. 163–164
   antennae, 33 (figs.), 35 (fig.)
   body, 39 (fig.)
   elytra, 7
noctural beetles, collecting, 18
Nosodendridae, 164
Nothopleurus, 238
   (formerly *Stenodontes*) *lobigenis*, 238
Notoxus, 224, 231, 232, 233
   calcaratus, 225, 233
Nutall's scrub oak, 114

Nyctoporis, 216
   carinata, 6 (fig.), 216, pl. 193

oak, 18, 93, 114, 127, 132, 150, 167, 168, 169, 193, 210, 212, 233, 239, 242, 244, 245, 246, 247, 249, 250, 251, 265, 271
   California black oak, 93
   canyon live oak, 99, 114, 239
   coast live oak, 52, 93, 124, 176, 179, 201, 212, 236, 239, 246, 276
   interior live oak, 180, 239
   live oak, 58
   Nutall's scrub oak, 114
   Oregon white oak, 162
   scrub oak, 124, 128
   shingle oak, 121
   tanbark-oak, 93
Oak Ambrosia Beetle, California, 276
Oak Cordwood Borer. *See* Nautical Borer
Oak Stag Beetle, 92–93, pl. 47
Oak Twig Girdler, Pacific, 133
Oblivious Tiger Beetle, 25
observing beetles, with binoculars, 23
Ochodaeidae, 101
Octinodes, 147
Ocypus olens, 33 (fig.), 90, pl. 45
Oedemeridae, **221–224**, pls. 207–209
   antennae, 33 (fig.), 41 (fig.)
   pronotum, 45 (fig.)
Ohlone Tiger Beetle, 25, 285
Olla v-nigrum, 200, 204, pl. 181
Olneya tesota, 117, 128
Omoglymmius hamatus, 53 (fig.), pl. 3
Omophron, 55, 56
   dentatus, 56, pl. 4
   tesselatus, 56
Omorgus, 95–96
   punctatus, 96
   suberosus, 95–96, pl. 50
Omus, 56, 58–59, 63
   auduoini, 58
   californicus, 58, pl. 9
   submetallicus, 58
Oncerus, 119
onion, wild, 179, 218

Onitis, 110
  alexis, 109–110
Onthophagus, 106, 107, 119
  gazella, 110, 119, pl. 61
  taurus, 110
Ophraella, 255
Ophryastes, 273
  desertus, 273, pl. 295
  geminatus, 273
Optioservus, 136–137
  divergens, 136
  quadrimaculatus, 137, pl. 102
  seriatus, 136–137
Opuntia, 187, 252
  basilaris, 250, 270
  bigelovii, 250
orange, 263
order, defined, 9
Oregon Fir Sawyer, 250, pl. 260
Oregon Stag Beetle, 93, pl. 48
Oregon white oak, 162
Orgyia pseudotsugata, 276
Ornate Checkered Beetle, 179–180, pl. 157,
Orphilus, 164
  subnitidus, 164, pl. 136
Ortholeptura, 237, 242–243
  insignis, 242
  valida, 243, pl. 238
Oryzaephilus, 188, 189
  mercator, 189
  surinamensis, 14, 189, pl. 165
Osmia, 165
Ostoma, 175, 176, 177
  pippingskoeldi, 6 (fig.), 35 (figs.), 39 (fig.), 176, pl. 151
Ostomidae (Ostomatidae), 174
Otiorhynchus (formerly *Brachyrhinus*), 273–274
  cribricollis, 273
  meridionalis, 273, pl. 296
  ovatus, 274
  rugostriatus, 273
  sulcatus, 273–274
owl's clover, 111
Oxacis
  bicolor, 223
  pallida, 223

Oxylaemus, 195
  californicus, 196, pl. 170

Pacific Coast Wireworm, 151
Pacific Dampwood Termite, 118
Pacific Flatheaded Borer, 133
Pacific madrone, 93, 127, 167, 239
Pacific Oak Twig Girdler, 133
Pacific Powder-post Beetle, 173
Pacificanthia, 159, 160, 162
  (formerly *Cantharis*) *consors*, 45 (figs.), 161, pl. 132
pack rat, 16, 80, 117
Paederus, 89
Palaeoxenus, 145
  dohrni, 33 (figs.), 144, 146, pl. 110
palafoxia, 226
Palafoxia arida, 127, 226
palm, California fan, 168
Palm Borer, Giant, 178, pl. 140
palo verde, 127, 128, 248, 249, 254
  blue, 116, 117, 128, 133, 249
Pan-American Big-headed Tiger Beetle, 58–59, pl. 10
pantry pests, 14
  See also specific beetles
Paracotalpa, 115
  deserta, 115
  granicollis, 115
  puncticollis, 115, pl. 73
  ursina, 115, pl. 72
Parandra, 237
  marginicollis, 226, 237
Paraneotermes simplicicornis, 120
Paranoplium, 244–245
  gracile, 244, pl. 244
  gracile gracile, 244
  gracile laticolle, 244
Parasite, Ladybird Beetle, 199–200
Paratyndaris
  knulli, 128
  olneyae, 128
parish mallow, 263
parthenogenesis, 267, 274
Pasture Wireworm. *See* Sugar Cane Wireworm
Pea Weevil, 263–264, pl. 284

peach, 239, 251, 263
  desert peach, 265, 267
  dooryard peach, 133
Peach Beetle. *See* Green Fig Beetle
pear, 98, 250, 251
pebble pincushion, 123
pectinate antennae, 6 (fig.)
Pectis papposa, 227
Pediacus, 190
  depressus, 191
pedicel, 5
Peltodytes, 66, 67, 68
  simplex, 31 (fig.), pl. 23
Pepsis, 249
Persea, 247
Peruvian pepper tree, 238
pesticides, 25
pests, beetles as, 14–15
Phaenops
  californica, 132, 275, pl. 96
  gentilis, 132
Phaleria, 218, 221
  rotundata, 218
Phellopsis, 210
  obcordata (formerly *porcata*), 210–211, pl. 185
Phengodidae, **153–156**, pl. 127
  antennae, 6 (fig.)
  elytra, 7
Phloeodes, 208, 209, 211–212
  diabolicus, 212
  plicatus, 212, pl. 188
  pustulosus, 212, pl. 187
Phloeosinus, 179
Phobetus, 114–115
  comatus, 114, pl. 70
  mojavus, 114–115
  saylori, 115
Phodaga
  alticeps, 227, pl. 214
  marmorata, 227
Pholiota, 193
Phoracantha, 236, 245
  recurva, 245, pl. 246
  semipunctata, 245
phoresy, 240
phoretic mite, 85, 148

Phrixothrix, 154
Phyllobaenus, 179, 181
  merkeli, 179
  scaber, 179, pl. 156
Phyllophaga, 106
Picea, 130, 241, 276
  engelmannii, 223
Pictured Rove Beetle, 90–91, pl. 46
pincushion, 226
  pebble pincushion, 123
pine, 18, 52, 83, 114, 128, 132, 148, 176, 191, 235, 238, 240, 241, 242, 243, 246, 252, 267, 276
  Coulter pine, 173
  gray pine, 58, 240, 252
  Jeffrey pine, 149, 150, 173, 179, 196, 211, 242, 275, 276
  lodgepole pine, 150, 276
  ponderosa pine, 94, 112, 130, 149, 150, 153, 173, 176, 240, 244, 252, 275, 276
  single-leaf pine, 211
  sugar pine, 146, 275, 276
  western white pine, 275, 276
pine beetles
  Jeffrey Pine Beetle, 275, 276
  Mountain Pine Beetle, 275, 276
  Western Pine Beetle, 275
pine borers
  Flatheaded Pine Borer, 132
  Hairy Pine Borer, 240, pl. 231
  Ponderous Pine-borer. *See* Pine Sawyer
  Ribbed Pine Borer, 243, pl. 239
  Sculptured Pine Borer, 128, pl. 90
Pine Engraver, 275, 276, pl. 300
pine sawyers
  Pine Sawyer, 238–239, pl. 228
  Spotted Pine Sawyer, 250–251, pl. 261
Pineapple Beetle, 187, pl. 163
Pink Glowworm, 158, pl. 128
Pinus, 18, 52, 83, 114, 128, 148, 176, 191, 198, 235, 267
  contorta, 150, 275
  coulteri, 173

edulis, 252
jeffreyi, 149, 173, 179, 196, 211, 242, 275
lambertiana, 146, 275
monophylla, 173, 211
monticola, 275
ponderosa, 94, 112, 130, 149, 153, 173, 176, 240, 252, 275
sabiniana, 58, 240, 252
pinyon
   Colorado pinyon, 252
   single-leaf pinyon, 173
Pittosporum, 247
pittosporum, 247
Pityobius murrayi, 151, pl. 124
Pityophagus, 185
plant beetles, soft-bodied, **119–121**
plants, as beetle habitat, 18
plasmodia, 54
plastron, defined, 138
Platanus, 133
   racemosa, 52, 167, 201, 237
Plateros lictor (formerly *californicus*), 153, pl. 126
Platyceroides, 92–93
   agassizi, 92–93, pl. 47
Platycerus, 92, 93
   oregonensis, 93, pl. 48
Platypodinae, 195, 196
pleasing fungus beetles, **191–194**
Plectrodes, 112
   pubescens, 112
Pleocoma, 96, 98
   australis, 35 (fig.), 99, pl. 52
   behrensi, 99
   conjugens, 100
   fimbriata, 99–100
   puncticollis, 99
Pleocomidae, **96–100**, pl. 52
   antennae, 35 (fig.)
   sensitive species, 286
Pleotomus, 159
   nigripennis, 6 (fig.), pl. 131
Pleuropasta, 227
   mirabilis, 227, pl. 215
Pleurota, 193
Plicate Ironclad Beetle, 212, pl. 188

Plionoma, 248–249
   rubens, 248
   suturalis, 248–249, pl. 255
Pluchea, 273
   odorata, 208
   sericea, 128
pluchea, 273
Podabrus, 159, 160, 162
   pruinosus, 162
Podabrus sp., pl. 134
Poecilonota salicis, 129–130
Pogonomyrmex, 118, 213
Poliaenus, 251–252
   californicus, 251, 252
   obscurus, 252
   obscurus albidus, 252
   obscurus ponderosae, 252, pl. 263
   obscurus schaefferi, 252
   oregonus, 251
pollen beetles. See *Meligethes*
Polycaon, 167
   stouti, 167–168, pl. 139
Polycesta, 127
   californica, 127, pl. 87
Polyphylla, 106, 112–113
   barbata, 25, 112–113, 285
   crinita, 112
   decemlineata, 112, pl. 66
   nigra, 112
Polyporaceae, 176
Polyporus, 176, 193
ponderosa pine, 94, 112, 130, 149, 150, 153, 173, 176, 240, 244, 252, 275, 276
Ponderous Pine-borer. See Pine Sawyer
poplar, 132
   white poplar, 93
Poplar Dicerca, 129
poppies, 124
   prickly poppy, 194
Populus, 82, 129, 212, 239
   alba, 93
   balsamifera trichocarpa, 242
   fremontii, 93, 115
Postelichus, 138, 139
   immsi, 8 (fig.), 139, pl. 105
   productus, 138, 139

powder-post beetles, 166, 170
　Pacific Powder-post Beetle, 173
　Southern Powder-post Beetle, 169, pl. 144
predaceous diving beetles, 12, 19, **70–75**
Priacma serrata, 52, pl. 2
prickly poppy, 194
primitive carrion beetles, **83–85**
　See also carrion beetles
Prionus, 239–240
　californicus, 6 (fig.), 41 (fig.), 236, 239–240, pl. 230
　lecontei, 240
prionus
　California Prionus, 236, 239–240, pl. 230
　LeConte's Prionus, 240
Procyon lotor, 55, 98
prognathous mouthparts, 4–5, 40, 41 (figs.)
Prolixocupes lobiceps, 52, pl. 1
pronotum, 6–7, 40, 41 (figs.), 44, 45 (figs.)
Prosopis, 18, 127, 168, 254
　glandulosa, 238
　glandulosa var. *torreyana*, 102, 117, 127, 248
　pubescens, 128
prothorax, 6
Prunus, 93, 98, 120, 239
　andersonii, 265
　persica, 133
Psephenidae, **142–144**, pl. 108
　body, 47 (fig.)
　collecting, 19
Psephenus falli, 143
Pseudococcis citri, 161
pseudoscorpions, 240
Pseudotsuga
　macrocarpa, 146
　menziesii, 52, 128, 149, 176, 221, 238
Psoa maculata, 169
Psorothamnus
　arborescens, 127
　emoryi, 127
　schottii, 123
psyllids, 203, 204
　Red Gum Eucalyptus Psyllid, 204

Pterostichus, 54, 60–61
　californicus, 61
　lama, 61, pl. 15
Pterotus, 157, 159
　obscuripennis, 158–159, pl. 130
Ptinus, 172
　californicus, 172
　clavipes, 172
　fur, 172, pl. 145
Punctate Blister Beetle, 228, pl. 217
Punctate Stag Beetle, 94
puncture vine, 227
purple loosestrife, 268
push ups, defined, 15, 101
Pyropyga, 159
　nigricans, 158
Pyrota, 226
　palpalis, 226, pl. 210
Pyrus, 98, 250

Quercus, 18, 58, 114, 127, 150, 167, 193, 210, 233, 265, 271
　agrifolia, 52, 93, 124, 176, 179, 201, 212, 236, 276
　berberidifolia, 124, 128
　chrysolepis, 99, 114, 239
　dumosa, 114
　garryana, 162
　imbricarius, 121
　kelloggii, 93
　wislezenii, 180, 239

rabbitbrush, 16, 18, 228, 229, 247, 248, 273
　rubber rabbitbrush, 124, 134, 260
Raccoons, 55, 98
rain beetles, **96–100**
　Behren's Rain Beetle, 99
　Black Rain Beetle, 99
　Fimbriate Rain Beetle, 99–100
　Santa Cruz Rain Beetle, 100
　Southern Rain Beetle, 99, pl. 52
raspberries, 265, 266, 274
Rathvon's Bumblebee Scarab, 105
rats
　kangaroo rat, 80, 116
　pack rat, 16, 80, 117
rearing beetles, 24–25

red alder, 93
Red and Yellow Leaf Beetle, 263, pl. 282
Red and Black Burying Beetle, 87, pl. 42
red cedar, western, 131, 223
red clover, 192
red elderberry, 243, 244
red fir, California, 247
Red Flat Bark Beetle, 190-191, pl. 166
red gum, 259
Red Gum Eucalyptus Psyllid, 204
Red Turpentine Beetle, 275, pl. 298
Red-bellied Clerid, 180–181
Red-legged Ham Beetle, 181, pl. 159
Red-shouldered Ham Beetle, 181
Red-shouldered Leaf Beetle, 263, pl. 283
redwood, coast, 94
reflex bleeding, 197, 198, 199
reproductive strategies, 9–10
  parthenogenesis, 267, 274
reticulated beetles, **51–52**
Rhagium, 237
  inquisitor, 243, pl. 239
Rhamnus, 243
  californica, 251
Rhantus, 73
  gutticollis, 73, pl. 27
Rhinoplatia
  mortivallicoa, 224
  ruficollis, 223–224, pl. 209
Rhipiceridae, **121–122**, pl. 84
  antennae, 6 (fig.)
Rhodobaenus, 270
  tredecimpunctatus, 270
Rhus
  integrifolia, 179, 249
  ovata, 249
Rhynchites, 267
Rhysodidae, **52–54**, pl. 3
Ribbed Pine Borer, 243, pl. 239
Rice Weevil, 14, 267, 269, pl. 287
Ricinus, 250
riffle beetles, 19, **134–137**
ringstem, 123
ripiphorid beetles, 7, **206–208**
Ripiphoridae, **206–208**, pls. 183–184
  antennae, 33 (fig.)
  elytra, 7

Ripiphorus, 206, 207, 208
  rex, 33 (fig.), 208, pl. 184
  smithi, 208
Rodolia cardinalis, 201, pl. 175
root borers
  Giant Root Borer, 239
  Willow Root Borer, 237, pl. 227
Root Weevil, Strawberry, 274
Rosa, 111, 266
Rosalia, 246
  funebris, 246, pl. 248
rose curculios, 265–266
  Eastern Rose Curculio, 266
  Western Rose Curculio, 266, pl. 286
Rose Weevil, Fuller's, 272, pl. 294
roses, 111, 114, 265
  wild rose, 266
round sand beetles, 55, 56
roundheaded borers, 234–235, 236
rove beetles, **88–91**
  collecting, 19, 20, 21
  elytra, 7
  Hairy Rove Beetle, 90, pl. 44
  Pictured Rove Beetle, 90–91, pl. 46
rubber rabbitbrush, 124, 134, 260
Rubus, 263, 266
Rudbeckia, 242
Rugose Stag Beetle, 93, pl. 49
Rumex, 259
rushes, 73
Russian thistle, 218, 223

sage
  burr sage, 115
  white sage, 169
sagebrush, 182, 227, 229, 260
  California sagebrush, 114, 259
  Great Basin sagebrush, 113, 122, 247, 248
saguaro, 270
St. John's wort. *See* Klamathweed
Salix, 93, 105, 111, 115, 127, 136, 150, 204, 233, 238, 261
Salsola tragus, 218, 223
salt grass, 218
salt marsh fleabane, 207–208
saltbush, 128, 217, 273
  big saltbush, 127, 133

Salvia apiana, 169
Sambucus, 179
   mexicana, 243
   racemosa, 243
San Joaquin Valley Tiger Beetle, 25
sand beetles, round, 55, 56
sand dunes, as beetle habitat, 17, 19–20
sand verbena, 57, 217
Sandalus, 121, 122
   californicus, 122
   cribricollis, 6 (fig.), 122, pl. 84
   niger, 121, 122
Santa Cruz Rain Beetle, 100
sap beetles, **184–188**
   collecting, 18
   elytra, 7
   Yellow-shouldered Sap Beetle. *See* Pineapple Beetle
Saperda, 252
   calcarata, 252, pl. 266
   horni, 252
Saprinus, 82
   lugens, 33 (fig.), 82, pl. 37
Sargent's cypress, 131, 133
Satin Carrion Beetle, 87
Saw-toothed Grain Beetle, 14, 189, pl. 165
sawyers
   Oregon Fir Sawyer, 250, pl. 260
   Pine Sawyer, 238–239, pl. 228
   Spotted Pine Sawyer, 250–251, pl. 261
   See also wood-boring beetles
Saxinis, 263
   saucia, 263, pl. 283
scale broom, 16–17, 18, 229
scale insects, 184
   Cottony Cushion Scale, 201
   Incense-cedar Scale, 175
   Sycamore Scale, 201
Scaphinotus, 57
   cristatus, 57
   punctatus, 57, pl. 7
   striatopunctatus, 57
   ventricosus, 57
scarab beetles, **105–119**

Bee Scarab, 105, pl. 56
beetles formerly classified as, 95, 100
bumblebee scarab beetles, **103–105**
collecting, 18
dung scarabs, 106, 107
dung-rolling scarabs, 109
earth-boring scarab beetles, 6–7, **100–103**
larvae, 80, 91, 106
Rathvon's Bumblebee Scarab, 105
White Sand Bear Scarab, 104–105
Scarabaeidae, **105–119**, pls. 57–81
   antennae, 6 (fig.), 35 (figs.)
   beetles formerly classified as, 95, 100
   body, 31 (figs.)
   collecting, 20
   endangered species, 25, 285
   as introduced beneficial, 15
   larvae, 80, 91
   mouthparts, 5
   sensitive species, 286–287
   tarsi, 8 (fig.)
Scarites subterraneus, 60, pl. 14
scavenger beetles, water, 19, 20, 62, **74–80**
Sceliphron, 165
Schinus molle, 238
Schizax senax, 249
Schizillus, 215–216
   laticeps, 216, pl. 192
schizopodid beetles, **123–125**
Schizopodidae, **123–125**, pls. 85–86
Schizopus, 123
   laetus, 123–124, pl. 85
   sallaei nigricans, 124
   sallei sallei, 124
Scirpus, 73, 271
scirpus, 73
Scobicia declivis, 169, pl. 143
Scoliidae, 207
Scolytinae, 176, 177, 179, 267
Scolytus, 180, 275–276
   ventralis, 275–276, pl. 299
Scotch broom, 256
Scotobaenus, 220
   parallelus, 220, pl. 204
Scraptiidae, 45 (fig.), 47 (fig.)
screw-bean, 128

scrub oak, 124, 128
    Nutall's scrub oak, 114
Sculptured Pine Borer, 128, pl. 90
Scyphophorus, 270
    acupunctatus, 270
    yuccae, 6 (fig.), 270, pl. 289
Searcher, Fiery, 57, pl. 6
seasons, role in beetle collecting, 15–17
seed beetles (formerly seed weevils), **254–264**
senna, spiny, 123
Senna armata, 123
Sennius sp., 31 (fig.)
sensitive species, 25, 67, 285–288
    See also endangered and threatened species
sequoia, giant, 131, 238
Sequoia sempervirens, 94, 238
Sequoiadendron giganteum, 131
Serica, 31 (fig.), 106, 111, 118
    perigonia, 111, pl. 63
service-berry, 242
Seven-spotted Lady Beetle, 203, pl. 178
Sheperdia, 242
sheperdia, 242
shingle oak, 121
Shining Black Turfgrass Ataenius, 108–109
Shiny Spider Beetle, 172
Short-circuit Beetle. *See* Lead Cable Borer
shot-hole borers. *See* bostrichid beetles
Shrike, Loggerhead, 115
Silis, 160
silk tassel bush, 127
Silphidae, **85–88**, pls. 41–43
    antennae, 6 (fig.), 35 (fig.)
    body, 37 (fig.)
    collecting, 20
silvanid flat bark beetles, **188–189**
Silvanidae, **188–189**, pl. 165
Silvanus, 188, 189
Silybum marianum, 250
single-leaf pine, 211
single-leaf pinyon, 173

Sinodendron, 92
    rugosum, 93, pl. 49
Sisal Weevil, 270
Sitophilus
    granarius, 267, 269–270
    oryzae, 14, 267, 269, pl. 287
    zeamais, 14, 269
skin beetles, **162–166**
    collecting, 20
    Skin Beetle, 165, pl. 138
skunks, 98
Sleeper's Elephant Beetle, 116–117, pl. 76
slime molds, 52–53, 54
Slug, Gray Garden, 58
smoke tree, 129
snail eaters, 57–58
snails, 158
    Brown Garden Snail, 87, 90, 90
snakeweed, 128, 229, 273
sneezeweed, 242
Snout Beetle, Eucalyptus, 271
snout beetles. *See* weevils
snowberries, 120
societies, for studying beetles, 289–290
soft-bodied plant beetles, **119–121**
soft-winged flower beetles, 7, **182–184**
Solanum, 258
soldier beetles, **159–162**
Solidago, 228
solitary bees, 165, 179, 225, 226
Sosylus, 195
    dentiger, 195
Southern Powder-post Beetle, 169, pl. 144
Southern Rain Beetle, 99, pl. 52
Southwestern Stump Borer, 238
Spanish broom, 268
Spanish fly. *See* blister beetles
Spanish needle, 127
species
    book's selection of, 2
    defined, 9
    extinct, 25
    introduced, 25
    sensitive, 25, 67, 285–288
    threatened and endangered, 14, 25, 285

Spectralia purpurascens, 130
Spermophilus, 80
   beecheyi, 109
Sphaeralcea, 127
   ambigua var. *rosacea*, 263
Sphaericus gibboides, 172–173
Sphaeridium scarabaeoides, 37 (fig.), 79–80, pl. 35
Sphecidae, 207
Sphenophorus, 270–271
   aequalis, 270–271
   aequalis discolor, 271
   aequalis ochrea, 271
   aequalis picta, 271, pl. 290
spider beetles, **170–174**
   American Spider Beetle, 171–172
   Brown Spider Beetle, 172
   parasites of, 194
   Shiny Spider Beetle, 172
   White-marked Spider Beetle, 172, pl. 145
Spilodiscus, 80
spiny senna, 123
Spiny Wood-borer. *See* Pine Sawyer
Spondylis upiformis, 240–241
Spotted Asparagus Beetle, 258
Spotted Cucumber Beetle, Western, 260–261, pl. 275
Spotted Dung Beetle, 79–80, pl. 35
Spotted Flower Buprestid, 127, pl. 88
Spotted Limb Borer, 169
Spotted Pine Sawyer, 250–251, pl. 261
Spotted Tree Borer, 251, pl. 262
spruce, 130, 241, 242, 243, 251, 276
   bigcone spruce, 146
   Englemann spruce, 223
Squamodera, 127
   barri, 127
   vanduzee, 127
squirrels
   California Ground Squirrel, 109
   ground squirrel, 16, 80
stag beetles, **91–94**
   Oak Stag Beetle, 92–93, pl. 47
   Oregon Stag Beetle, 93, pl. 48
   Punctate Stag Beetle, 94
   Rugose Stag Beetle, 93, pl. 49
   Striated Stag Beetle, 94

Staphylinidae, **88–91**, pls. 44–46
   elytra, 7, 33 (fig.)
starthistle, 268
Stegobium paniceum, 170, 174, pl. 148
Stem Borer, Clover, 192, 194, pl. 168
Stenomorpha, 217
   costipennis, 217
   mckittricki, 217
Stenostrophia, 237, 243
   tribalteata sierrae, 243, pl. 240
   tribalteata tribalteata, 243
Stictotarsus, 72
   striatellus, 72, pl. 25
stink beetles, 218, 220
Stomacoccus platani, 201
Stout's Hardwood Borer, 167–168, pl. 139
strawberries, 98, 263, 274
Strawberry Root Weevil, 274
streams, as beetle habitat, 19
Striated Hister Beetle, 83
Striated Stag Beetle, 94
stridulation, 76, 85–86, 95, 101
Striped Cucumber Beetle, Western, 261, pl. 276
studying beetles, resources for, 289–290
Stump Borer, Southwestern, 238
subterranean dry-rot termite, 120
sugar bush, 249
Sugar Cane Wireworm, 149, pl. 119
sugar pine, 146, 275, 276
Sugar-beet Wireworm, 150–151, pl. 123
sumac, laurel, 250
Sunburst Diving Beetle, 73–74, pl. 29
sunflower family, 18, 179, 227, 259
   desert-sunflower, 123, 226, 227
Super Mealworm, 220
sycamore, 133, 246
   western sycamore, 52, 167, 201, 237
Sycamore Scale, 201
Symphoricarpos, 120
Synaphaeta guexi, 251, pl. 262

tachinid flies, 216, 217
tamarisk, 211
Tamarix, 211

Tanarthrus, 231, 232–233
  alutaceus, 233, pl. 224
tanbark-oak, 93
Tanystoma maculicolle, 61, pl. 19
Taphrocerus, 125–126
tarantula wasps, 249
tarsal formulas, 8
tarsi, morphology, 7–8, 8 (fig.)
tarweed, 228, 275
taxonomy, 9
tea
  desert tea, 123
  Mormon tea, 128
tecate cypress, 133
teddy bear cholla, 250
Tegrodera, 227–228
  aloga, 228
  erosa, 228, pl. 216
  latecincta, 227–228
Temnoscheila (formerly *Temnochila*), 175, 176, 177
  chlorodia, 39 (fig.), 176, pl. 153
Tenebrio, 213
  molitor, 219–220, pl. 203
Tenebrionidae, **212–221**, pls. 189–206
  antennae, 6 (fig.)
  body, 39 (fig.)
  distribution patterns, 12
  head, 37 (fig.)
  mouthparts, 41 (fig.)
  pronotum, 41 (fig.)
  sensitive species, 287
Tenebroides, 175, 177, pl. 154
  corticalis, 177
  crassicornis, 41 (fig.)
  mauritanicus, 175, 177
Ten-lined June Beetle, 112, pl. 66
termites
  Pacific Dampwood Termite, 118
  subterranean dry-rot termite, 120
Tetracha, 56
  carolina, 58–59, pl. 10
Tetraopes, 234, 253
  basalis, 253
  femoratus, 253, pl. 267
  sublaevis, 253
Thanatophilus, 86
  lapponicus, 87

Therevid flies, 217
Thermonectus, 73–74
  intermedius (formerly *basillaris*), 74
  marmoratus, 73–74, pl. 29
Thinopinus pictus, 90–91, pl. 46
thistle, 268
  milk thistle, 250
  Russian thistle, 218, 223
  starthistle, 268
thorax, morphology, 3, 4, 5–7
threatened species. *See* endangered and threatened species
Thuja plicata, 131, 223
tiger beetles, **54–63**
  Auduoin's Night-stalking Tiger Beetle, 58
  Boreal Long-lipped Tiger Beetle, 60
  California Night-stalking Tiger Beetle, 58, pl. 9
  California Tiger Beetle, 59–60, pl. 12
  collecting, 19
  feeding habits, 233
  Lustrous Night-stalking Tiger Beetle, 58
  Mojave Giant Tiger Beetle, 58, pl. 8
  Oblivious Tiger Beetle, 25
  Ohlone Tiger Beetle, 25, 285
  Pan-American Big-headed Tiger Beetle, 58–59, pl. 10
  San Joaquin Valley Tiger Beetle, 25
  Western Tiger Beetle, 59, pl. 11
  White-striped Tiger Beetle, 60, pl. 13
Tiphiidae, 207
Tiquilia (formerly *Coldenia*), 226
  palmeri, 227
  plicata, 227
tiquilia, 226, 227
Tomarus (formerly *Ligyrus*) *gibbosus obsoletus*, 116, pl. 75
tortoise beetles, 255
  Eggplant Tortoise Beetle, 258, pl. 269
  Golden Tortoise Beetle, 258–259
  Green Tortoise Beetle. *See* Eggplant Tortoise Beetle
toyon, 93, 120, 249, 250, 263

Trachyderes (formerly *Dendrobias*)
 mandibularis mandibularis, 249
 mandibularis reductus, 249, pl. 256
Trachykele, 131
 blondeli, 131
 hartmanni, 131
 nimbosa, 131, pl. 95
 opulenta, 131
Trachymela sloanei, 259, pl. 270
Tragidion, 249
 armatum, 249, pl. 257
 californicum, 249
 peninsulare, 249
Tragosoma
 depsarius, 240, pl. 231
 pilosicornis, 45 (fig.), 240
Tribolium confusum, 14
Tribulus terrestris, 227
Trichodes, 177–178, 179–180, 181
 ornatus, 179–180
 ornatus douglasianus, 180, pl. 157
 ornatus tenuosus, 180
Trifolium pratense, 192
Trigonogenius, 172
 globulus, 172
Trigonoscuta, 272
 morroensis, 272, pl. 293
Triplax, 191, 193–194
 californica antica, 194
 californica californica, 193–194
Trirhabda, 259–260
 confusa, 260, pl. 273
 diducta, 260
 geminata, 260, pl. 272
 luteocincta, 259
 nitidicollis, 259–260
Trogidae, 20, **94–96**, pls. 50–51
Trogoderma, 165
 simplex, 165, pl. 138
Trogossitidae, **174–177**, pls. 151–154
 antennae, 6 (fig.), 35 (figs.)
 body, 39 (figs.)
 mouthparts, 41 (fig.)
Tropicus, 141–142
 pusillus, 141–142
Tropisternus, 79
 californicus, 79
 ellipticus, 39 (fig.), 79, pl. 33

trout-stream beetles, **68–69**, 79, pl. 24
Trox, 95, 96
 fascifer, 96
 gemmulatus, 96, pl. 51
Tsuga, 176
 mertensiana, 94, 131
tule, 271
Tule Beetle, 61, pl. 19
Tule Billbug, 270–271, pl. 290
tumbling flower beetles, 18, 182, **204–206**
turf pests, 103, 106, 108–109, 261–262, 272
Turpentine Beetle, Red, 275, pl. 298
Tussock Moth, Douglas-fir, 276
twig borers, 7, 166
 Western Twig Borer, 168, pl. 141
Twig Girdler, Pacific Oak, 133
Two-spotted Lady Beetle, 203, pl. 177
Typha, 73

Uleiota, 188, 189
Ulex europaea, 268
Ulmus, 52, 260
 americanus, 121
Ulochaetes, 244
 leoninus, 234, 244, pl. 243
Uloma longula, 39 (fig.), 41 (fig.)
Umbellularia californica, 93, 167, 246
Uropoda, 148
Usechimorpha barberi, 210
Usechus, 210
 lacerta, 210
 nucleatus, 210
Utah juniper, 124

Vacusus, 233
Valgus, 118
 californicus, 118, pl. 81
Valley Elderberry Longhorn Borer, 25, 244, 285
Varied Carpet Beetle, 165, pl. 137
variegated mud-loving beetles, 19, **140–142**
Vedalia, 201, pl. 175
Vegetable Weevil, 271–272, pl. 292
Velvet Beetle, Yellow, 242, pl. 236
verbena, sand, 57, 217

Vespidae, 207
Vetch Bruchid, 264
Vine Weevil, Black, 273–274
Vitis, 111, 169
Vulpes, 98

wallflowers, 132
walnut, 247, 250, 251
  English walnut, 237
Washingtonia filifera, 168
wasps, 178, 206, 207
  braconid wasp, 216
  Ladybird Beetle Parasite, 199–200
  mud dauber wasp, 165
  tarantula wasp, 249
watching beetles, with binoculars, 23
water beetles
  crawling water beetles, 7, **65–68**
  Giant Black Water Beetle, 79, pl. 34
  Giant Green Water Beetle, 73, pl. 28
  long-toed water beetles, 19, **137–140**
  See also diving beetles
water birch, 93
water penny beetles, 19, **142–144**
  Western Water Penny Beetle, 143
water scavenger beetles, 19, 20, 62, **74–80**
web sites, for studying beetles, 289–290
weeds, biological controls for, 15, 256, 263, 268
weevils, **267–277**
  Alfalfa Weevil, 268, 274
  Bean Weevil, 264, pl. 285
  as biological controls, 15
  Black Vine Weevil, 273–274
  Boll Weevil, 267
  Broadbean Weevil, 264
  Cactus Weevil, 270, pl. 288
  California Acorn Weevil, 271, pl. 291
  California Filbert Weevil. *See* California Acorn Weevil
  Clover Leaf Weevil, 274
  collecting, 19
  distribution patterns, 13
  Fuller's Rose Weevil, 272, pl. 294
  Grainary Weevil, 267, 269–270
  leaf-rolling weevils, 254, **264–267**
  mouthparts, 5
  parasites of, 194
  Pea Weevil, 263–264, pl. 284
  predators of, 80
  Rice Weevil, 14, 267, 269, pl. 287
  Sisal Weevil, 270
  Strawberry Root Weevil, 274
  Vegetable Weevil, 271–272, pl. 292
  Yucca Weevil, 270, pl. 290
Western Banded Glowworm, pl. 127
Western Carrot Beetle, 116, pl. 75
Western Cedar Borer, 131
Western Eyed Click Beetle, 149–150, pl. 120
Western Pine Beetle, 275
western red cedar, 131, 223
Western Rose Curculio, 266, pl. 286
Western Spotted Cucumber Beetle, 260–261, pl. 275
Western Striped Cucumber Beetle, 261, pl. 276
western sycamore, 52, 167, 201, 237
Western Tiger Beetle, 59, pl. 11
Western Twig Borer, 168, pl. 141
Western Water Penny Beetle, 143
western white pine, 275, 276
Wharf Borer, 222, pl. 207
whirligig beetles, **63–65**
  collecting, 19
  mouthparts, 5
white alder, 130, 237
white fir, 52, 94, 176, 247
White Fringed Beetle, 272–273
white oak, Oregon, 162
white poplar, 93
white sage, 169
White Sand Bear Scarab, 104–105
White-lined June Beetle, 112
White-lined Sphinx Moth, 57
White-marked Spider Beetle, 172, pl. 145
White-striped Tiger Beetle, 60, pl. 13
wild nightshade, 258
wild onion, 179, 218
wild roses, 266

willow, 93, 105, 111, 114, 115, 127, 136, 150, 204, 233, 238, 246, 247, 250, 251, 252, 261, 262, 263
Willow Agrilus, Common, 133–134
Willow Root Borer, 237, pl. 227
wireworms, 147, 150–151
   false wireworms, 213
   Pacific Coast Wireworm, 151
   Pasture Wireworm. *See* Sugar Cane Wireworm
   Sugar Cane Wireworm, 149, pl. 119
   Sugar-beet Wireworm, 150–151, pl. 123
Wisteria, 251
wisteria, 251
wood (dead and stumps), as beetle habitat, 18–19
wood-boring beetles
   as forest pests, 14
   mouthparts, 5
   parasites of, 194–195
   predators of, 206
   See also borers; metallic wood-boring beetles; sawyers
Wooly Darkling Beetle, 219, pl. 200
wounded-tree beetles, 164
wrinkled bark beetles, **52–54**
   Wrinkled Bark Beetle, 53, pl. 3

Xanthochroa, 222–223
   californica, 222–223
Xanthogaleruca luteola, 260, pl. 274
Xeranobium, 173–174, pl. 147
Xerosaprinus sp., 82, pl. 36
Xylococculus macrocarpae, 175
Xylocrius agassizi, 236
Xylotrechus, 247
   albonotatus, 247
   insignis, 247
   nauticus, 247, pl. 252

yarrow, 105, 242
Yellow Mealworm. *See* Common Mealworm
Yellow Velvet Beetle, 242, pl. 236
Yellow-bellied Marmot, 109
Yellow-shouldered Sap Beetle. *See* Pineapple Beetle
Yellow-spotted Diving Beetle. *See* Sunburst Diving Beetle
yerba santa, 111, 260
Yucca, 82, 179, 212
   brevifolia, 269
   whipplei, 249, 270
yucca, 82, 179, 212
   chaparral yucca, 249, 270
Yucca Weevil, 270, pl. 289
Yuccaborus frontalis, 269

Zabrotes subfasciatus, 33 (fig.)
Zaitzevia, 136, 137
   parvula, 137, pl. 103
   posthonia, 137
Zarhipis, 154, 155
   integripennis, 6 (fig.), 155, 156, pl. 127
   tiemanni, 156
   truncaticeps, 155–156
Zootermopsis angusticollis, 118
zopherid beetles, **208–212**
Zopheridae, **208–212**, pls. 185–188
Zopherus, 208, 209, 211, 212
   chilensis, 208–209
   granicollis granicollis, 211, pl. 186
   granicollis ventriosus, 211
   opacus, 211
   sanctaehelenae, 211
   tristis, 211
   uteanus, 211
Zophobas, 213
   morio, 220

ABOUT THE AUTHORS

Arthur V. Evans (left) is a research associate in the Department of Entomology at the National Museum of Natural History, at the Smithsonian Institution, and in the Department of Recent Invertebrates at the Virginia Museum of Natural History. He is coauthor, with Charles L. Bellamy, of *An Inordinate Fondness for Beetles* (UC Press, 2000).

James N. Hogue (right) is the manager of biological collections in the Department of Biology at California State University, Northridge, and a research associate at the Natural History Museum of Los Angeles County. Together, Art and Jim wrote *Introduction to California Beetles* (UC Press, 2004).

Series Design:	Barbara Jellow
Design Enhancements:	Beth Hansen
Design Development:	Jane Tenenbaum
Composition:	Jane Rundell
Indexer:	Jean Mann
Text:	9/10.5 Minion
Display:	ITC Franklin Gothic Book and Demi
Printer and binder:	Golden Cup Printing Company Limited

Field Guides

Sharks, Rays, and Chimaeras of California, by David A. Ebert, illustrated by Mathew D. Squillante

Field Guide to Beetles of California, by Arthur V. Evans and James N. Hogue

Geology of the Sierra Nevada, Revised Edition, by Mary Hill

Mammals of California, Revised Edition, by E.W. Jameson, Jr., and Hans J. Peeters

Field Guide to Amphibians and Reptiles of the San Diego Region, by Jeffrey M. Lemm

Dragonflies and Damselflies of California, by Tim Manolis

Field Guide to Freshwater Fishes of California, Revised Edition, by Samuel M. McGinnis, illustrated by Doris Alcorn

Raptors of California, by Hans J. Peeters and Pam Peeters

Geology of the San Francisco Bay Region, by Doris Sloan

Trees and Shrubs of California, by John D. Stuart and John O. Sawyer

Pests of the Native California Conifers, by David L. Wood, Thomas W. Koerber, Robert F. Scharpf, and Andrew J. Storer

Introductory Guides

Introduction to Air in California, by David Carle

Introduction to Water in California, by David Carle

Introduction to California Beetles, by Arthur V. Evans and James N. Hogue

Introduction to California Birdlife, by Jules Evens and Ian C. Tait

Weather of the San Francisco Bay Region, Second Edition, by Harold Gilliam

Introduction to Trees of the San Francisco Bay Region, by Glenn Keator

Introduction to California Soils and Plants: Serpentine, Vernal Pools, and Other Geobotanical Wonders, by Arthur R. Kruckeberg

Introduction to Birds of the Southern California Coast, by Joan Easton Lentz

Introduction to California Mountain Wildflowers, Revised Edition, by Philip A. Munz, edited by Dianne Lake and Phyllis M. Faber

Introduction to California Spring Wildflowers of the Foothills, Valleys, and Coast, Revised Edition, by Philip A. Munz, edited by Dianne Lake and Phyllis M. Faber

Introduction to Shore Wildflowers of California, Oregon, and Washington, Revised Edition, by Philip A. Munz, edited by Dianne Lake and Phyllis Faber

Introduction to California Desert Wildflowers, Revised Edition, by Philip A. Munz, edited by Diane L. Renshaw and Phyllis M. Faber

Introduction to California Plant Life, Revised Edition, by Robert Ornduff, Phyllis M. Faber, and Todd Keeler-Wolf

Introduction to California Chaparral, by Ronald D. Quinn and Sterling C. Keeley, with line drawings by Marianne Wallace

Introduction to the Plant Life of Southern California: Coast to Foothills, by Philip W. Rundel and Robert Gustafson

Introduction to Horned Lizards of North America, by Wade C. Sherbrooke

Introduction to the California Condor, by Noel F. R. Snyder and Helen A. Snyder

Regional Guides

Sierra Nevada Natural History, Revised Edition, by Tracy I. Storer, Robert L. Usinger, and David Lukas

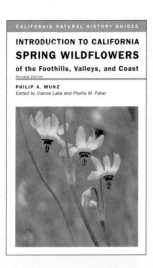

INTRODUCTION TO CALIFORNIA
SPRING WILDFLOWERS
of the Foothills, Valleys, and Coast
Revised Edition

PHILIP A. MUNZ
Edited by Dianne Lake and Phyllis M. Faber

RAPTORS
OF CALIFORNIA

HANS PEETERS and PAM PEETERS

The **CALIFORNIA NATURAL HISTORY GUIDES** are the most authoritative resource on the state's flora and fauna. These short, inexpensive, and easy-to-use books help outdoor enthusiasts make the most of California's abundant natural resources. The series is divided into two groups: **INTRODUCTIONS** for beginners and **FIELD GUIDES** for more experienced naturalists. Please visit our web site for announcements, a regular natural history column, and the most up-to-date list of books. To hear about new guides through UC Press E-News, fill out and return this card, or sign up online at www.californianaturalhistory.com.*

Name _____

Address _____

City/State/Zip _____

Email _____

Which book did this card come from? _____

Where did you buy this book? _____

What is your profession? _____

Comments _____

WE'D LOVE TO HEAR FROM YOU!

* UC Press will not share your information with any other organization.

POST OFFICE WILL NOT DELIVER WITHOUT POSTAGE

Return to:

University of California Press
Attn: Natural History Editor
2120 Berkeley Way
Berkeley, California 94704-1012